Microbial Bioinformatics in the Oil and Gas Industry

Microbes, Materials, and the Engineered Environment

Series Editor
Torben Lund Skovhus and Richard B. Eckert

Microbial Bioinformatics in the Oil and Gas Industry: Applications to Reservoirs and Processes
Kenneth Wunch, Marko Stipaničev, and Max Frenzel

Failure Analysis of Microbiologically Influenced Corrosion
Torben Lund Skovhus and Richard B. Eckert

For more information about this series, please visit:
https://www.routledge.com/Microbes-Materials-and-the-Engineered-Environment/book-series/CRCMMEE

Microbial Bioinformatics in the Oil and Gas Industry

Applications to Reservoirs and Processes

Edited by
Kenneth Wunch
Marko Stipaničev
Max Frenzel

CRC Press is an imprint of the
Taylor & Francis Group, an **informa** business

First edition published 2022
by CRC Press
6000 Broken Sound Parkway NW, Suite 300, Boca Raton, FL 33487-2742

and by CRC Press
2 Park Square, Milton Park, Abingdon, Oxon, OX14 4RN

© 2022 Taylor & Francis Group, LLC

CRC Press is an imprint of Taylor & Francis Group, LLC

Reasonable efforts have been made to publish reliable data and information, but the author and publisher cannot assume responsibility for the validity of all materials or the consequences of their use. The authors and publishers have attempted to trace the copyright holders of all material reproduced in this publication and apologize to copyright holders if permission to publish in this form has not been obtained. If any copyright material has not been acknowledged, please write and let us know so we may rectify in any future reprint.

Except as permitted under U.S. Copyright Law, no part of this book may be reprinted, reproduced, transmitted, or utilized in any form by any electronic, mechanical, or other means, now known or hereafter invented, including photocopying, microfilming, and recording, or in any information storage or retrieval system, without written permission from the publishers.

For permission to photocopy or use material electronically from this work, access www.copyright.com or contact the Copyright Clearance Center, Inc. (CCC), 222 Rosewood Drive, Danvers, MA 01923, 978-750-8400. For works that are not available on CCC, please contact mpkbookspermissions@tandf.co.uk

Trademark notice: Product or corporate names may be trademarks or registered trademarks and are used only for identification and explanation without intent to infringe.

Library of Congress Cataloging-in-Publication Data
Names: International Symposium on Applied Microbiology and Molecular Biology in Oil Systems (7th : 2019 : Halifax, N.S.) | Wunch, Kenneth, a editor. | Stipanicev, Marko, editor. | Frenzel, Max (Microbiologist), editor.
Title: Microbial bioinformatics in the oil and gas industry : applications to reservoirs and processes / edited by Kenneth Wunch, Marko Stipanicev, and Max Frenzel.
Description: First edition. | Boca Raton, FL : CRC Press, 2021. | Papers selected from the Seventh International Symposium on Applied Microbiology and Molecular Biology in Oil Systems held in Halifax, Nova Scotia, June 18-21, 2019. | Includes bibliographical references and index. |
Summary: "This book brings together contributions from leading scientists, academics, and experts from the oil and gas industry to discuss microbial-mediated problems faced by the industry and how bioinformatics and an interdisciplinary scientific approach can address these challenges. This book presents the major industrial problems caused by microbes as well as the beneficial activities. It provides microbial ecologists, molecular biologists, operators, engineers, chemists and academics involved in the sector an improved understanding of the significance of microcosmos in oil and gas detection, production, and degradation, leading to better solutions impacting on operations and profitability"– Provided by publisher.
Identifiers: LCCN 2021004514 | ISBN 9780367900939 (hbk) |
ISBN 9781032039749 (pbk) | ISBN 9781003023395 (ebk)
Subjects: LCSH: Petroleum–Microbiology–Congresses. |
Molecular microbiology–Congresses.
Classification: LCC QR53.5.P48 I58 2019 | DDC 547/.83–dc23
LC record available at https://lccn.loc.gov/2021004514

ISBN: 978-0-367-90093-9 (hbk)
ISBN: 978-1-032-03974-9 (pbk)
ISBN: 978-1-003-02339-5 (ebk)

Typeset in Times
by SPi Global, India

Contents

Foreword vii
Preface ix
Editors xi
Contributors xiii

Chapter 1 Bioinformatics and Genomics Breakthroughs That Enabled the Microbiome Revolution 1

David A. Selinger

Chapter 2 Unravelling the Oil and Gas Microbiome Using Metagenomics 15

Nicolas Tsesmetzis

Chapter 3 A Case for Molecular Biology: How This Information Can Help in Optimizing Petroleum Top Side Facility Operations 41

Geert M. van der Kraan and Nora Eibergen

Chapter 4 Using a Holistic Monitoring Approach to Set Effective Control Strategies: How Data Can Lead to Information and Subsequently to "Wisdom" 69

Heike Hoffmann

Chapter 5 Influence of Chemical Treatments and Topside Processes on the Dominant Microbial Communities at Conventional Oilfields 81

Matheus Paschoalino

Chapter 6 Microbial Control during Hydraulic Fracking Operations: Challenges, Options, and Outcomes 99

Renato De Paula, Irwan Yunus, and Conor Pierce

Chapter 7 The Application of Bioinformatics as a Diagnostic Tool for Microbiologically Influenced Corrosion 119

Nora Eibergen, Geert M. van der Kraan, and Kenneth Wunch

Chapter 8 Methanogens and MIC: Leveraging Bioinformatics to Expose an Underappreciated Corrosive Threat to the Oil and Gas Industry 137

Timothy J. Tidwell and Zachary R. Broussard

Chapter 9 Molecular Methods for Assessing Microbial Corrosion and Souring Potential in Oilfield Operations 169

Gloria N. Okpala, Rita Eresia-Eke, and Lisa M. Gieg

Chapter 10 Microbial Reservoir Souring: Communities Relating to the Initiation, Propagation, and Remediation of Souring 207

Matthew Streets and Leanne Walker

Chapter 11 Quantitative PCR Approaches for Predicting Anaerobic Hydrocarbon Biodegradation 227

Courtney R. A. Toth, Gurpreet Kharey, and Lisa M. Gieg

Chapter 12 Leveraging Bioinformatics to Elucidate the Genetic and Physiological Adaptation of *Pseudomonas aeruginosa* to Hydrocarbon-Rich Jet Fuel 249

Thusitha S. Gunasekera

Index 273

Foreword

ISMOS^7 Takeaways

The Seventh International Symposium on Applied Microbiology and Molecular Biology in Oil Systems (ISMOS^7) held in Halifax, Nova Scotia, Canada from June 18 to 21, 2019 was both enjoyable and informative. The meeting offered 39 contributed talks and 69 posters. Thinking back of the meeting and looking out into an eerily-quiet COVID19-dominated world, it is clear that we have seen profound changes since the meeting. Profound changes that no one saw coming and which have wreaked havoc on the oil and gas industry with which our symposium was, is, and will be so intimately intertwined. Planes are not flying and there are few cars on the road. As a result, there is little demand for oil and prices have dropped to historic lows. Authors are to be forgiven if they are feeling a little blue while completing their promised chapters. Yet there is also light at the end of this tunnel, the length of which we do as yet not know. There always is. Also, most of us now have time on our hands, a commodity that was until recently in such short supply in our lives. So, this offers an incentive to get going on promises made toward completion of this book.

Following a workshop on "Science-based oilfield management – From the lab to the field" the symposium started out with a plenary from the Nova Scotia Department of Energy and Mines in which the stakes the province has in the topic of this meeting were clearly explained. Revenues from offshore oil and gas production have been important to the province. Offshore exploration costs are high and the province would like to contribute to de-risk new entries into their offshore. De-risking could include incorporating genotyping sub-seafloor bacteria as a new tool to identify subsurface hydrocarbon reserves. Quite a few of the presented papers dealt with this issue and indicated that the presence of hydrocarbon in cores is associated with a distinct microbial community composition and the presence of high concentrations of intact polar lipids. Thermophilic spore-forming bacteria, potentially associated with oil seeps, can also leave chemically detectable markers. This area will likely see more activity in the future.

As in previous ISMOS meetings, we obtained a lot of information on microbial community compositions in oilfield systems and on the application of molecular microbial methods (MMMs). I got a chuckle out of learning that this research is also pursued by the 3M Company. Separating the signal from the noise in microbial community composition data from oil fields is complicated. When we follow the flow lines of injected and produced water below and above the surface in our minds, we can envisage a large succession of environments with different physical and chemical properties like temperature, salinity, and available electron acceptors. Hence, in addition to their specific roles, it may be difficult to pinpoint identified taxa present in a sample of produced water to specific locations. Bioinformatics techniques like co-occurrence analysis may help in this regard. Of course, localization problems may not complicate the interpretation of microbial community data of more homogeneous environments such as storage tanks, hydrocarbon-containing seafloor cores, or laboratory cultures. Analysis of the occurrence of specific taxa by qPCR is now used in a wide-range of hydrocarbon resource environments to identify those potentially involved in microbially influenced corrosion (MIC), in degradation of biodiesel, in hydraulic fracturing and so on.

MIC was a major topic at ISMOS7. Electrical MIC (EMIC) organisms, which are thought to be capable of directly accepting cathodic electrons from iron and channeling these into their energy metabolism, continued to be of interest. Genes, encoding proteins specific for this kind of metabolism like conductive nanowires, have not yet been definitively identified. Indeed, a well-understood theoretical framework for this kind of metabolism is still lacking. Case studies of MIC in offshore pipelines and water handling systems as well as novel methods for its electrochemical detection and novel agents for its inhibition were also presented.

Further study of hydrocarbon degradation included the mapping of composition and catalytic potential of microbial communities in marine sediments in the Barents Sea and along the coasts of Canada. Remediation of oil spills is known to be accelerated by the application of chemical surfactants. These create small emulsion droplets, which remain suspended in the water column increasing bioavailability in regions away from the upper oil-water interface. In a previous ISMOS meeting, we saw a video of hydrocarbon-degrading bacteria attached to and growing at the oil-water interface, creating increasingly smaller oil droplets as growth continued and more surface area was needed. It is an unresolved question whether cells producing surfactants or cells being surfactants are concomitant, alternative, or competing strategies for the biodegradation of spills or for the microbial metabolism of oil in water-injected reservoirs. Hydrocarbon toxicity, as experienced by cells growing in tanks with low molecular weight hydrocarbons such as jet fuel, was shown to have a profound effect on their gene expression resembling that of heat shock response.

In microbially enhanced energy recovery, the water-mediated anaerobic metabolism of hydrocarbons into methane and CO_2 offers a route to produce energy from stranded hydrocarbons in water-injected reservoirs. Interestingly, yeast extract (essentially dead biomass) proved to be a superior nutrient amendment to stimulate methane formation. Bioaugmentation of reservoirs lacking the requisite microbial activities by injecting produced waters from active reservoirs was being contemplated. Bioaugmentation was also a viable strategy for degradation of BTEX benzene in environments where natural biodegradation is often slow.

Many other interesting topics were presented and discussed at ISMOS7. We will necessarily differ in the topics that caught our attention and in the take-home messages that we derived from these. However, we can all agree that we enjoyed a great meeting in a beautiful environment and that we can and will learn more from future ISMOS meetings.

Gerrit Voordouw
University of Calgary, Canada

Preface

This new book overviews novel applications of the rapidly evolving field of bioinformatics to the conservative and traditional world of oil and gas production. Poignantly, this work was created as the COVID-19 pandemic impacted the world and ravaged the oil and gas industry to where on April 20, 2020, the price for a barrel of West Texas Intermediate (WTI) crude was an unprecedented negative $37. However, this impact paled in comparison to the loss of millions of lives and exponentially more livelihoods. This book is a testament to the leading scientists, academics and experts from the oil and gas industry that were able to navigate the hardships of the pandemic to deliver 12 peer-reviewed chapters with contributors sharing their field experiences and research findings.

The Dutch botanist and microbiologist Lourens Baas Becking developed the Baas Becking hypothesis that "*Everything is everywhere, but the environment selects*". The oil and gas industry inattentively believed for decades that microorganisms could not survive in most petroleum reservoirs, pipelines, equipment, and processes as these environments, due to extreme temperatures, pressures, and other physio-chemical parameters, were not conducive for life. However, as ambient water is introduced into oil and gas processes for activities such as hydrotesting, secondary recovery, and hydraulic fracturing, so are microbes. For example, seawater is commonly injected by offshore platforms into the petroleum reservoir to maintain pressure for hydrocarbon recovery. It is not uncommon for an offshore platform to transport 10 million liters of seawater a day through its topside processes and subsea injection system into the reservoir. For perspective, a single liter of seawater contains about one billion bacteria, representing thousands of different species, resulting in 10,000,000,000,000,000 (10^{16} or 10 quadrillion) microbial cells being injected per day! Clearly, *everything is everywhere* in these operations, but the key question is what will survive and what are the consequences of its survival.

The oil and gas industry has been plagued by corrosion, souring (production of H_2S), and fouling since its inception but, until recently, rarely have these been attributed to microbial growth. Microbially Influenced Corrosion (MIC) occurs as biofilms are established on metallurgical surfaces resulting in metabolically driven corrosion mechanisms that can lead to failures in equipment and pipelines. Reservoir or systematic souring ensues when sulfate- or thiosulfate-reducing microbes "breathe in" sulfate or thiosulfate and release the highly toxic and corrosive molecule H_2S. Finally, biofouling occurs when microbial biofilms form on various equipment or geological surfaces impeding the flow of water and hydrocarbons, impacting extraction and separation and exacerbating MIC and souring. All of these microbiological activities cost the oil and gas industry billions of dollars annually along with threating the health and safety of workers and the inadvertent release of hydrocarbons into the environment.

For many years, any knowledge of microorganisms in oil and gas systems and reservoirs was largely based on culture-based methods. However, the recent convergence of computational biology and DNA sequencing has led to significant advances in bioinformatics allowing researchers to correlate taxonomy and genetic functionality to the community dynamics of oilfield microbial populations. Development and increased utilization of an interdisciplinary scientific approach allowed improved understanding of relations between ever-changing microbial communities and oilfield environments revealing some mysteries of oil and gas microbiome-related challenges. The synthesis of these data has provided a deeper understanding of key problems and opportunities facing the oil and gas industry.

This book strives to elaborate how the technological advances in bioinformatics over the last few years are being utilized by the oil industry to address the key issues faced by the sector. Specifically, The *Oil and Gas Biome* section of this book overviews the evolution of bioinformatics and its eventual application into the oil and gas industry. *Oilfield Process Management* discusses how bioinformatic data is collected, interpreted, and actioned in the field where *Microbial Control Strategies* focuses on the chemical treatments used to control microbial contamination in conventional and hydraulic fracturing operations. The subsequent two sections, *Microbiologically Influenced Corrosion (MIC)* and *Reservoir Souring*, outline the major industrial problems caused by microbes along with the application of bioinformatic techniques to diagnose and assess MIC and souring communities. Finally, *Emerging Innovation and Applications in Oilfield Microbiology* takes the reader out of the conventional upstream operation and into the arenas of anaerobic hydrocarbon degradation and bacterial adaption to jet fuel.

The editors wish to thank all of the authors and reviewers who contributed their time, knowledge, and expertise to this book, making it an invaluable resource for many years to come. The editors hope this book will stimulate further research, discussions, and developments in the field of oilfield bioinformatics and its importance to the oil and gas industry.

Kenneth Wunch
DuPont
United States

Marko Stipaničev
Schlumberger
Norway

Max Frenzel
Crown Technology
United Kingdom

Editors

Kenneth Wunch, PhD, holds the position of Energy Technology Advisor at DuPont in Houston responsible for business development, technology transfer, and shaping the innovation pipeline and strategy for global oil and gas applications. He earned a PhD in environmental microbiology at Tulane University followed by an Exxon-funded post-doc working on bioremediation of the *Valdez* spill in Prince William Sound. He then accepted a professorship at the Texas Research Institute for Bioenvironmental Studies and became Director of the SHSU Disease Vector Program for the Air Force Border Health and Environmental Threats Initiative. Dr. Wunch moved into petroleum microbiology at Baker Hughes with responsibilities in oilfield production chemistry and development and application of technologies associated with oilfield microbiology, microbially influenced corrosion (MIC), sulfide control, and corrosion inhibition. He later moved to BP as a production microbiologist with responsibilities in development and implementation of R&D strategies for reservoir souring and MIC before his current role at DuPont. Dr. Wunch is the author of more than 40 publications and patents and has chaired several committees, including SPE/NACE Deepwater Field Life Corrosion Prevention, Detection, Control and Remediation, International Symposium on Applied Microbiology and Molecular Biology in Oil Systems (ISMOS), NACE Control of Problematic Microorganisms in the Oil and Gas Industry, and the Energy Institute and Reservoir Microbiology Forum.

Marko Stipaničev, PhD, is Technical Expert at Schlumberger (Sandsli, Norway). He earned a master's degree in chemical engineering and technology at the University of Zagreb, Croatia, in 2009. In 2013, he earned a PhD in environmental process at the Laboratoire de Génie Chimique (LGC), Université de Toulouse, and Institut National Polytechnique de Toulouse (INPT), France. In 2010, he was employed at DNV (Det Norske Veritas) at the Department for Material Technology and Corrosion, where he was responsible for the development project targeting biocorrosion management, and in addition, he provided consultancy activities for the oil and gas industry. In a role of research engineer, Dr. Stipaničev was expanding DNV Bergen laboratory capacities to microbiology and advanced electrochemistry. During this period, he was a member of the BIOCOR ITN network. Thereafter, Dr. Stipaničev worked as Corrosion Specialist and Corrosion Discipline Lead at Schlumberger Production Technologies Scandinavia, where currently he is Technical Expert and Subject Matter Expert (SME) in the domain of oilfield corrosion, microbiology, and H_2S treatment. Dr. Stipaničev is a member of an International Symposium on Applied Microbiology and Molecular Biology in Oil Systems (ISMOS) TSC. He is an international scientific reviewer and the author of more than 20 technical and scientific papers and conference proceedings related to oilfield corrosion, microbiology, scale, and H_2S treatment.

Max Frenzel, PhD, is Senior Microbiologist at Crown Technology UK responsible for microbiology and analytical chemistry. In 2007, he earned a PhD in oilfield microbiology on persistence and bioremediation of unresolved complex mixtures (UCMs) of hydrocarbons in the environment at the University of Exeter and Plymouth University, UK. Between 2007 and 2008, he was employed as a postdoctoral researcher in microbiology at the University of Exeter. In 2008, he was employed as a researcher at the Department of Marine Environmental Technology at SINTEF Materials and Chemistry, Trondheim, Norway. In this role, he was focused on the fate and effects of oil spills in the environment, improving heavy oil recovery by MEOR, and the impact of pollution nanoparticles on the marine environment. Between 2010 and 2020, he was employed at Oil Plus Ltd where he was the focal point for consultancy projects in oilfield microbiology and production chemistry. During this time, he was subject matter expert in MIC, reservoir souring, and calcium naphthenate formation. Dr. Frenzel is the author of numerous technical papers and conference proceedings and a past member of the International Symposium on Applied Microbiology and Molecular Biology in Oil Systems (ISMOS) TSC.

Contributors

Zachary R. Broussard
Cemvita Factory Inc.
Houston, Texas, United States

Renato De Paula
ChampionX – Water Solutions RD&E and Marketing
Sugar Land, Texas, United States

Nora Eibergen
DuPont Microbial Control
Wilmington, Delaware, United States

Rita Eresia-Eke
Department of Biological Sciences
University of Calgary
Calgary, Alberta, Canada

Lisa M. Gieg
Department of Biological Sciences
University of Calgary
Calgary, Alberta, Canada

Thusitha S. Gunasekera
Fuels and Energy Branch
Aerospace Systems Directorate
Air Force Research Laboratory
Wright-Patterson Air Force Base, Ohio, United States

Heike Hoffmann
Intertek
Aberdeen, United Kingdom

Gurpreet Kharey
McGill University
Montreal, Quebec, Canada

Gloria N. Okpala
Department of Biological Sciences
University of Calgary
Calgary, Alberta, Canada

Matheus Paschoalino
DuPont Microbial Control
Wilmington, Delaware, United States

Conor Pierce
ChampionX – Water Solutions RD&E and Marketing
Sugar Land, Texas, United States

David A. Selinger
DuPont Industrial Biosciences
Wilmington, Delaware, United States

Matthew Streets
Scientific Division
Rawwater Engineering Company Limited
Cheshire, United Kingdom

Timothy J. Tidwell
ChampionX – Water Solutions RD&E and Marketing
Sugar Land, Texas, United States

Courtney R. A. Toth
University of Toronto
Toronto, Ontario, Canada

Nicolas Tsesmetzis
New Energies Research and Technology Group
Shell International Exploration and Production Inc.
Houston, Texas, USA

Geert M. van der Kraan
DuPont Microbial Control
Leiden, Netherlands

Leanne Walker
Scientific Division
Rawwater Engineering Company Limited
Cheshire, United Kingdom

Kenneth Wunch
DuPont Microbial Control
Houston, Texas, United States

Irwan Yunus
ChampionX – Water Solutions RD&E and Marketing
Sugar Land, Texas, United States

1 Bioinformatics and Genomics Breakthroughs That Enabled the Microbiome Revolution

David A. Selinger

CONTENTS

1.1 Early DNA Sequencing 2
1.2 Automated DNA Sequencing 3
1.3 Next-Generation DNA Sequencing 3
1.4 Microbiome Sequencing 5
1.5 Data Analysis 7
References 12

The notion that advances in DNA sequencing technology and bioinformatics underlie the advances in the elucidation of microbial ecology is hardly surprising. What maybe more interesting is what that journey can tell us about the limits of today's tools and where tomorrow's advancements may take us. For those of us who entered the field of Molecular Biology in the last three decades, it may be hard to appreciate how difficult it was to obtain a single DNA sequence let alone the millions or billions of sequences we routinely generate in a single run today. Prior to 1977, only a few DNA sequences had been determined and those were relatively short.

Advances in both DNA sequencing and the Bioinformatics tools needed to make sense of these sequences has equipped microbial ecologists of today with capabilities that seemed impossible to most scientists in the 1980s. Perhaps the most commonly used and powerful capability is to rapidly and inexpensively determine the microbes present in an environmental sample. This capability has powered the explosion of interest and knowledge in microbiomes. The capability to identify organisms based only on sequence data without having to culture them greatly extends the ability of Microbiologists to study microbes that live in environments that are difficult to replicate in a lab. Lastly, the availability of low-cost, whole genome sequencing for bacteria and fungi has revolutionized microbial taxonomies and changed how we think about the definition of microbial species. In this review, I will explore the development of DNA sequencing technologies and discuss the corresponding data analytical methods that have enabled us to make use of sequence data.

1.1 EARLY DNA SEQUENCING

Practical DNA sequencing began in 1977 with the publication of "A new method for sequencing DNA" by Maxam and Gilbert followed a few months later by Sanger, Nicklin and Coulson's paper "DNA sequencing with chain-terminating inhibitors" (Maxam and Gilbert 1977; Sanger, Nicklen and Coulson 1977b). Initially, Maxam and Gilbert's method was more popular owing to availability of reagents. Within a few years, the chain terminating method of Sanger et al became the main method because it was easier to perform and avoided toxic chemicals. While a small number of DNA sequences had been determined by previous methods, these two methods ushered in a step change in the acquisition of DNA sequences.

The first evidence of this step change occurred earlier in 1977, when Sanger and colleagues published the first whole genome sequence, that of the bacteriophage phiX174. They had sequenced the entire 5,375 bp genome using a precursor to the dideoxy method published later in 1977 (Sanger et al. 1977a). While tiny by today's standards, the sequence of this genome was far beyond what earlier DNA sequencing efforts had produced and foreshadowed whole genome sequencing of first bacteria in 1997 and then the first eukaryote in 1999. Along with these new sequencing methods and consequently new sequences, came the need for databases to hold them. Protein sequence databases existed starting in the 1960s, but DNA databases were not formally organized until the early 1980s with GenBank and EMBL databases emerging in the early 1980s (reviewed in Smith 1990).

The original Sanger methodology used radioactively labeled nucleotides to visualize the different sized fragments via gel electrophoresis. X-ray films and strips of tape applied along the side of each column of four lanes, one for each dideoxy-terminator, and a ball point pen were the means of reading the sequence. This method sufficed until the development of automated sequencing machines using fluorescent dye labeled dideoxy-terminators first reported in 1986 (Smith et al. 1986). With the introduction of a new generation of ABI machines in 1996 that replaced slab gels, which needed to be manually loaded, with capillary electrophoresis gels that could be automatically replaced and loaded, these second-generation machines achieved much greater output. These fully automated Sanger sequencing machines opened the capability to sequence larger genomes and various sequencing factories appeared to do that. This was the technology used to sequence the human genome, along with a growing list of microbial genomes. A good detailed review of these early days of automated DNA sequencing is (Hutchison 2007).

Even before the advancements in DNA sequencing technology that Maxam, Gilbert, and Sanger introduced, nucleic acid sequences were revolutionizing microbiology. Coincidentally in 1977, Carl Woese and George Fox proposed that prokaryotes should more properly be separated into two kingdoms, eubacteria and archaebacteria (later simply archaea) based on grouping 16S and 18S ribosomal RNA sequences by nucleotide similarity (Woese and Fox 1977). In looking back at this article, it is amazing how little data they had available to draw this conclusion compared to the massive amounts of sequence data we look at today. In a sense, though, we have not moved too far beyond what they did as 16S rRNA continues to be a highly useful molecule for evaluating the microbial diversity of samples.

1.2 AUTOMATED DNA SEQUENCING

As the more advanced and automated Sanger sequencing methods came into general use, what we now refer to as microbiome profiling became practical. The use of DNA sequencing technology to profile environmental samples was first reported in 1990 by Bateson and colleagues. They selectively cloned 16S rRNA genes from a hotspring microbial mat community in Yellowstone National Park and reported the identification of novel sequences that did not match microbes that had been isolated from this mat community (Ward, Weller and Bateson 1990). In 1991, two groups reported similar surveys of marine environmental samples. Britschgi and Giovannoni used PCR to amplify conserved regions of 16S rRNA from DNA collected from the Sargasso Sea and reported the characterization of 51 clones (Britschgi and Giovannoni 1991). Pace and colleagues reported the sequence of 38 clones of 16S rRNA derived from DNA extracted from a Pacific Ocean environmental sample (Schmidt, DeLong and Pace 1991).

With the rapidly decreasing cost of Sanger sequencing, Banfield and colleagues were able to completely sequence multiple microbes from an extreme environment (Tyson et al. 2004). By sequencing over 100,000 clones and generating 76 Mbp of sequence, they ushered in the field of metagenomics. Beyond the technology tour de force, they demonstrated the practicality of using sequencing alone to characterize a difficult to culture microbial assemblage. However, we should note that the system they chose was quite simple and dominated by a single taxon. Metagenomic sequencing of more complex samples required further advances in technology and even today it is not trivial to assemble whole genomes from metagenomic sequencing. While not trivial, it is often feasible to recover whole genome sequences from reasonably abundant organisms using today's technologies. These approaches will be further discussed in the Data Analysis section that follows.

1.3 NEXT-GENERATION DNA SEQUENCING

In late 2005, the beginning of the Next-Generation Sequencing (NGS) era dawned with the release of the first 454 Pyrosequencing Instrument. The initial instrument was capable of around 1 million reads of around 100 bp in length. Improved versions were eventually capable of 3–4 million reads of 300–400 basepairs in length. Approximately one year later, in 2006, the Solexa Genome Analyzer was launched. Solexa's machine offered substantially more reads that the 454 technology, but read lengths were substantially shorter, initially around 30 basepairs. The Solexa technology was the forerunner of today's Illumina NGS technology. Figure 1.1 illustrates the reduction in sequencing costs from 2001 to 2019. As these two technologies gained wider usage and improved, the curve in Figure 1.1 of cost per megabase of sequence takes a dramatic dive toward the end of 2007. The 454 and Solexa technologies were the two most successful of several similar short-read technologies to reach the

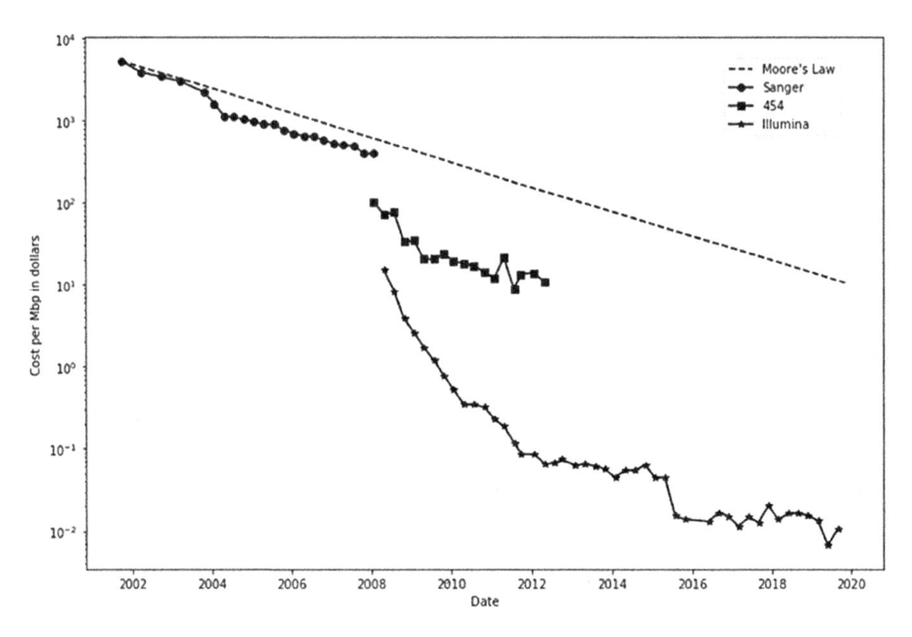

FIGURE 1.1 Reduction in DNA sequencing costs as compiled by NIH. Data replotted from NCBI (Wetterstrand 2020). The three sets of points track the cost of a million basepairs of DNA sequence by three different sequence technologies. These are compared to the dashed line of Moore's law, which was coined from the observation by Gordon Moore of Intel that the density of transistors in integrated circuits doubled approximately every two years. As transistor density is a proxy for computational power, Moore's law provides a useful reference for the increase in sequence data requiring this power. As noted in the text, the Sanger dideoxy sequencing method dates back to 1977, but the data here only goes back to 2001 as large-scale sequencing centers scaled up to sequence the human genome, which was completed in 2003.

market in the 2005–2008 timeframe. A good review of these technologies is found in (Metzker 2010). In today's marketplace, Illumina's technology has cleared the field of other short-read competitors and Illumina has continued to release improved versions, mostly in cost per gigabase (Gbp), which was reported to be as low as $7.00/Gbp with their NovaSeq machine and latest chemistry. In contrast, the 454 technology has succumbed to competition, first from the increasing read length of Illumina reads and finally from newer long-read technologies.

While the 454 technology has faded from the field, it was the Next-gen sequencing technology that opened the door to high-throughput microbiome studies. The 454 technology was almost immediately applied to microbial ecology. In 2006, the first paper describing the use of 454 sequencing, that of a mammoth sample using the 454 sequencing technology, reported the identification of microbial sequences from environmental contaminants of the sample (Poinar et al. 2006). It was shortly followed by the first intentional study of a microbiome using Next-generation sequencing (Edwards et al. 2006). This study compared two samples from an iron mine using a shotgun metagenomics approach. Like the microbial community studied by Banfield and colleagues, these communities were the result of manmade disruption and had low complexity. These two early studies highlighted the potential of Next-Generation Sequencing technologies to truly revolutionize microbial ecology.

In 2009, the first long-read NGS technology was reported by scientists at Pacific Biosciences (Eid et al. 2009) and released to market in 2011. This technology is fundamentally different from the short-read technologies and for genome sequences effectively complements the short-read data. More recently, in 2014, a competing nanopore-based technology from Oxford Nanotechnology (ONT) has produced similar long-reads (reviewed in van Dijk, Jaszczyszyn, Naquin and Thermes 2018). While the two technologies have similar inherent accuracies (85–90% depending on the sequence), the Pacific Biosciences (PacBio) platform can achieve much higher accuracy on shorter (<10 kbp) sequences because the same sequence can be read several times from the same molecule in a single run, resulting in base calling accuracies of 99%. The predominate errors in both platforms are insertion or deletion errors at homopolymer stretches caused by the difficulty of defining the length of the stretch from the length of base call signal in real time. Despite the accuracy advantage of PacBio on shorter sequences, in our experience, the two platforms are comparable for whole genome sequencing as the errors can be corrected by multiple reads and the longer length of ONT sequences has advantages in bridging repetitive regions. From a microbiome perspective, the long-read technologies offer the promise of much more precise strain level identification. To date, these technologies have been limited in microbiome applications because they do not produce reliable quantitative data on microbial taxa abundance, require higher quality DNA, and are more expensive than short-read technologies. A recent comparison of the PacBio and ONT platforms indicated utility of PacBio for longer amplicon sequencing (i.e. entire 16S rRNA, long fungal ITS amplicons) using agricultural samples (Loit et al. 2019). However, the authors indicate that the longer PacBio amplicons probably do not have enough advantage over Illumina short read amplicons to induce a shift in technology platforms. They also found that the ONT technology was not suitable for amplicon sequencing due to its higher error rate on short sequences. They did find the ONT platform to have potential for rapid identification of agricultural pathogens. In a clinical setting, ONT has been used for rapid identification of antibiotic resistance in patient isolates (Břinda et al. 2020). The field of long-read NGS is continuously advancing and there are many interesting microbial applications where the two technologies are being tried.

1.4 MICROBIOME SEQUENCING

Today, there are two main approaches to obtain sequence information from a microbiome sample: amplicon sequencing and shotgun sequencing. Amplicon sequencing relies on conserved primers that amplify a variable region in between. The variable region provides the sequence differences to resolve different organisms and the conserved primer locations allow the amplicon to sample a diverse group of organisms. For prokaryotes (Eubacteria and Archaea), the gene of choice is the 16S rRNA gene that was originally used by Woese and Fox to document the existence of the archaea. For fungi, the Internal Transcribed Spacer (ITS) regions from the eukaryotic rRNA gene complex are commonly used. There are two ITS regions in the eukaryotic

18S-5.8S-25/28S gene, one on either side of the 5.8S rRNA coding region, and typically the first of these is used for amplicon sequencing. Amplicon sequencing is efficient as each sequence approximately corresponds to an individual cell and thus 10,000 sequences can give an accurate measure of the microbiome of a sample. It does suffer from two drawbacks. The first is that the "universal" primers are not truly universal, no matter which primer sets are used, there are a small percentage of taxa that have variation in the conserved regions used for primers and thus are missed in amplicon sequencing studies. Second, the resolution of amplicons is limited. A substantial number of species share amplicon sequences; thus, they cannot be resolved by amplicon sequencing. A third, more minor issue is that the copy number of 16S rRNA genes in different species and strains varies, so quantitation is not exact. However, methods have been developed to correct for copy number variation (Kembel, Wu, Eisen and Green 2012).

The alternative or in many cases, complementary, approach is shotgun sequencing, either of genomic DNA (metagenomics) or mRNA (metatranscriptomics). Shotgun sequencing has the theoretical advantage of capturing everything and can resolve taxa to species and even to strains. It is not efficient, requiring millions of sequences per sample, but as sequencing costs have fallen, so have costs per sample. Metagenomics or metatranscriptomics data are less simple to process, but several packages have been released to convert these data into microbial profiles. These data also hold information about the metabolic pathways that were either encoded by the microbes (metagenomics) or expressed by the microbes (metatranscriptomics). The ability to directly measure expressed pathways has been cited as an advantage for metatranscriptomics over metagenomics. However, since the most abundant microbes dominate the signals in both and the most highly expressed pathways dominate for metatranscriptomics, it is not clear that the theoretical advantage translates into a practical one unless one sequences more deeply than in most studies (reviewed in Migun, Lo and Chain 2019). In practice, metagenomic shotgun sequencing is often used on a subset of samples after amplicon sequencing. While this approach has advantages in cost, the two approaches can best be considered as having different biases as will be explored in more detail in the following section on data analysis.

A drawback for both methods is that contamination by host cells, in the case of samples from animals or plants, can consume a substantial amount of sequence space. Because an animal cell has 500–1,000 times as much DNA as a prokaryotic cell, even if the prokaryotes are more numerous, at the DNA level, the eukaryotic cells dominate and can make up more than 90% of the DNA in the sample. For example, stool samples from the Human Microbiome Project had less than 10% of human DNA, whereas other sample types contained more than 90% human-aligned reads (The Human Microbiome Project Consortium et al. 2012; Lloyd-Price et al. 2017). For this reason, shotgun sequencing is more effective on environmental samples and those animal samples, like feces, that have few host cells. Deeper sequencing can partially overcome host cell contamination along with use of software tools to filter the data (Pereira-Marques et al. 2019; McArdle and Kaforou. 2020). Fortunately for Oil and Gas samples, higher eukaryotic cell contamination should be minimal.

1.5 DATA ANALYSIS

Alongside the advances in sequencing technology came the need for computational tools to make sense of the data. In 1977, a table showing relatedness metrics for 13 organisms was sufficient to make the case for a fundamental restructuring of microbial phylogeny. This small table represented at least four years of data gathering and included data that was part of seven previous publications. Suffice it to say that analysis needed for this data set could be managed with a scientific calculator. As hundreds of sequences became available, computer algorithms were needed. Each advancement in sequencing technology has been closely followed by a wave of algorithm development. For example, the BLAST and FASTA algorithms, which are heuristics on the general dynamic programming algorithm, were developed to deal with the increasing size of Genbank and other sequence databases (Pearson and Lipman 1988; Altschul et al. 1990). As short-read sequencing technologies entered the marketplace, new algorithms were required to align and assemble the shorter, less information-rich sequences. The current aligners and assemblers are based on different families of string-matching algorithms than BLAST and FASTA were. An example is the popular BWA tool based on the Burroughs-Wheeler Transform algorithm (Li and Durbin 2009).

In the microbiome space, the most common approach to data analysis starts with aligning sequences to reference sequences to identify and quantify the microbes present. Alignment is straightforward, and a number of good tools have emerged to translate metagenomic data into lists of taxa with counts of occurrence in the sample. The list of taxa and abundances within a single sample is often referred to as alpha diversity, and the comparison between samples is referred to as beta diversity after the ecology terms originally proposed by R. H. Whittaker (Whittaker 1972). In the case of 16S Amplicon data, an early approach, which is still used today by several packages, was to assign reads to Operational Taxonomic Units (OTUs). The concept behind OTUs dates back to early microbiome studies the 1990s using Sanger sequencing and involves comparing the reads to a refence database of 16S rRNA sequences (Bond, Hugenholtz, Keller and Blackall 1995). Clustering by OTU became popular with the introduction of software tools to cluster the larger numbers of sequences generated by NGS platforms (Schloss and Handelsman 2005). The concept is that sequences with less than 3% nucleotide difference most likely represent the same or closely related species. Conveniently, this criterion encompasses the typical error rate of single pass Sanger reads and 454 reads. In practice, sequences clustered using OTUs are resolved to the level of genera and not species. Several popular packages including QIIME and MOTHUR use OTU clustering (Schloss et al. 2009; Caporaso et al. 2010). In addition to QIIME and MOTHUR, there are quite a number of other tools to cluster sequences and generate OTUs. All of them rely on curated ribosomal RNA sequence databases such as Silva and RDP (Quast et al. 2013; Cole et al. 2014), both of which were recently updated as of the writing of this article.

More recently, an alternative clustering concept, Amplicon Sequence Variant (ASV) has been introduced to take advantage of shorter, but highly accurate Illumina

Amplicon sequences (Callahan, McMurdie, Rosen, Han, Johnson and Holmes 2016). ASV has the advantage of resolving sequences that differ by a single nucleotide. While ASVs, in many cases represent species, depending on the amplicon used, around 20%–30% of prokaryotic species share the same 16S amplicon sequence. The newer version of QIIME, QIIME 2 can use ASVs computed by DADA2 and other packages (Bolyen et al. 2019). A recent comparison of multiple packages for 16S amplicon data indicated that different packages return different measures of alpha and beta diversity from the same input samples, highlighting the need to analyze a set of samples to be compared with the same platform (Marizzoni et al. 2020).

Alignment of shotgun metagenomic reads to get alpha and beta diversity is more complex than aligning 16S amplicon reads. Sequence reads from a genome come in two basic types, those that are unique to a species or strain and those that are conserved across a wider range of taxa. Unique short reads to a species are less common than those that are shared by related species and for strains the unique reads are even less common. Additionally, there is not a neat nesting of related reads. For example, one species may share reads with different related species or strains and even relatively unrelated ones due to horizontal gene transfer. Furthermore, genome sizes vary across prokaryotes, so the quantitation of taxa needs to take that into account. Because Shotgun metagenomics did not become practical until the higher throughput Illumina sequencing machines pushed the cost per Mbp into the $1 range, around 2011, there remains active development of algorithms and tools in this space. MetaPhlAn is one of the commonly used tools for shotgun metagenomics data and relies on a curated database of reference sequences where the sequence regions in common between higher level taxonomic groups are masked. (Segata, Waldron, Ballarini, Narasimhan, Jousson and Huttenhower 2012). Focusing on the unique sequences enables high resolution in shotgun metagenomics studies but greatly reduces the amount of information actually used to determine alpha and beta diversity. Consequently, taxa with similar reported abundance may be quantified using a different number of raw sequences. Because the variance of quantification depends on the number of sequences counted, the statistical significance values may not be the same for taxa with seemingly similar levels in samples. Alternative methods that do not rely on unique marker sequence databases are available. A good review of tools and methods of analysis for metagenomic sequences is found in Breitwieser, Lu and Salzberg (2019). In addition to the variety of open source tools, there are public servers with integrated tools, like MG-RAST (Meyer et al. 2008; Meyer et al. 2019), and proprietary software packages like Geneious and CLC Genomics. A comprehensive review of metagenomic sequencing and data analysis methods can be found in Pérez-Cobas, Gomez-Valero and Buchrieser (2020).

While shotgun metagenomic sequencing has been cited as more accurate than 16S amplicon sequencing, a direct comparison shows that the two methods have different biases. Figure 1.2 shows the application of 16S amplicon and shotgun Metagenomic sequencing of the same sample (unpublished data). One takeaway from this figure is that control experiments are valuable when setting up any large metagenomic experiment. A number of life science vendors offer targeted sequencing panels made up of known microbes and known amounts which can be useful in evaluating alternative steps in an experimental plan.

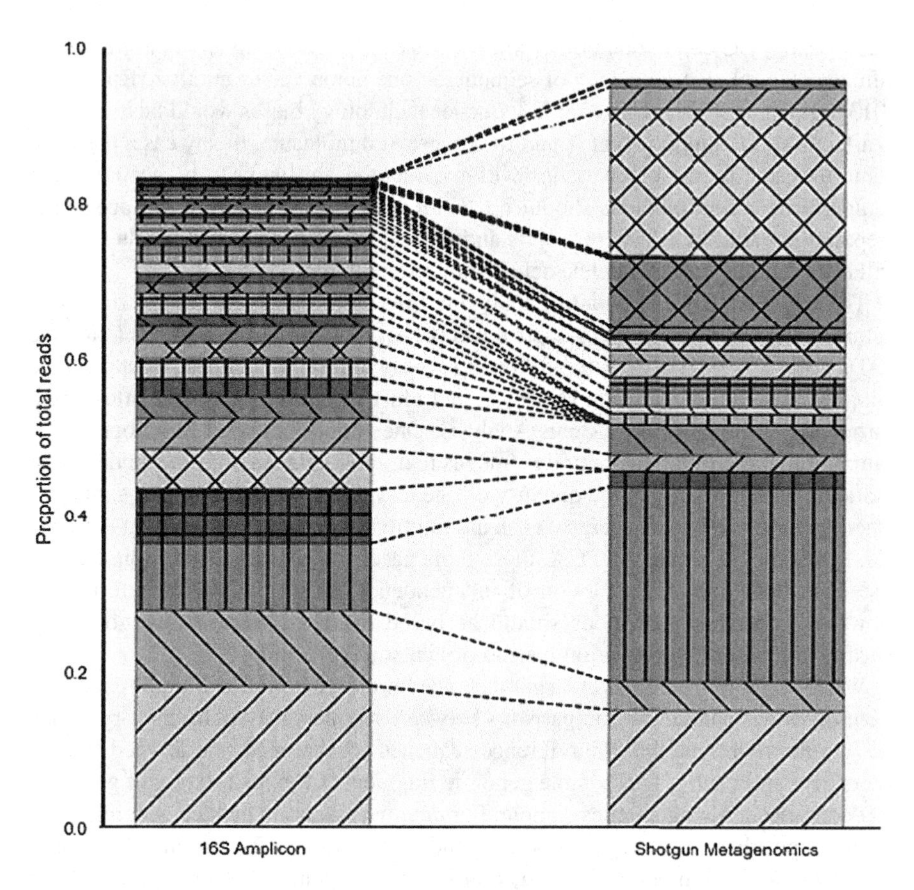

FIGURE 1.2 Comparison of Amplicon and Shotgun Metagenomics Sequencing. DNA from the same human fecal sample was used for 16S amplicon sequencing, generating approximately 60,000 sequences, and shotgun Metagenomics sequencing, generating approximately 40 Million sequences (approximately 100 Gbp of sequence; Unpublished data). Taxa assignments to genera were done using the RDP database of 16S sequences for the 16S data and MetaPhlAn for the shotgun sequences. Organisms in both bars are ordered by abundance in 16S Amplicon data and the dashed lines show the connections between corresponding taxa. Out of 49 genera detected, 6 genera were not detected by 16S Amplicon sequencing and 25 genera were not detected by shotgun Metagenomic sequencing.

Based on the preceding discussion of amplicon and metagenomic shotgun sequencing, it is not surprising that 16S amplicon sequencing might miss taxa due to mismatches in the conserved primer regions, or that shotgun metagenomic sequencing might miss low abundance taxa due to sampling depth issues. However, the comparison shows that quantitation varies substantially between the techniques and that there are abundant taxa in each set that are not detected in the other. It could be that differences in copy number of 16S rRNA are artificially inflating the quantitation of some taxa as rRNA operon copy number varies both between and within species; see the rrnDB for details on rRNA operon copy number variation (Stoddard et al. 2015).

Additionally, GC content and other known biases of sequencing technologies could influence the relative efficiency of sequencing one taxon versus another (Ross et al. 2013; Sato et al. 2019). These same sequence technology biases would be less likely to affect a short amplicon that is part of a conserved molecule. In any case, the take-home message is that comparisons within a method are likely to be accurate, but comparisons across methods should be avoided. The same advice goes for sample preparation methods as different lysis and nucleic acid extraction protocols and kits will differentially extract nucleic acids from different taxa.

The statistical considerations of analyzing Microbiome data are beyond the scope of this introduction and more in-depth coverage of this topic can be found in (Ramette 2007; Calle 2019). However, the following are some brief guidelines. With multiple samples, statistical dimension reduction and clustering approaches are often used, particularly Principal Components Analysis. One should note that the proportional nature of the microbial diversity data means that these data are not independent. For example, within a sample, the quantity of one taxon is dependent on the quantity of other taxa. Because independence is an assumption of parametric statistical methods like Pearson correlation and PCA, these tools need to be used with caution. PCA is less dependent on the assumption of independence, so it is relatively safe to use. However, comparison methods should be based on non-parametric methods, e.g. Spearman rank order correlation instead of Pearson correlation.

While alignment methods are a good choice for most Microbiome studies and lend themselves to quantitative comparisons between samples, they will miss organisms that are not represented in the reference sequence set that reads are aligned to. The alternative approach is to assemble genomic fragments from the mixture of genomes that comprise a Metagenomics sample. For *de novo* assembly of Metagenomic data, several algorithms have been introduced and tools released. The main challenge in this space is discriminating the reads that belong to a single organism from related species or strains. With Illumina reads, even the best algorithms will assemble multiple strains of a species into a single assembly, or more likely into multiple chimeric contigs. Assembly is also computationally much more complex, so the analysis takes longer and requires more computing horsepower, especially access to servers with large amounts of shared memory. These assemblies can be useful if looking at the collective metabolic capability of a microbiome but recovering complete genomes from these is difficult. Realistically, complete genomes can be recovered only for very abundant species in a sample. Long-read technologies offer advantages in this application as the long reads can be more reliably assembled and each read can capture a significant fraction of the genome. While data on recovery of complete genomes of bacteria as a function of abundance in the sample across different sequencing technologies are lacking, publications of viral genome recovery have been made. Using the MinIon technology from Oxford Nanotech and Illumina's technology, researchers reported obtaining complete genomes of the 11,826 bp chikungunya virus using the MinION from samples having as little as 4% chikungunya reads. For Illumina, 10x coverage of the viral genome dropped below 95% when viral reads dropped below 10%, suggesting a significant advantage for long-read sequencing in recovering complete or nearly complete genomes from metagenomic samples (Kafetzopoulou et al.

2018). For much larger bacterial genomes, 4% would be at best a lower limit for recovery of a complete genome sequence. As discussed previously in the Next-Gen DNA Sequencing section, long-read technologies do require higher quality DNA and generally more DNA than the short-read Illumina technology and are thus not yet routine in the Microbiome space. Looking forward, these technologies will hopefully improve to the point of more routine use as long reads would provide the capability of reliably discriminating between different strains of a single species in a sample.

Returning to the question at the beginning of this chapter, "what does the past tell us about the future of genomics and Bioinformatics in this space?" Our look at the past for DNA sequencing indicates that each major technology introduction has led to a rapid improvement phase followed by refinement and increased adoption. These phases can be seen in Figure 1.1 for both the 454 and Illumina technologies. The short-read NGS technology space is now quite mature and the Illumina technology is both dominant in this space and probably close to optimum in terms of scale and cost. Algorithms for this kind of data are also mature and relatively little improvement can be expected. In the long-read technology space, we are probably near the end of the rapid improvement phase as many of the early hurdles have been surmounted for both the Pacific Biosciences and Oxford Nanopore technologies. There is still plenty of room for improvement in these technologies and application to new areas. As these technologies continue to improve in scale and cost, we can expect their use in microbiome studies to become more common and to enable new types of analyses. We can also expect that the sequencing technology space will continue to develop. Single cell sequencing is now practical for human cells, including single cell expression analysis. These techniques are not as easy to apply to microbes, but history suggests that new solutions will be found sooner rather than later.

From a societal perspective, the genomics and bioinformatics technologies described in this chapter have fundamentally changed many aspects of our lives. One example of this fundamental change is the use of DNA in criminal investigations. By using modern DNA sequencing and bioinformatics technologies, criminals have been captured based on comparing DNA sequences from crime scenes with DNA profiles of siblings uploaded to ancestry databases. The application of these technologies to microbial ecology has enabled the booming interest in probiotic consumption for health. To wit, there are now companies that offer personalized dietary advice based on metagenomic sequencing. Genomic and bioinformatic technologies have only reached into the Oil and Gas industry in the last few years. Further development of the long-read technologies may be especially relevant given the ability of the Oxford Nanotech to rapidly identify microbes in field settings. Further reductions in cost of the competing PacBio platform along with its advantage in sequence accuracy may transform amplicon sequencing from the short reads and limited taxonomic resolution of 16S rRNA-based microbiome profiles to a longer read, higher resolution data set with the potential to give a better diagnosis of what species and strains are actually present. If other areas are any indication, advances in metagenomic sequencing and analysis have the potential to transform aspects of the Oil and Gas industry from discovery and production through processing and infrastructure maintenance.

REFERENCES

Altschul, S. F., W. Gish, W. Miller, E. W. Myers and D. J. Lipman. 1990. Basic local alignment search tool. *Journal of Molecular Biology* 215(3):403–410.

Bolyen, E., J. R. Rideout, M. R. Dillon, et al. 2019. Reproducible, interactive, scalable and extensible microbiome data science using QIIME 2 [published correction appears in *Nature Biotechnology* 2019 37: 1091] *Nature Biotechnology* 37: 852–857.

Bond, P. L., P. Hugenholtz, J. Keller and L. L. Blackall. 1995. Bacterial community structures of phosphate-removing and non-phosphate-removing activated sludges from sequencing batch reactors. *Applied and Environmental Microbiology* 61: 1910–1916.

Breitwieser, F. P., J. Lu and S. L. Salzberg. 2019. A review of methods and databases for metagenomic classification and assembly. *Briefings in Bioinformatics* 20: 1125–1136.

Britschgi, T. B. and S. J. Giovannoni. 1991. Phylogenetic analysis of a natural marine bacterioplankton population by rRNA gene cloning and sequencing. *Applied and Environmental Microbiology* 57: 1707–1713.

Břinda, K., A. Callendrello, K. C. Ma, et al. 2020. Rapid inference of antibiotic resistance and susceptibility by genomic neighbour typing. *Nature Microbiology* 5: 455–464.

Callahan, B. J., P. J. McMurdie, M. J. Rosen, A. W. Han, A. J. Johnson and S. P. Holmes. 2016. DADA2: high-resolution sample inference from Illumina amplicon data. *Nature Methods* 13: 581–583.

Calle, M. L. 2019. Statistical analysis of metagenomics data. *Genomics & Informatics* 17: e6. doi:10.5808/GI.2019.17.1.e6

Caporaso, J. G., J. Kuczynski, J. Stombaugh, et al. 2010. QIIME allows analysis of high-throughput community sequencing data. *Nature Methods* 7: 335–336.

Cole, J. R., Q. Wang, J. A. Fish, et al. 2014. Ribosomal database project: data and tools for high throughput rRNA analysis. *Nucleic Acids Research* 42: D633–D642.

Edwards, R. A., B. Rodriguez-Brito, L. Wegley, et al. 2006. Using pyrosequencing to shed light on deep mine microbial ecology. *BMC Genomics* 7: 57. doi:10.1186/1471-2164-7-57.

Eid, J., A. Fehr, J. Gray, et al. 2009. Real-time DNA sequencing from single polymerase molecules. *Science* 323: 133–138.

Hutchison, C. A. 2007. DNA sequencing: bench to bedside and beyond. *Nucleic Acids Research* 35: 6227–6237.

Kafetzopoulou, L. E., K. Efthymiadis, K. Lewandowski et al. 2018. Assessment of metagenomic Nanopore and Illumina sequencing for recovering whole genome sequences of chikungunya and dengue viruses directly from clinical samples. *Euro Surveillance* 23(50):pii=1800228. doi:10.2807/1560-7917.ES.2018.23.50.1800228

Kembel, S. W., M. Wu, J. A. Eisen and J. L. Green. 2012. Incorporating 16S gene copy number information improves estimates of microbial diversity and abundance. *PLoS Computational Biology* 8(10): e1002743. doi:10.1371/journal.pcbi.1002743

Li, H. and R. Durbin (2009) Fast and accurate short read alignment with Burrows-Wheeler Transform. *Bioinformatics* 25:1754–1760.

Lloyd-Price, J., A. Mahurkar, G. Rahnavard, et al. 2017. Strains, functions and dynamics in the expanded Human Microbiome Project. *Nature* 550: 61–66.

Loit, K., K. Adamson, M. Bahram, et al. 2019. Relative performance of MinION (Oxford Nanopore Technologies) versus sequel (Pacific Biosciences) third-generation sequencing instruments in identification of agricultural and forest fungal pathogens. *Applied Environmental Microbiology* 85: e01368–e01319. doi:10.1128/AEM.01368-19

Marizzoni, M., T. Gurry, S. Provasi, et al. 2020. Comparison of bioinformatics pipelines and operating systems for the analyses of 16S rRNA gene amplicon sequences in human fecal samples. *Frontiers in Microbiology*, 11: 1262. doi:10.3389/fmicb.2020.01262

Maxam, A. M. and W. Gilbert. 1977. A new method for sequencing DNA. *Proceedings of the National Academy of Sciences USA* 74: 560–564.

McArdle, A. J. and M. Kaforou. 2020. Sensitivity of shotgun metagenomics to host DNA: abundance estimates depend on bioinformatic tools and contamination is the main issue. *Access Microbiology* 2: e000104. doi:10.1099/acmi.0.000104

Metzker, M. 2010. Sequencing technologies—the next generation. *Natural Review Genetics* 11: 31–46.

Meyer, F., D. Paarmann, M. D'Souza et al. 2008. The metagenomics RAST server—a public resource for the automatic phylogenetic and functional analysis of metagenomes. *BMC Bioinformatics* 9: 386.

Meyer, F., B. Saurabh, S. Chaterji et al. 2019. MG-RAST version 4—lessons learned from a decade of low-budget ultra-high-throughput metagenome analysis. *Briefings in Bioinformatics* 20: 1151–1159.

Migun, S., C.-C. Lo, P. S. G. Chain. 2019. Advances and challenges in metatranscriptomic analysis. *Frontiers in Genetics* 10: 904 https://www.frontiersin.org/article/10.3389/fgene.2019.00904

Pearson, W. R. and D. J. Lipman. 1988. Improved tools for biological sequence comparison. *Proceedings of the National Academy of Sciences of the United States of America* 85: 2444–2448.

Pereira-Marques, J., A. Hout, R. M. Ferreira, et al. 2019. Impact of host DNA and sequencing depth on the taxonomic resolution of whole metagenome sequencing for microbiome analysis. *Frontiers in Microbiology* 10: 1277 https://www.frontiersin.org/article/10.3389/fmicb.2019.01277

Pérez-Cobas, A. E., L. Gomez-Valero and C. Buchrieser. 2020. Metagenomic approaches in microbial ecology: an update on whole-genome and marker gene sequencing analyses. *Microbial Genomics* 6: mgen000409. doi:10.1099/mgen.0.000409

Poinar, H. N., C. Schwarz, J. Qi, et al. 2006. Metagenomics to paleogenomics: large-scale sequencing of mammoth DNA. *Science* 311: 392–394.

Quast, C., E. Pruesse, P. Yilmaz, et al. 2013. The SILVA ribosomal RNA gene database project: improved data processing and web-based tools. *Nucleic Acids Research* 41: D590–D596.

Ramette, A. 2007. Multivariate analyses in microbial ecology. *FEMS Microbiology Ecology* 62: 142–160.

Ross, M. G., C. Russ, M. Costello et al. 2013. Characterizing and measuring bias in sequence data. *Genome Biology* 14: R51. doi:10.1186/gb-2013-14-5-r51

Sanger, F., G. Air, B. Barrell et al. 1977a. Nucleotide sequence of bacteriophage φX174 DNA. *Nature* 265: 687–695.

Sanger, F., S. Nicklen and A. R. Coulson. 1977b. DNA sequencing with chain-terminating inhibitors. *Proceedings of the National Academy of Sciences USA* 74: 5463–5467.

Sato, M. P., Y. Ogura, K. Nakamura et al. 2019. Comparison of the sequencing bias of currently available library preparation kits for Illumina sequencing of bacterial genomes and metagenomes. *DNA Research* 26: 391–398.

Schloss, P. D. and J. Handelsman. 2005. Introducing DOTUR, a computer program for defining operational taxonomic units and estimating species richness. *Applied and Environmental Microbiology* 71: 1501–1506.

Schloss, P. D., S. L. Westcott, T. Ryabin, et al. 2009. Introducing mothur: open-source, platform-independent, community-supported software for describing and comparing microbial communities. *Applied and Environmental Microbiology* 75: 7537

Schmidt, T. M., E. F. DeLong and N. R. Pace. 1991. Analysis of a marine picoplankton community by 16S rRNA gene cloning and sequencing. *Journal of Bacteriology* 173: 4371–4378.

Segata, N., L. Waldron, A. Ballarini, V. Narasimhan, O. Jousson and C. Huttenhower. 2012. Metagenomic microbial community profiling using unique clade-specific marker genes. *Nature Methods* 9: 811–814.

Smith, L. M., J. Z. Sanders, R. J. Kaiser et al. 1986. Fluorescence detection in automated DNA sequence analysis. *Nature* 321: 674–679.

Smith, T. F. 1990. The history of the genetic sequence databases. *Genomics* 6:701–707.
Stoddard, S. F., B. J. Smith, R. Hein, B. R. K. Roller and T. M. Schmidt. 2015. rrnDB: improved tools for interpreting rRNA gene abundance in bacteria and archaea and a new foundation for future development. *Nucleic Acids Research* 43: D593–D598.
The Human Microbiome Project Consortium., C. Huttenhower, D. Gevers, et al. 2012. Structure, function and diversity of the healthy human microbiome. *Nature* 486: 207–214.
Tyson, G. W., J. Chapman, P. Hugenholtz, et al. 2004. Community structure and metabolism through reconstruction of microbial genomes from the environment. *Nature* 428: 37–43.
van Dijk, E. L., Y. Jaszczyszyn, D. Naquin and C. Thermes. 2018. The third revolution in sequencing technology. *Trends in Genetics* 34: 666–681.
Ward, D. M., R. Weller and M. M. Bateson 1990. 16S rRNA sequences reveal numerous uncultured microorganisms in a natural community. *Nature* 345: 63–65.
Wetterstrand, K. A. 2020. DNA sequencing costs: data from the NHGRI Genome Sequencing Program (GSP) https://www.genome.gov/sequencingcostsdata (accessed March 25, 2020).
Whittaker, R.H. 1972. Evolution and measurement of species diversity. *Taxon* 21: 213–251.
Woese, C. R. and G. E. Fox. 1977. Phylogenetic structure of the prokaryotic domain: the primary kingdoms. *Proceedings of the National Academy of Sciences USA* 74: 5088–5090.

2 Unravelling the Oil and Gas Microbiome Using Metagenomics

Nicolas Tsesmetzis

CONTENTS

2.1 Introduction 15
2.1.1 Sampling and Lab Processing Considerations 17
2.2 Downstream Processing Considerations 18
2.3 Single Marker Gene Sequencing 19
2.4 Single Marker Gene Analysis 20
2.5 Whole Metagenome Sequencing 22
2.6 Metagenomic Data Analysis 22
2.7 Metagenomic Applications to Oil and Gas 26
2.7.1 Exploration 26
2.7.2 Production Allocation 27
2.7.3 Souring and Microbiologically Influenced Corrosion (MIC) Monitoring 27
2.7.4 Baseline Environmental Assessment and Monitoring 28
2.7.5 Biofouling and Contamination Control 28
2.7.6 Novel Biocatalysts for Enzymatic Hydrolysis of Biomass 28
2.7.7 Biofuel Production 28
2.8 Concluding Remarks 29
References 29

2.1 INTRODUCTION

Recent microbiological studies have revealed that microbial activity, abundance, and diversity in oil reservoirs and associated oil and gas installations is much more evident than originally thought. More importantly, it is now widely accepted that detrimental microbial processes like reservoir souring, biofouling, and microbiologically influenced corrosion (MIC) pose significant financial and environmental threats. Mitigating strategies include chemical treatment and microbial monitoring to assess their effectiveness. These new insights in microbial processes in conjunction with the recent technological advances in DNA sequencing and bioinformatics are revolutionizing the way we understand, monitor, and manage these processes and their effects.

Sequencing of the human genome is a contemporary example that demonstrates the impact of the new DNA sequencing technologies. The publicly funded Human Genome Project (Lander et al., 2001), which was solely based on the Sanger sequencing platform, cost US$2.7Bn and required 13 years for its completion (1990–2003) (Wetterstrand, 2019). In 2007, the diploid genome of James Watson was sequenced using the now obsolete 454™ sequencing platform in four months and a total cost of US$1.5M (Wheeler et al., 2008). In 2014, Illumina announced the first sub-US$1000 human genome sequenced in less than a day and, as of February 2020, service providers offer personalized genome sequencing services for US$299 (Nebula Genomics).

Lower DNA sequencing cost has opened new opportunities in genomic research and biological sciences. Among them, metagenomics (aka environmental genomics) is one of the biological fields which benefited the most from such developments. Metagenomics aims at identifying the members of microbial communities present in environmental samples and the elucidation of their biochemical capabilities. One of the biggest advantages of metagenomics over traditional methods, such as MPN (most probable number) counts, is that it is culture independent and completely bypasses the need for time-consuming culturing these microorganisms in the lab. The fact that more than 99% of the microorganisms from an environmental sample cannot be cultured in the lab (Schloss and Handelsman, 2005) puts metagenomic approaches in a very strong position as a robust, high throughput and high resolution, screening tool for environmental samples.

These advances led to an exponential increase of microbiome studies including those pertaining to hydrocarbon-rich environments as shown in Figure 2.1. At the same time, numerus bioinformatics tools are being developed for the handling and analysis of gigabases (Gb) of data produced from these Next Generation Sequencing (NGS) platforms. Moreover, this ever-increasing data accumulation necessitates appropriate cataloguing in a way that facilitates the comparison and better contextualization of the underlying investigations. To this extent, a robust experimental design and the capturing of contextual information about the samples and the environment of origin are crucial. Emerging standards and repositories for sample collection, preservation, and lab processing of hydrocarbon-related samples can greatly facilitate downstream analysis of the data, improve the power of statistical analysis and enable comparison across datasets from these environments (Tsesmetzis et al., 2017, Tsesmetzis et al., 2016; Marks et al., 2018; NACE TM0212, 2018; NACE TM0106, 2016; Mitraka, 2019; Manuel, 2020). In addition to standard operating procedures for sampling and lab processing, bioinformatics analysis can also influence the analysis outcome. This is mainly due to the inherent tradeoff between precision and performance of the underling algorithm as well as the theoretical or statistical framework that they are based on (for example, Naive Bayes versus Random Forests classification methods). Furthermore, bioinformatics algorithms are often trained and optimized using synthetic data or data that are unrelated to oil and gas microbiomes (e.g. human gut) and therefore might not perform optimally on those unrelated datasets. It is therefore important that metadata, such as software versions and run parameters, are reported in the public repositories along with the raw sequencing data. Toward that goal, data provenance tracking is

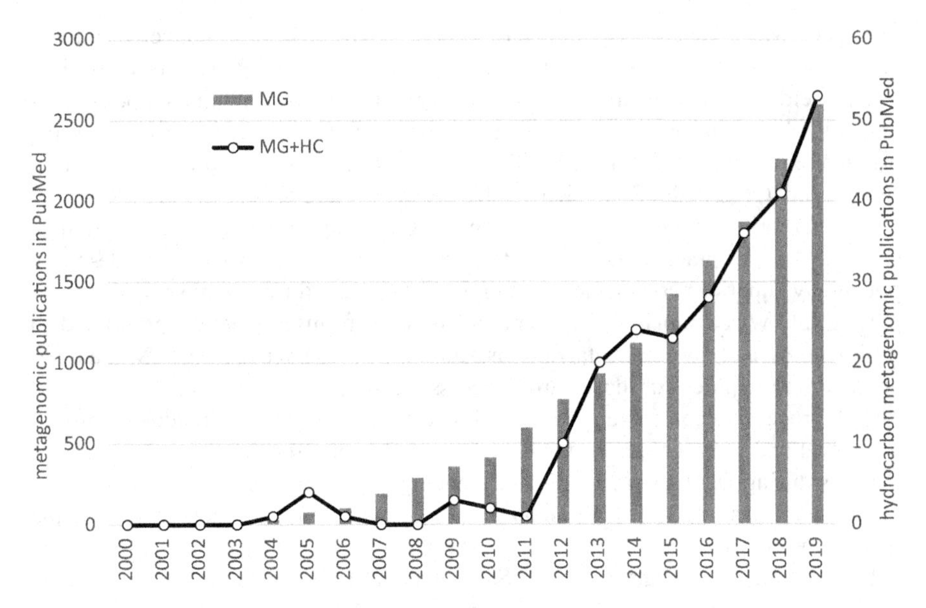

FIGURE 2.1 Growth of metagenomic (MG) and hydrocarbon-related metagenomic (MG+HC) publications in PubMed over the last 2 decades.

becoming increasingly popular and several widely used standalone and online tools such as QIIME 2 (Bolyen et al., 2019), Galaxy (Afgan et al., 2018), KBase (Arkin et al., 2018), Qiita (Gonzalez et al., 2018) have already integrated such functionalities which keeps track of all the previous processing steps and parameters used in data analysis up to that point.

This chapter provides a series of recommendations for an effective and reliable analysis of DNA sequences derived from oil and gas associated samples such as reservoir cores and produced fluid samples, pigging debris, and corrosion coupons from pipelines, fracking fluids, etc. These procedures are making use of publicly available bioinformatics tools and form the basis for best practices for single gene and metagenome data analysis. Adherence to these best practices should enable more reliable comparison across different studies and research groups.

2.1.1 Sampling and Lab Processing Considerations

Drawing meaningful conclusions from metagenome analysis necessitates uncompromised sample and data integrity. Loss of integrity can be introduced at any stage of the sampling, transport, or analysis processes and can lead to technical variation. Over the last decade, considerable effort was made toward investigating various causes for technical variation in environmental samples, particularly from oil and gas operations.

Past and current joint industry programs (e.g. NSERC IRC Petroleum Microbiology, geno-MIC) have made considerable efforts to address technical variation in oil and gas samples and develop Standard Operating Procedures (SOPs) for sampling and shipment of liquid and solid samples as well as swabs from corrosion coupons (e.g. geno-MIC Standard Operating Procedures). Through these collective efforts, information on chemical preservation (e.g. ethanol, RNAlater™, isopropyl alcohol, nucleic acid preservation buffer - NAP) and temperature preservation methods (ambient, chilled, frozen) have been investigated for their effectiveness in preserving the original microbial community and inhibiting any microbial growth during transport (Rachel and Gieg, 2020; Song et al., 2016; Menke et al., 2017).

Considerations regarding logistical challenges which might impact adoption of such methods in practice also needed to be accounted for. For example, despite a potentially superior performance as sample preservative for onshore samples, transporting 95% v/v ethanol by helicopter to offshore platforms is generally prohibited on health, safety, and environmental reasons. Alternative preservatives (e.g. RNALater™) are therefore required for those hard to access locations.

Lab processing considerations to minimize technical variation include the use of biological and technical replicates, and the use of appropriate positive and negative controls during DNA extraction and sequencing (e.g. sample spiking with a known organism) (Chase et al., 2016). Such precautionary measures can help assess sample heterogeneity, variation between technical replicates, and possible sources of external contamination (Sinha et al., 2017; Salter et al., 2014; Knights et al., 2011; Minich et al., 2019). Furthermore, commercial or custom-made mock communities can be a valuable resource for standardizing the lab procedures and parameters (e.g. PCR primers, cycling conditions, reagents, additives, etc.) prior to use with actual samples (Gieg, personal communication; Sui et al., 2020; Walker et al., 2015, Bonnet et al., 2002).

2.2 DOWNSTREAM PROCESSING CONSIDERATIONS

Reliable microbial monitoring of oil and gas infrastructure necessitates the taxonomic identification of the *in-situ* microbial community i.e.: *Which microorganisms are present?, How many of them are there?*, and finally, *What are their role in these environments?* (Figure 2.1). A fourth question could be added to those three regarding the state of microbial growth, i.e. *How active they are* or *what are they actually doing?* Comparative analysis of oil and gas microbiome data derived from different sampling locations or over different time points could reveal trends reflecting changes in the microbial communities and their metabolic potential. For example, the comparison of spatiotemporal metagenomic data derived from an oil reservoir injected with nitrate to prevent reservoir souring could reveal valuable information relating to nitrate treatment efficacy. Nitrate is thermodynamically a more favorable electron acceptor than sulfate and can therefore control the population of sulphate reducing microorganisms (SRP, sulphate reducing prokaryotes) preventing them from producing hydrogen sulfide. Monitoring the trends of the nitrate reducing and sulphate reducing populations could therefore provide production chemists with actionable information regarding status of souring control and nitrate dosage efficacy (Vigneron et al., 2017b; Voordouw et al., 2009).

Addressing the four questions above (i.e. who is there, how many they are, what are their functions, and what they are *actually* doing) requires different research methodologies to be followed as shown in Figure 2.2. These include single marker

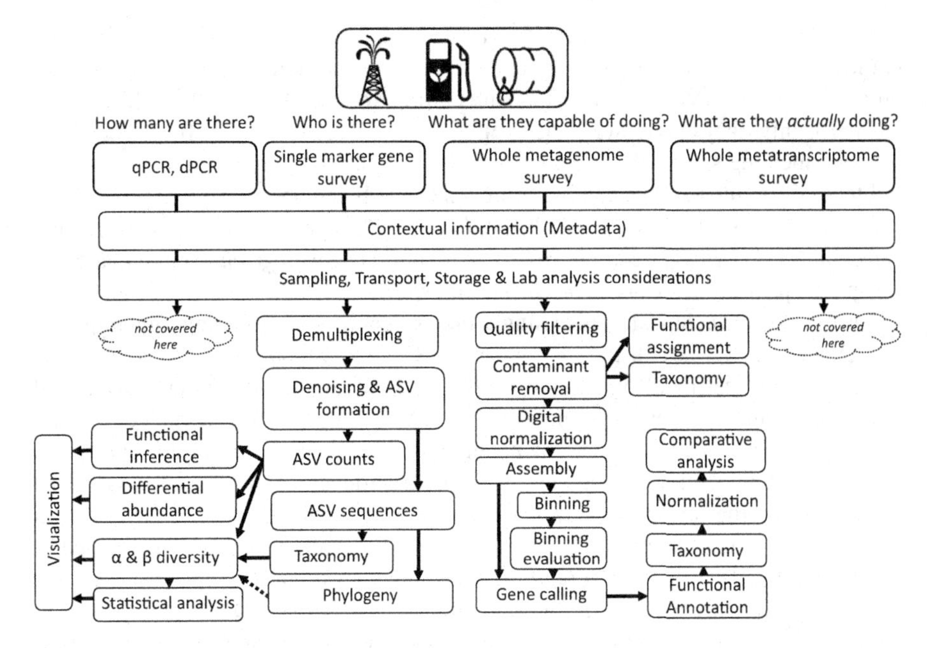

FIGURE 2.2 Flow diagram showing the different types of analysis and processing steps that can be performed on oil and gas samples.

gene sequencing for high-level microbial taxonomic identification, quantitative PCR (qPCR) for microbial enumeration, whole metagenome sequencing for in-depth information into molecular function and metatranscriptome sequencing for gene expression analysis. Single marker gene sequencing and whole metagenome sequencing are covered in the sections below but for brevity, real-time PCR (qPCR) and metatranscriptome analysis are reviewed in detail in other literature (Bonk et al., 2018; Bashiardes et al., 2016; Shakya et al., 2019).

2.3 SINGLE MARKER GENE SEQUENCING

Single marker gene sequencing involves the use of PCR primers to amplify a region of a target gene of interest. This could be either a taxonomy-relevant gene such as the 16S rRNA for prokaryotes (Klindworth et al., 2013) and the internal transcribed spacer (ITS) for fungi (Schoch et al., 2012) or a functional gene such as the *mcr* or *dsr* for the identification of taxa capable of methanogenesis or sulfate reduction respectively (Vigneron et al., 2018). The nucleotide differences of the amplified region in the various microorganisms are then utilized for distinguishing them from each other and for identifying their closest matches in reference databases such as SILVA (Quast et al., 2013), RDP (Cole et al., 2014) or Greengenes (DeSantis et al., 2006). Depending on the extent of the similarity between the query sequence and the database entries, an assignment is made at the appropriate taxonomic level (i.e. species, genus, family, etc.).

Despite the usefulness as a high-level microbial community profiling tool, single marker gene-based diversity surveys can skew the representation of certain taxa in the produced microbial community. The main culprits for such bias are the PCR primers and the affinity they exert toward the template DNA (Walters et al., 2016). Taxa whose primer annealing sites lack sufficient complementarity to the used PCR primers will fail to amplify during PCR and therefore appear as being absent from these samples (Sharma et al., 2020; Walker et al., 2015). *In-silico* PCR primer design and evaluation can certainly address some issues related to taxonomic coverage but cannot guarantee acceptable primer performance *in-vitro* (Klindworth et al., 2013; Walters et al., 2011). Other sources of technical bias in single marker gene surveys include the amplicon size (Walker et al., 2015), the number of PCR cycles which appears to impact the alpha-diversity metrics (Bonnet et al., 2002), and the amount of template DNA used in the PCR with low biomass samples being more susceptible to external contamination (Salter et al., 2014). Despite the potential pitfalls, single marker gene analysis is a relatively quick and low-cost way of getting high-level taxonomic information on the organisms that are present in a sample. Primer optimization can, to a certain extent, address amplification bias leading to good overall agreement with the taxonomic profiles derived from shotgun metagenomes. Moreover, single marker gene surveys are typically a good starting point to assess a sample's DNA quality prior to endeavoring to more expensive, data-rich, labor intensive metagenome or metatranscriptome analysis. As this field is constantly developing, techniques like full-length 16S rRNA gene sequencing by combining short reads (Callahan et al., 2020) and synthetic spike-in standards which can be used for evaluating data quality and for determining the absolute microbial abundances in a sample bypassing the need for qPCR (Tourlousse et al., 2017) are holding great promise.

2.4 SINGLE MARKER GENE ANALYSIS

For the analysis of single marker gene sequencing data, open-source microbiome bioinformatics platforms such as QIIME 2 (Bolyen et al., 2019), mothur (Schloss et al., 2009), and phyloseq (McMurdie and Holmes, 2013) have been developed. These are mostly community-driven platforms that provide a comprehensive array of tools for the handling, analysis, and visualization of marker-gene data. Among the three, QIIME 2 appears to be the most popular mainly due to its interactive visualizations and its plugin architecture that allows third party plugins to extend its functionality (Bolyen et al., 2019).

Typically, a single marker gene analysis starts with a quality check of the produced sequencing reads to remove sequencing errors and chimeras as these would artificially inflate the microbial diversity of the sample. Until recently, error-free sequencing reads were clustered into Operational Taxonomic Units (OTUs) based on sequence similarity (typically 97% and above). Nowadays, the OTU clustering approach was superseded by a more accurate method which allows discrimination between closely related but distinct taxa termed amplicon sequence variants (ASVs) (Eren et al., 2013). Tools like DADA2 (Callahan et al., 2016) and Deblur (Amir et al., 2017) can perform error correction, chimera removal, and formation of ASVs which correspond to the exact sequence features rather than OTU groups. A table that reports the count of each ASV found in each sample and a list of all the identified ASVs along with their corresponding nucleotide sequences are the main outputs from these two tools. The latter output (ASVs with their corresponding nucleotide sequences) can be subsequently used for the construction of a phylogenetic tree and for the taxonomic assignments of the ASVs. Taxonomic assignment is performed against a reference database such as SILVA, RDP or Greengenes, using a naïve Bayesian classifier (Bolyen et al., 2019). Among these three reference databases, SILVA (Release 138, 2019) is the most up to date while RDP (Release 11, 2016) and Greengenes (Release 13_8, 2013) lagging with their updates. The table containing the counts of each unique ASV in each sample in the dataset and optionally the produced phylogenetic tree can then be used for the computation of alpha and beta diversity metrics. Alpha-diversity indices provide quantitative and qualitative measures of community richness (e.g. Shannon, Faith's PD, Observed Features), or evenness (e.g. Pielou's Evenness) with or without the incorporation of phylogenetic relationships between the features (Faith's PD and Shannon respectively) (Faith, 1992; Shannon, 1948). Typically, the higher the value on the diversity index (e.g. Shannon), the richer the sample's microbial community. Beta diversity, on the other hand, determines the community dissimilarity between samples. Beta diversity can be qualitative, i.e. presence/absence of an ASV (e.g. Jaccard distance) or quantitative where the relative abundance of each ASV is taken into consideration in the calculation (e.g. Bray-Curtis distance). Where phylogenetic relationships between the features (ASVs) are incorporated in the calculation of community dissimilarity, phylogenetic distance matrices can be derived such the unweighted (qualitative) and weighted (quantitative) UniFrac matrices (Lozupone et al., 2011). To determine whether any of the metadata correlate with the alpha diversity metrics, Spearman or Pearson correlation coefficients can be calculated (Pearson, 1895; Spearman, 1904). In the case of beta diversity matrices, testing whether distances between samples within a group are more similar to each other than they are to samples from other groups, statistical tests such as PERMANOVA and ANOSIM can be performed (Anderson, 2001; Clarke, 1993). Where identification of features (ASVs) that are overrepresented or underrepresented across sample groups (e.g. treated vs. untreated) is wanted, differential abundance testing can be performed using tools like gneiss (Morton et al., 2017) and ANCOM (Mandal et al., 2015). Finally, despite the limitation of single marker gene sequencing lacking information on the functional composition of the sampled communities, tools like Picrust2 (Douglas et al., 2020), Tax4Fun2 (Wemheuer et al., 2020) and Piphillin (Iwai et al., 2016) can make functional inferences. These tools exhibit better performance with taxa that are not too phylogenetically distant to the bacterial and archaeal genomes that were used as references. Evaluation of these metagenome prediction tools on datasets from different environments which include human, non-human animal, and soil samples showed that they had more consistent assessment from human datasets than non-human animals or environmental datasets (Sun et al., 2020). It is likely that these differences reflect the bias of genome databases toward human-associated microorganisms. The use of oil and gas-centric microbiome reference databases for the training of these tools should improve the accuracy of their functional inference on oil and gas samples as observed in other cases (Wilkinson et al., 2018).

2.5 WHOLE METAGENOME SEQUENCING

Moving beyond simple taxonomic profiling toward what microbial functions can be performed in a given environment, whole metagenome analysis can lead the way. Whole metagenome sequencing sequences the extracted DNA in its totality, thus avoiding bias introduced by PCR, primer choice, or target region. Moreover, it captures not only prokaryotic (i.e. bacterial and archaeal) genetic information but any DNA molecule that is isolated during the extraction process (i.e. viral, prokaryotic, and eukaryotic). This broader and unbiased coverage of metagenomic information can provide a higher resolution and thus more accurate representation of the biological processes taking place in an environmental sample. For example, metagenomic analysis on produced fluid samples from oil reservoirs provided novel insights into the dominant microbial metabolic processes in relation to other environmental parameters such as reservoir temperature, seawater and nitrate injections, and level of biodegradation (Christman et al., 2020; Hu et al., 2016; Sierra-Garcia et al., 2020; Vigneron et al., 2017b). Furthermore, it unearthed the pivotal role of microbial controlling factors like bacteriophages in these environments (Cai et al., 2015; Daly et al., 2019; Pannekens et al., 2019). Metagenome assembled genomes (MAGs) from metagenomic data can provide additional information on the role of individual community members, allowing the association of MAG-derived taxa with previously unassigned biochemical processes like hydrocarbon degradation or even entirely novel biosynthetic pathways.

While metagenomic studies are more informative than single gene marker surveys, they tend to be more expensive, and their analysis is more complex and labor intensive. As the sequencing data are not restricted to a certain region of interest, a larger number of sequencing reads are required for sufficient coverage of the metagenome under investigation. For example, where 50,000 MiSeq reads per sample might be adequate for a single marker gene survey, metagenomic samples would require at least an order of magnitude more reads (i.e. >500,000) per sample for comparable taxonomic resolution (Hillmann et al., 2018). Lastly, as standard metagenome sequencing methodologies do not discriminate on the origin of DNA, a large proportion of the producing sequencing data could be from an unwanted source (i.e. contaminant) such as plants or humans. Typically, this is not an issue for samples derived from oil and gas environments, but it is more prominent on samples with low microbial load such as reservoir cores and cuttings.

2.6 METAGENOMIC DATA ANALYSIS

The first two steps in the analysis of metagenomic data are quality filtering and contaminant removal. For sequencing reads produced by the Illumina platform, the former step can be achieved using trimming tools such as Trimmomatic (Bolger et al., 2014). Following trimming, contaminant sequences, such as from PhiX or host/human DNA, can be removed using bowtie2 by mapping them against the reference genomes of the contaminants, while preserving the unmapped ones (Langmead and Salzberg, 2012). Quality-filtered and contaminant-free sequencing reads can be subsequently analyzed directly, or they can be assembled into contigs. Taxonomic assignments of the unassembled sequencing reads can be performed by comparing them against reference databases or subsets of those (e.g. unique clade-specific marker genes or proteins). For this comparison, alignment or k-mer based metagenome classifiers have been developed such as ganon (Piro et al., 2020), KRAKEN2 (Wood et al., 2019), Kaiju (Menzel et al., 2016), MetaPhlAn2 (Truong et al., 2015) and PhyloFlash (Gruber-Vodicka et al., 2019) among others (McIntyre et al., 2017). When functional information is required in addition to taxonomic information, tools like MEGAN6 (Huson et al., 2016) and PROMMenade (Utro et al., 2020) can be used instead.

For the reconstruction of genomes from metagenomic data (i.e. MAGs), assembly of the sequencing reads is typically required. The more abundant microorganisms are in a sample, the higher their representation in the whole metagenome sequencing dataset is going to be. Metagenome assemblers can effectively deal with this unevenness in genomic representation expressed as differences in sequencing depth. Several metagenome assemblers have been developed with MetaSPAdes (Nurk et al., 2017) and MEGAHIT (Li et al., 2015) being among the most widely used (Forouzan et al., 2018). In some cases, assembly of a complex metagenomic dataset can be too challenging and computationally intensive. In these cases, digital normalization can be performed where highly repetitive and/or single k-mers can be removed to lower the sizes of the input files, allowing for the assembly to proceed within the available amount of RAM. Digital normalization can be done using the tools like khmer (Crusoe et al., 2015), Bignorm (Wedemeyer et al., 2017), and ORNA (Durai and Schulz, 2019).

Following assembly, the produced contigs can be grouped into draft genomes (MAGs) using metagenome binning (Kang et al., 2015). Binning methods can be either supervised where individual contigs are first assigned taxonomic information before grouping of those contigs with the same taxonomy, or unsupervised where statistical methods are applied on contigs to uncover similarities that would group contigs together (e.g. GC content, tetranucleotide frequency, etc.). Popular binning tools include MetaBAT2 (Kang et al., 2019), MaxBin 2.0 (Wu, Simmons & Singer, 2016), CONCOCT (Alneberg et al., 2014), and more recently VAMB a binning tool that uses deep learning (Nissen et al., 2018). To assess the quality of the produced MAGs, tools like CheckM (Parks et al., 2015) which use single-copy gene profiling to determine the completeness and contamination of each MAG can be used. In addition to CheckM, other post-binning polishing methods (Chen et al., 2020) and tools (Wang et al., 2017) can be deployed to further improve the quality, accuracy, and completeness of the produced MAGs.

Assembled metagenomic data (i.e. contigs and MAGs) can subsequently be taken forward for annotation. Metagenome annotation refers to the identification of genomic features such as protein coding genes (CDS) as well as ribosomal RNA (rRNA) and transport RNA (tRNA) genes on the assembled metagenomic data. Protein coding genes can be predicted with tools like Prodigal (Hyatt et al., 2010), whereas barrnap (Seemann, 2013) and RNAmmer (Lagesen et al., 2007) can identify rRNA genes. For the identification of transport RNA genes, tRNAscan (Chan and Lowe, 2019) or ARAGORN (Laslett and Canback, 2004) can be used instead. Functional annotation of the predicted protein coding genes can be performed against databases such as KEGG, UniProt, NCBI, Pfam, and InterPro where functions from orthologous proteins and functional motifs and domains can be inferred. Standalone tools such as PROKA (Seemann, 2014), DFAST (Tanizawa et al., 2018) and MetaErg (Dong and

Strous, 2019) combine the gene finding and functional annotation processes described above. In addition, web-based bioinformatics pipelines like IMG/M (Chen et al., 2019) and MG-RAST (Meyer et al., 2008) can provide gene finding and functional annotation capability as well as other advanced metagenome analysis tools.

Metagenomic analysis pipelines that provide end-to-end solutions for metagenomic data analysis by integrating multiple processing steps such as quality filtering, assembly, functional annotation, taxonomic assignments, MAG creation, and metagenomic comparisons can be found either as standalone tools like Anvi'o (Eren et al., 2015), ATLAS (Kieser et al., 2020), IMP (Narayanasamy et al., 2016), SqeezeMeta (Tamames and Puente-Sanchez, 2018), or as web-based platforms such as MG-RAST (Meyer et al., 2008) and KBase (Arkin et al., 2018).

Metagenomes are differently sized due to sample, sequencing, and assembly differences, but it is crucial that they are normalized for fair comparative analyses. Use of simple proportions or rarefying the counts of metabolic genes are not considered statistically appropriate methods and therefore may obscure the biological interpretation of the data (McMurdie and Holmes, 2014). Therefore, more suitable methods have been proposed including Trimmed Mean of M-values (TMM) (Robinson and Oshlack, 2010), the Relative Log Expression (RLE) (Love et al., 2014), and Cumulative Sum Scaling (CSS) (Paulson et al., 2013) (for comparison of these methods, see Pereira et al., 2018). MUSiCC, an alternative metagenome normalization method which combines universal single-copy genes with machine learning has also been proposed (Manor and Borenstein, 2015). MUSiCC can perform abundance corrections on KEGG Orthology Groups (KO) and Clusters of Orthologous Groups (COGs) gene abundance profiles. In addition to data-driven normalization approaches, lab-based methods for the normalization of metagenomic studies is an active area of research (Hardwick et al., 2018; Reis et al., 2020). In terms of identification of differentially abundant genes in normalized metagenomic datasets, a number of comparative methods have been proposed. A comprehensive statistical evaluation of the most common comparative methods has been previously reported (Jonsson et al., 2016).

The bioinformatics tools that are mentioned in this chapter are summarized in Table 2.1.

TABLE 2.1
Bioinformatics Tools Mentioned in This Chapter in the Order They Appear

Tool	Description	Reference
MetaHCR	Web-enabled metagenome data management system for hydrocarbon resources	Marks et al., 2018
Quorem	Database for cataloguing and analyzing MIC related metagenomic data	Manuel 2020
QIIME 2	Microbiome analysis platform	Bolyen et al., 2019
Galaxy	Web-based metagenome analysis platform	Afgan et al., 2018
KBase	Web-based metagenome analysis platform	Arkin et al., 2018
Qiita	Web-based metagenome analysis platform	Gonzalez et al., 2018
phyloseq	Microbiome analysis platform	McMurdie and Holmes, 2013
mothur	Microbiome analysis platform	Schloss et al., 2009

(Continued)

TABLE 2.1 (Continued)
Bioinformatics Tools Mentioned in This Chapter in the Order They Appear

Tool	Description	Reference
DADA2	Error correction, chimera removal and formation of ASVs	Callahan et al., 2016
Deblur	Error correction, chimera removal and formation of ASVs	Amir et al., 2017
gneiss	Differential abundance testing	Morton et al., 2017
ANCOM	Differential abundance testing	Mandal et al., 2015
Picrust2	Functional inference from taxonomic information	Douglas et al., 2020
Tax4Fun2	Functional inference from taxonomic information	Wemheuer et al., 2020
Piphillin	Functional inference from taxonomic information	Iwai et al., 2016
Trimmomatic	Quality filtering of metagenomic data	Bolger et al., 2014
bowtie2	Tool for contaminant removal by mapping reads against the contaminants reference genomes	Langmead and Salzberg, 2012
ganon	Tool for taxonomic assignment of metagenomic data	Piro et al., 2020
KRAKEN2	Tool for taxonomic assignment of metagenomic data	Wood et al., 2019
Kaiju	Tool for taxonomic assignment of metagenomic data	Menzel et al., 2016
MetaPhlAn2	Tool for taxonomic assignment of metagenomic data	Truong et al., 2015
PhyloFlash	Tool for taxonomic assignment of metagenomic data	Gruber-Vodicka et al., 2019
MEGAN6	Tool for taxonomic and functional assignment of metagenomic data	Huson et al., 2016
PROMMenade	Tool for taxonomic and functional assignment of metagenomic data	Utro et al., 2020
MetaSPAdes	Metagenome assembler	Nurk et al., 2017
MEGAHIT	Metagenome assembler	Li et al., 2015
khmer	Digital normalization tool	Crusoe et al., 2015
Bignorm	Digital normalization tool	Wedemeyer et al., 2017
ORNA	Digital normalization tool	Durai and Schulz, 2019
MetaBAT2	Metagenome binning tool	Kang et al., 2019
MaxBin 2.0	Metagenome binning tool	Wu, Simmons & Singer, 2016
CONCOCT	Metagenome binning tool	Alneberg et al., 2014
VAMB	Metagenome binning tool	Nissen et al., 2018
CheckM	Tool for determining completeness and contamination of MAGs	Parks et al., 2015
Prodigal	Prediction of protein coding genes	Hyatt et al., 2010
barrnap	Prediction of rRNA genes	Seemann 2013
RNAmmer	Prediction of rRNA genes	Lagesen et al., 2007
tRNAscan	Prediction of tRNA genes	Chan and Lowe 2019
ARAGORN	Prediction of tRNA genes	Laslett and Canback 2004
PROKA	Gene finding and functional annotation tool	Seemann, 2014
DFAST	Gene finding and functional annotation tool	Tanizawa et al., 2018
MetaErg	Gene finding and functional annotation tool	Dong and Strous, 2019
IMG/M	Web-based metagenome analysis platform	Chen et al., 2019
MG-RAST	Web-based metagenome analysis platform	Meyer et al., 2008
Anvi'o	Standalone metagenomic analysis pipeline	Eren et al., 2015
ATLAS	Standalone metagenomic analysis pipeline	Kieser et al., 2020
IMP	Standalone metagenomic analysis pipeline	Narayanasamy et al., 2016
SqeezeMeta	Standalone metagenomic analysis pipeline	Tamames and Puente-Sanchez, 2018
MUSiCC	Metagenome normalization method using universal single-copy genes	Manor and Borenstein, 2015

2.7 METAGENOMIC APPLICATIONS TO OIL AND GAS

The field of environmental genomics has seen an enormous growth over the last decade and it continues to grow exponentially. Many of the scientific advances are tightly linked to human microbiome studies driven by large research initiatives such as the human microbiome project (HMP) (Integrative HMP (iHMP) Research Network Consortium). The oil and gas industry has been slower compared to other industries in adopting these new technologies but nonetheless is making considerable efforts toward benefiting from these new developments. More specifically, novel applications leveraging the power of microbial profiling and metagenomics have started emerging in various parts of the oil and gas industry. Some of these application areas are shown in Figure 2.3 and briefly described below.

2.7.1 Exploration

The tight relationship between food availability and feeders can be exploited for the discovery of new hydrocarbon-rich environments by identifying the microorganisms that are feeding on them (Schumacher, 1996). Recent studies claim they can identify sweet spot areas for drilling by combining microbial profiling at the soil surface with machine learning (Te Stroet et al., 2017). Similarly, microbial profiling of piston cores from offshore exploratory campaigns led to the identification of microbial indicators that strongly correlate with hydrocarbon presence. Combining these microbial indicators with machine learning can assist the detection of hydrocarbon-altered sediments from newly acquired samples based on their microbial profiles (Miranda et al., 2019). Another microbial-based offshore exploration technique uses thermospores as proxies for hydrocarbon seepage locations and have shown promising results (Chakraborty et al., 2018; Chakraborty et al., 2020). Moving beyond hydrocarbon prospecting and more toward cold seep characterization, considerable progress has been made in distinguishing between seepage types (gas *vs.* hydrocarbons) (Vigneron et al., 2017); identifying key microbial players and processes in these unique environments (Vigneron et al., 2019; Zhao et al., 2020); and revealing that actual *in-situ* microbial activity (metatranscriptome) might significantly differ from what would have been predicted based on the metagenomic data alone (Vigneron et al., 2019).

In general, metagenomics can help us better understand the relationship between microorganisms and hydrocarbon seeps. Yet, metagenomics is not a silver bullet for

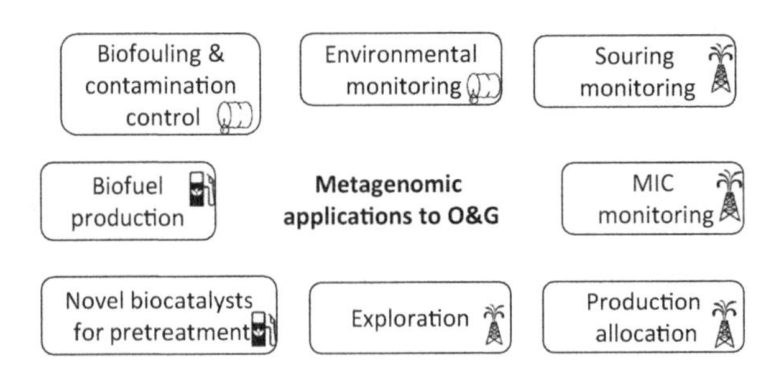

FIGURE 2.3 Applications of metagenomics in oil and gas range from upstream to downstream as well as for Health, Safety and Environment (HSE).

hydrocarbon exploration but should be deployed as part of an integrated charge evaluation workflow, incorporating sub-surface fluid modeling and 3-D seismic. Where metagenomics might be able to provide valuable insights is on the notion of microseepage (i.e. near vertical upward migration of hydrocarbons toward the surface) where there has been a lot of debate among the geosciences community on this elusive for some scientists, hydrogeological phenomenon.

2.7.2 Production Allocation

Production allocation aims to quantify the proportional contribution of each of the hydrocarbon baring zones to the comingled hydrocarbon fluid recovered at the surface. Such information can help determine the vertical and horizontal spacing of the wells to maximize drainage of the targeted zones as well as monitor the draining process over time. Production allocation is typically based on geochemical parameters, but microbial DNA-based methods have started to appear. These methods leverage the changes in microbial diversity and abundance as a function of depth. Microbial DNA isolated from the well cuttings during drilling is used for the creation of a biostratigraphic profile. End member mixing analysis is then used for the identification and quantification of the contributing depth intervals to the comingled microbial signal measured in the produced fluids (Percak-Dennett et al., 2019).

2.7.3 Souring and Microbiologically Influenced Corrosion (MIC) Monitoring

As mentioned in the previous section, injection of seawater during secondary oil recovery can stimulate the growth of sulphate reducing prokaryotes (SRP) and lead to hydrogen sulfide formation, a process better known as reservoir souring (Machel, 2001; Johnson et al., 2017; Khatib and Salanitro, 1997). Given the toxicity and corrosive nature of the produced sulfide, prophylactic measures against the rise of the SRP populations is of paramount importance for the production chemists and the HSE personnel. Nitrate injections during waterflooding can control reservoir souring by promoting the growth of nitrate reducing prokaryotes (NRP) which outcompete the SRP population (Voordouw et al., 2007). Regular microbial monitoring of the production facilities should reveal trends of the nitrate and sulphate reducing populations and provide actionable information regarding the health status of the facilities and the efficacy of the nitrate dosage (Vigneron et al., 2017b; Voordouw et al., 2009). Baseline metagenomic studies on reservoir cores to assess souring potential prior to seawater flooding (Tsesmetzis et al., 2018) as well as reservoir souring simulation tools like REVEAL™ and SourSimRL™ can further assist with souring prediction (Johnson et al., 2017).

In addition to reservoir souring, SRP along with methanogenic archaea and organic acid producers can be directly or indirectly involved in pitting corrosion on metal surfaces a process better known as microbiologically influenced corrosion (MIC) (Vigneron et al., 2018b; Skovhus et al., 2017). Using metagenomics, MIC-related microorganisms and processes can be identified and quantified (Vigneron et al., 2016; Lomans et al., 2016; Sharma et al., 2020b).

2.7.4 Baseline Environmental Assessment and Monitoring

Metagenomics can be used to monitor microbial community changes to determine whether they are linked to natural or anthropogenic activities such as oil and gas operations (Chapelle et al., 2002). Groundwater and aquifers, for example, harbor an extensive microbial ecosystem supported by the presence of carbon substrates including methane which occasionally can represent more than 20% of the total carbon (Barker and Fritz, 1981). Microbial monitoring of natural and hydrocarbon-impacted aquifers using metagenomic tools can help determine the various sources (e.g. biogenic *vs.* thermogenic methane), track the in-situ conversion processes (e.g. methanogenesis and methane oxidation), and the coupling of these processes to other biogeochemical cycles (Vigneron et al., 2017c). Such improved environmental surveillance methods should improve leak detection and monitoring of oil and gas operations providing another layer of information toward environmentally conscious asset management and regulatory compliance.

2.7.5 Biofouling and Contamination Control

Biofilm formation in oil and gas processing facilities such as refineries can lead to blocked pipelines and clogging of cooling towers (Wang et al., 2013). Chemicals with biocidal properties are typically used for microbial control but due to the layered nature of the formed biofilms, these biocides are not entirely effective in penetrating the deeper biofilm layers (Balamurugan et al., 2011; Wang et al., 2013). Metagenomic analysis has been proven valuable in identifying key microbial perpetrators in these systems, while providing comprehensive microbial monitoring capability for the asset (Di Gregorio et al., 2017; Wang et al., 2013). Furthermore, metagenomic analysis can assist with the screening process for determining the optimal biocide formulation and dosage for effective microbial control of such facilities.

2.7.6 Novel Biocatalysts for Enzymatic Hydrolysis of Biomass

Enzymatic hydrolysis is a process used for the conversion of cellulosic and hemi-cellulosic material into fermentable sugars using cellulytic enzymes known as cellulases. The cost of cellulases comprises a significant proportion of the overall second-generation ethanol cost and any improvements on the enzyme's stability and reaction kinetics would greatly benefit the process economics. To that extent, mining of environmental metagenomic datasets can lead to the discovery of novel enzymes that can improve the enzymatic hydrolysis process like in the case of the novel endoxylanase family GH10 (Alvarez et al., 2013; Ferraz Junior et al., 2017).

2.7.7 Biofuel Production

Certain biofuel production processes could greatly benefit from the use of metagenomics. For example, temporal metagenomic analysis on ensiled sweet sorghum inoculated with an ethanologenic yeast was able to shed light on the underlying microbial processes during bioethanol production and the microbial response to the various treatments (Gallagher et al., 2018). Also, during industrial fermentation of sugars into ethanol, bacterial and wild-yeast contamination can be a major hindrance leading to lower ethanol yield and profitability. There are multiple entry points for these contaminants including the reactors, the pipelines, the nutrients, and through yeast recycling. Surveying the various stages of the fermentation process using metagenomics can locate the culprit's entry points and alert the operators before they reach critical levels which would significantly impact the fermentation efficiency (Khullar et al., 2013; Skinner and Leathers, 2004). Finally, metagenomics can provide valuable insights into the biogas production process where operating parameters such as temperature can greatly impact the microbial community composition and process stability. Several key microorganisms have been identified and the microbial community composition of several anaerobic digesters have been elucidated using metagenomics (Campanaro et al., 2016; Maus et al., 2016; for review see Junemann et al., 2017).

2.8 CONCLUDING REMARKS

Microorganisms are ubiquitous in the environment and they can greatly impact oil and gas operations. Microbial processes can play a pivotal role as the oil and gas industry strives for energy transition towards a lower-carbon future while meeting the rising energy demand. Metagenomics is a rapidly evolving technology that can help us better understand, monitor, and even manipulate those microorganisms for our benefit. This is the start of a promising era for the oil and gas industry.

REFERENCES

Afgan, E., D. Baker, B. Batut, M. van den Beek, D. Bouvier, M. Cech, J. Chilton, D. Clements, N. Coraor, B. A. Gruning, A. Guerler, J. Hillman-Jackson, S. Hiltemann, V. Jalili, H. Rasche, N. Soranzo, J. Goecks, J. Taylor, A. Nekrutenko, and D. Blankenberg. 2018. The Galaxy platform for accessible, reproducible and collaborative biomedical analyses: 2018 update. *Nucleic Acids Res* 46 (W1):W537–W544. doi: 10.1093/nar/gky379.

Alneberg, J., B. S. Bjarnason, I. de Bruijn, M. Schirmer, J. Quick, U. Z. Ijaz, L. Lahti, N. J. Loman, A. F. Andersson, and C. Quince. 2014. Binning metagenomic contigs by coverage and composition. *Nat Methods* 11 (11):1144–1146. doi: 10.1038/nmeth.3103.

Alvarez, T. M., R. Goldbeck, C. R. dos Santos, D. A. Paixao, T. A. Goncalves, J. P. Franco Cairo, R. F. Almeida, I. de Oliveira Pereira, G. Jackson, J. Cota, F. Buchli, A. P. Citadini, R. Ruller, C. C. Polo, M. de Oliveira Neto, M. T. Murakami, and F. M. Squina. 2013. Development and biotechnological application of a novel endoxylanase family GH10 identified from sugarcane soil metagenome. *PLoS One* 8 (7):e70014. doi: 10.1371/journal.pone.0070014.

Amir, A., D. McDonald, J. A. Navas-Molina, E. Kopylova, J. T. Morton, Z. Zech Xu, E. P. Kightley, L. R. Thompson, E. R. Hyde, A. Gonzalez, and R. Knight. 2017. Deblur rapidly resolves single-nucleotide community sequence patterns. *mSystems* 2 (2). doi: 10.1128/mSystems.00191-16.

Anderson, Marti J. 2001. A new method for non-parametric multivariate analysis of variance. *Austral Ecology* 26 (1):32–46. doi: 10.1111/j.1442-9993.2001.01070.pp.x.

Arkin, A. P., R. W. Cottingham, C. S. Henry, N. L. Harris, R. L. Stevens, S. Maslov, P. Dehal, D. Ware, F. Perez, S. Canon, M. W. Sneddon, M. L. Henderson, W. J. Riehl, D. Murphy-Olson, S. Y. Chan, R. T. Kamimura, S. Kumari, M. M. Drake, T. S. Brettin, E. M. Glass, D. Chivian, D. Gunter, D. J. Weston, B. H. Allen, J. Baumohl, A. A. Best, B. Bowen,

S. E. Brenner, C. C. Bun, J. M. Chandonia, J. M. Chia, R. Colasanti, N. Conrad, J. J. Davis, B. H. Davison, M. DeJongh, S. Devoid, E. Dietrich, I. Dubchak, J. N. Edirisinghe, G. Fang, J. P. Faria, P. M. Frybarger, W. Gerlach, M. Gerstein, A. Greiner, J. Gurtowski, H. L. Haun, F. He, R. Jain, M. P. Joachimiak, K. P. Keegan, S. Kondo, V. Kumar, M. L. Land, F. Meyer, M. Mills, P. S. Novichkov, T. Oh, G. J. Olsen, R. Olson, B. Parrello, S. Pasternak, E. Pearson, S. S. Poon, G. A. Price, S. Ramakrishnan, P. Ranjan, P. C. Ronald, M. C. Schatz, S. M. D. Seaver, M. Shukla, R. A. Sutormin, M. H. Syed, J. Thomason, N. L. Tintle, D. Wang, F. Xia, H. Yoo, S. Yoo, and D. Yu. 2018. KBase: The United States Department of Energy Systems Biology Knowledgebase. *Nat Biotechnol* 36 (7):566–569. doi: 10.1038/nbt.4163.

Balamurugan, P., M. H. Joshi, and T. S. Rao. 2011. Microbial fouling community analysis of the cooling water system of a nuclear test reactor with emphasis on sulphate reducing bacteria. *Biofouling* 27 (9):967–978. doi: 10.1080/08927014.2011.618636.

Barker, J., Fritz, P. Carbon isotope fractionation during microbial methane oxidation. *Nature* 293, 289–291 (1981). https://doi.org/10.1038/293289a0.

Bashiardes, S., G. Zilberman-Schapira, and E. Elinav. 2016. Use of metatranscriptomics in microbiome research. *Bioinform Biol Insights* 10:19–25. doi: 10.4137/BBI.S34610.

Bolger, A. M., M. Lohse, and B. Usadel. 2014. Trimmomatic: a flexible trimmer for Illumina sequence data. *Bioinformatics* 30 (15):2114–2120. doi: 10.1093/bioinformatics/btu170.

Bolyen, E., J. R. Rideout, M. R. Dillon, N. A. Bokulich, C. C. Abnet, G. A. Al-Ghalith, H. Alexander, E. J. Alm, M. Arumugam, F. Asnicar, Y. Bai, J. E. Bisanz, K. Bittinger, A. Brejnrod, C. J. Brislawn, C. T. Brown, B. J. Callahan, A. M. Caraballo-Rodriguez, J. Chase, E. K. Cope, R. Da Silva, C. Diener, P. C. Dorrestein, G. M. Douglas, D. M. Durall, C. Duvallet, C. F. Edwardson, M. Ernst, M. Estaki, J. Fouquier, J. M. Gauglitz, S. M. Gibbons, D. L. Gibson, A. Gonzalez, K. Gorlick, J. Guo, B. Hillmann, S. Holmes, H. Holste, C. Huttenhower, G. A. Huttley, S. Janssen, A. K. Jarmusch, L. Jiang, B. D. Kaehler, K. B. Kang, C. R. Keefe, P. Keim, S. T. Kelley, D. Knights, I. Koester, T. Kosciolek, J. Kreps, M. G. I. Langille, J. Lee, R. Ley, Y. X. Liu, E. Loftfield, C. Lozupone, M. Maher, C. Marotz, B. D. Martin, D. McDonald, L. J. McIver, A. V. Melnik, J. L. Metcalf, S. C. Morgan, J. T. Morton, A. T. Naimey, J. A. Navas-Molina, L. F. Nothias, S. B. Orchanian, T. Pearson, S. L. Peoples, D. Petras, M. L. Preuss, E. Pruesse, L. B. Rasmussen, A. Rivers, M. S. Robeson, 2nd, P. Rosenthal, N. Segata, M. Shaffer, A. Shiffer, R. Sinha, S. J. Song, J. R. Spear, A. D. Swafford, L. R. Thompson, P. J. Torres, P. Trinh, A. Tripathi, P. J. Turnbaugh, S. Ul-Hasan, J. J. J. van der Hooft, F. Vargas, Y. Vazquez-Baeza, E. Vogtmann, M. von Hippel, W. Walters, Y. Wan, M. Wang, J. Warren, K. C. Weber, C. H. D. Williamson, A. D. Willis, Z. Z. Xu, J. R. Zaneveld, Y. Zhang, Q. Zhu, R. Knight, and J. G. Caporaso. 2019. Reproducible, interactive, scalable and extensible microbiome data science using QIIME 2. *Nat Biotechnol* 37 (8):852–857. doi: 10.1038/s41587-019-0209-9.

Bonk, F., D. Popp, H. Harms, and F. Centler. 2018. PCR-based quantification of taxa-specific abundances in microbial communities: Quantifying and avoiding common pitfalls. *J Microbiol Methods* 153:139–147. doi: 10.1016/j.mimet.2018.09.015.

Bonnet, R., A. Suau, J. Dore, G. R. Gibson, and M. D. Collins. 2002. Differences in rDNA libraries of faecal bacteria derived from 10- and 25-cycle PCRs. *Int J Syst Evol Microbiol* 52 (Pt 3):757–763. doi: 10.1099/00207713-52-3-757.

Cai, M., Y. Nie, C. Q. Chi, Y. Q. Tang, Y. Li, X. B. Wang, Z. S. Liu, Y. Yang, J. Zhou, and X. L. Wu. 2015. Crude oil as a microbial seed bank with unexpected functional potentials. *Sci Rep* 5:16057. doi: 10.1038/srep16057.

Callahan, Benjamin J, Dmitry Grinevich, Siddhartha Thakur, Michael A Balamotis, and Tuval Ben Yehezkel. 2020. Ultra-accurate microbial amplicon sequencing directly from complex samples with synthetic long reads. *bioRxiv*:2020.07.07.192286. doi: 10.1101/2020.07.07.192286.

Callahan, B. J., P. J. McMurdie, M. J. Rosen, A. W. Han, A. J. Johnson, and S. P. Holmes. 2016. DADA2: High-resolution sample inference from Illumina amplicon data. *Nat Methods* 13 (7):581–583. doi: 10.1038/nmeth.3869.

Campanaro, S., L. Treu, P. G. Kougias, D. De Francisci, G. Valle, and I. Angelidaki. 2016. Metagenomic analysis and functional characterization of the biogas microbiome using high throughput shotgun sequencing and a novel binning strategy. *Biotechnol Biofuels* 9:26. doi: 10.1186/s13068-016-0441-1.

Chakraborty, A., E. Ellefson, C. Li, D. Gittins, J. M. Brooks, B. B. Bernard, and C. R. J. Hubert. 2018. Thermophilic endospores associated with migrated thermogenic hydrocarbons in deep Gulf of Mexico marine sediments. *ISME J* 12 (8):1895–1906. doi: 10.1038/s41396-018-0108-y.

Chakraborty, A., S. E. Ruff, X. Dong, E. D. Ellefson, C. Li, J. M. Brooks, J. McBee, B. B. Bernard, and C. R. J. Hubert. 2020. Hydrocarbon seepage in the deep seabed links subsurface and seafloor biospheres. *Proc Natl Acad Sci U S A* 117 (20):11029–11037. doi: 10.1073/pnas.2002289117.

Chan, P. P., and T. M. Lowe. 2019. tRNAscan-SE: Searching for tRNA genes in genomic sequences. *Methods Mol Biol* 1962:1–14. doi: 10.1007/978-1-4939-9173-0_1.

Chapelle, F. H., P. M. Bradley, D. R. Lovley, K. O'Neill, and J. E. Landmeyer. 2002. Rapid evolution of redox processes in a petroleum hydrocarbon-contaminated aquifer. *Ground Water* 40 (4):353–360. doi: 10.1111/j.1745-6584.2002.tb02513.x.

Chase, J., J. Fouquier, M. Zare, D. L. Sonderegger, R. Knight, S. T. Kelley, J. Siegel, and J. G. Caporaso. 2016. Geography and location are the primary drivers of office microbiome composition. *mSystems* 1 (2). doi: 10.1128/mSystems.00022-16.

Chen, I. A., K. Chu, K. Palaniappan, M. Pillay, A. Ratner, J. Huang, M. Huntemann, N. Varghese, J. R. White, R. Seshadri, T. Smirnova, E. Kirton, S. P. Jungbluth, T. Woyke, E. A. Eloe-Fadrosh, N. N. Ivanova, and N. C. Kyrpides. 2019. IMG/M v.5.0: an integrated data management and comparative analysis system for microbial genomes and microbiomes. *Nucleic Acids Res* 47 (D1):D666–D677. doi: 10.1093/nar/gky901.

Chen, L. X., K. Anantharaman, A. Shaiber, A. M. Eren, and J. F. Banfield. 2020. Accurate and complete genomes from metagenomes. *Genome Res* 30 (3):315–333. doi: 10.1101/gr.258640.119.

Christman, G. D., R. I. Leon-Zayas, R. Zhao, Z. M. Summers, and J. F. Biddle. 2020. Novel clostridial lineages recovered from metagenomes of a hot oil reservoir. *Sci Rep* 10 (1):8048. doi: 10.1038/s41598-020-64904-6.

Clarke, K. R. 1993. Non-parametric multivariate analyses of changes in community structure. *Australian Ecol* 18 (1):117–143. doi: 10.1111/j.1442-9993.1993.tb00438.x.

Cole, J. R., Q. Wang, J. A. Fish, B. Chai, D. M. McGarrell, Y. Sun, C. T. Brown, A. Porras-Alfaro, C. R. Kuske, and J. M. Tiedje. 2014. Ribosomal Database Project: data and tools for high throughput rRNA analysis. *Nucleic Acids Res* 42 (Database issue):D633–D642. doi: 10.1093/nar/gkt1244.

Crusoe, M. R., H. F. Alameldin, S. Awad, E. Boucher, A. Caldwell, R. Cartwright, A. Charbonneau, B. Constantinides, G. Edvenson, S. Fay, J. Fenton, T. Fenzl, J. Fish, L. Garcia-Gutierrez, P. Garland, J. Gluck, I. Gonzalez, S. Guermond, J. Guo, A. Gupta, J. R. Herr, A. Howe, A. Hyer, A. Harpfer, L. Irber, R. Kidd, D. Lin, J. Lippi, T. Mansour, P. McA'Nulty, E. McDonald, J. Mizzi, K. D. Murray, J. R. Nahum, K. Nanlohy, A. J. Nederbragt, H. Ortiz-Zuazaga, J. Ory, J. Pell, C. Pepe-Ranney, Z. N. Russ, E. Schwarz, C. Scott, J. Seaman, S. Sievert, J. Simpson, C. T. Skennerton, J. Spencer, R. Srinivasan, D. Standage, J. A. Stapleton, S. R. Steinman, J. Stein, B. Taylor, W. Trimble, H. L. Wiencko, M. Wright, B. Wyss, Q. Zhang, E. Zyme, and C. T. Brown. 2015. The khmer software package: enabling efficient nucleotide sequence analysis. *F1000Res* 4:900. doi: 10.12688/f1000research.6924.1.

Daly, R. A., S. Roux, M. A. Borton, D. M. Morgan, M. D. Johnston, A. E. Booker, D. W. Hoyt, T. Meulia, R. A. Wolfe, A. J. Hanson, P. J. Mouser, J. D. Moore, K. Wunch, M. B. Sullivan, K. C. Wrighton, and M. J. Wilkins. 2019. Viruses control dominant bacteria colonizing the terrestrial deep biosphere after hydraulic fracturing. *Nat Microbiol* 4 (2):352–361. doi: 10.1038/s41564-018-0312-6.

DeSantis, T. Z., P. Hugenholtz, N. Larsen, M. Rojas, E. L. Brodie, K. Keller, T. Huber, D. Dalevi, P. Hu, and G. L. Andersen. 2006. Greengenes, a chimera-checked 16S rRNA gene database and workbench compatible with ARB. *Appl Environ Microbiol* 72 (7):5069–5072. doi: 10.1128/AEM.03006-05.

Di Gregorio, L., V. Tandoi, R. Congestri, S. Rossetti, and F. Di Pippo. 2017. Unravelling the core microbiome of biofilms in cooling tower systems. *Biofouling* 33 (10):793–806. doi: 10.1080/08927014.2017.1367386.

Dong, X., and M. Strous. 2019. An integrated pipeline for annotation and visualization of metagenomic contigs. *Front Genet* 10:999. doi: 10.3389/fgene.2019.00999.

Douglas, G. M., V. J. Maffei, J. R. Zaneveld, S. N. Yurgel, J. R. Brown, C. M. Taylor, C. Huttenhower, and M. G. I. Langille. 2020. PICRUSt2 for prediction of metagenome functions. *Nat Biotechnol* 38 (6):685–688. doi: 10.1038/s41587-020-0548-6.

Durai, D. A., and M. H. Schulz. 2019. Improving in-silico normalization using read weights. *Sci Rep* 9 (1):5133. doi: 10.1038/s41598-019-41502-9.

Eren, A. M., O. C. Esen, C. Quince, J. H. Vineis, H. G. Morrison, M. L. Sogin, and T. O. Delmont. 2015. Anvi'o: an advanced analysis and visualization platform for 'omics data. *PeerJ* 3:e1319. doi: 10.7717/peerj.1319.

Eren, A. M., L. Maignien, W. J. Sul, L. G. Murphy, S. L. Grim, H. G. Morrison, and M. L. Sogin. 2013. Oligotyping: Differentiating between closely related microbial taxa using 16S rRNA gene data. *Methods Ecol Evol* 4 (12). doi: 10.1111/2041-210X.12114.

Faith, D. P. 1992. Conservation evaluation and phylogenetic diversity. *Biological Conservation* 61 (1):1–10. doi: 10.1016/0006-3207(92)91201-3.

Ferraz, J., A. D. Nunes, A. R. L. Damásio, D. A. A. Paixão, T. M. Alvarez, and F. M. Squina. 2017. Applied metagenomics for biofuel development and environmental sustainability. In *Advances of Basic Science for Second Generation Bioethanol from Sugarcane*, edited by Marcos S. Buckeridge and Amanda P. De Souza, 107–129. Cham: Springer International Publishing.

Forouzan, E., P. Shariati, M. S. Mousavi Maleki, A. A. Karkhane, and B. Yakhchali. 2018. Practical evaluation of 11 de novo assemblers in metagenome assembly. *J Microbiol Methods* 151:99–105. doi: 10.1016/j.mimet.2018.06.007.

Gallagher, D., D. Parker, D. J. Allen, and N. Tsesmetzis. 2018. Dynamic bacterial and fungal microbiomes during sweet sorghum ensiling impact bioethanol production. *Bioresour Technol* 264:163–173. doi: 10.1016/j.biortech.2018.05.053.

geno-MIC program. 2020. https://wpsites.ucalgary.ca/microbial-corrosion/. Accessed 30 July 2020. https://wpsites.ucalgary.ca/microbial-corrosion/

geno-MIC Standard Operating Procedures. 2020. https://wpsites.ucalgary.ca/microbial-corrosion/sops2/. Accessed 30 July 2020. https://wpsites.ucalgary.ca/microbial-corrosion/sops2/

Genomics, Nebula. 2020. https://nebula.org/whole-genome-sequencing/. Accessed 30 July 2020. https://nebula.org/whole-genome-sequencing/.

Gonzalez, A., J. A. Navas-Molina, T. Kosciolek, D. McDonald, Y. Vazquez-Baeza, G. Ackermann, J. DeReus, S. Janssen, A. D. Swafford, S. B. Orchanian, J. G. Sanders, J. Shorenstein, H. Holste, S. Petrus, A. Robbins-Pianka, C. J. Brislawn, M. Wang, J. R. Rideout, E. Bolyen, M. Dillon, J. G. Caporaso, P. C. Dorrestein, and R. Knight. 2018. Qiita: rapid, web-enabled microbiome meta-analysis. *Nat Methods* 15 (10):796–798. doi: 10.1038/s41592-018-0141-9.

Gruber-Vodicka, H. R., B. K. B. Seah, and E. Pruesse. 2019. phyloFlash – Rapid SSU rRNA profiling and targeted assembly from metagenomes. *bioRxiv*:521922. doi: 10.1101/521922.

Hardwick, S. A., W. Y. Chen, T. Wong, B. S. Kanakamedala, I. W. Deveson, S. E. Ongley, N. S. Santini, E. Marcellin, M. A. Smith, L. K. Nielsen, C. E. Lovelock, B. A. Neilan, and T. R. Mercer. 2018. Synthetic microbe communities provide internal reference standards for metagenome sequencing and analysis. *Nat Commun* 9 (1):3096. doi: 10.1038/s41467-018-05555-0.

Hillmann, B., G. A. Al-Ghalith, R. R. Shields-Cutler, Q. Zhu, D. M. Gohl, K. B. Beckman, R. Knight, and D. Knights. 2018. Evaluating the information content of shallow shotgun metagenomics. *mSystems* 3 (6). doi: 10.1128/mSystems.00069-18.

Hu, P., L. Tom, A. Singh, B. C. Thomas, B. J. Baker, Y. M. Piceno, G. L. Andersen, and J. F. Banfield. 2016. Genome-resolved metagenomic analysis reveals roles for candidate phyla and other microbial community members in biogeochemical transformations in oil reservoirs. *mBio* 7 (1):e01669–e01615. doi: 10.1128/mBio.01669-15.

Huson, D. H., S. Beier, I. Flade, A. Gorska, M. El-Hadidi, S. Mitra, H. J. Ruscheweyh, and R. Tappu. 2016. MEGAN community edition - interactive exploration and analysis of large-scale microbiome sequencing data. *PLoS Comput Biol* 12 (6):e1004957. doi: 10.1371/journal.pcbi.1004957.

Hyatt, D., G. L. Chen, P. F. Locascio, M. L. Land, F. W. Larimer, and L. J. Hauser. 2010. Prodigal: prokaryotic gene recognition and translation initiation site identification. *BMC Bioinformatics* 11:119. doi: 10.1186/1471-2105-11-119.

Integrative, H. M. P. Research NETWORK CONSORTIUM. 2019. The integrative human microbiome project. *Nature* 569 (7758):641–648. doi: 10.1038/s41586-019-1238-8.

Iwai, S., T. Weinmaier, B. L. Schmidt, D. G. Albertson, N. J. Poloso, K. Dabbagh, and T. Z. DeSantis. 2016. Piphillin: Improved prediction of metagenomic content by direct inference from human microbiomes. *PLoS One* 11 (11):e0166104. doi: 10.1371/journal.pone.0166104.

Johnson, R. J., B. D. Folwell, A. Wirekoh, M. Frenzel, and T. L. Skovhus. 2017. Reservoir souring - latest developments for application and mitigation. *J Biotechnol* 256:57–67. doi: 10.1016/j.jbiotec.2017.04.003.

Jonsson, V., T. Osterlund, O. Nerman, and E. Kristiansson. 2016. Statistical evaluation of methods for identification of differentially abundant genes in comparative metagenomics. *BMC Genomics* 17:78. doi: 10.1186/s12864-016-2386-y.

Junemann, S., N. Kleinbolting, S. Jaenicke, C. Henke, J. Hassa, J. Nelkner, Y. Stolze, S. P. Albaum, A. Schluter, A. Goesmann, A. Sczyrba, and J. Stoye. 2017. Bioinformatics for NGS-based metagenomics and the application to biogas research. *J Biotechnol* 261:10–23. doi: 10.1016/j.jbiotec.2017.08.012.

Kang, D. D., J. Froula, R. Egan, and Z. Wang. 2015. MetaBAT, an efficient tool for accurately reconstructing single genomes from complex microbial communities. *PeerJ* 3:e1165. doi: 10.7717/peerj.1165.

Kang, D. D., F. Li, E. Kirton, A. Thomas, R. Egan, H. An, and Z. Wang. 2019. MetaBAT 2: an adaptive binning algorithm for robust and efficient genome reconstruction from metagenome assemblies. *PeerJ* 7:e7359. doi: 10.7717/peerj.7359.

Khatib, Z. I., and J. R. Salanitro. 1997. *Reservoir souring: Analysis of surveys and experience in sour waterfloods. SPE Annual Technical Conference and Exhibition*, San Antonio, Texas, 1997/1/1/.

Khullar, E., A. D. Kent, T. D. Leathers, K. M. Bischoff, K. D. Rausch, M. E. Tumbleson, and V. Singh. 2013. Contamination issues in a continuous ethanol production corn wet milling facility. *World J Microbiol Biotechnol* 29 (5):891–898. doi: 10.1007/s11274-012-1244-6.

Kieser, S., J. Brown, E. M. Zdobnov, M. Trajkovski, and L. A. McCue. 2020. ATLAS: a Snakemake workflow for assembly, annotation, and genomic binning of metagenome sequence data. *BMC Bioinf* 21 (1):257. doi: 10.1186/s12859-020-03585-4.

Klindworth, A., E. Pruesse, T. Schweer, J. Peplies, C. Quast, M. Horn, and F. O. Glockner. 2013. Evaluation of general 16S ribosomal RNA gene PCR primers for classical and next-generation sequencing-based diversity studies. *Nucleic Acids Res* 41 (1):e1. doi: 10.1093/nar/gks808.

Knights, D., J. Kuczynski, E. S. Charlson, J. Zaneveld, M. C. Mozer, R. G. Collman, F. D. Bushman, R. Knight, and S. T. Kelley. 2011. Bayesian community-wide culture-independent microbial source tracking. *Nat Methods* 8 (9):761–763. doi: 10.1038/nmeth.1650.

Lagesen, K., P. Hallin, E. A. Rødland, H. H. Stærfeldt, T. Rognes, and D. W. Ussery. 2007. RNAmmer: consistent and rapid annotation of ribosomal RNA genes. *Nucleic Acids Res* 35 (9):3100–3108. doi: 10.1093/nar/gkm160.

Lander, E. S., L. M. Linton, B. Birren, C. Nusbaum, M. C. Zody, J. Baldwin, K. Devon, K. Dewar, M. Doyle, W. FitzHugh, R. Funke, D. Gage, K. Harris, A. Heaford, J. Howland, L. Kann, J. Lehoczky, R. LeVine, P. McEwan, K. McKernan, J. Meldrim, J. P. Mesirov, C. Miranda, W. Morris, J. Naylor, C. Raymond, M. Rosetti, R. Santos, A. Sheridan, C. Sougnez, Y. Stange-Thomann, N. Stojanovic, A. Subramanian, D. Wyman, J. Rogers, J. Sulston, R. Ainscough, S. Beck, D. Bentley, J. Burton, C. Clee, N. Carter, A. Coulson, R. Deadman, P. Deloukas, A. Dunham, I. Dunham, R. Durbin, L. French, D. Grafham, S. Gregory, T. Hubbard, S. Humphray, A. Hunt, M. Jones, C. Lloyd, A. McMurray, L. Matthews, S. Mercer, S. Milne, J. C. Mullikin, A. Mungall, R. Plumb, M. Ross, R. Shownkeen, S. Sims, R. H. Waterston, R. K. Wilson, L. W. Hillier, J. D. McPherson, M. A. Marra, E. R. Mardis, L. A. Fulton, A. T. Chinwalla, K. H. Pepin, W. R. Gish, S. L. Chissoe, M. C. Wendl, K. D. Delehaunty, T. L. Miner, A. Delehaunty, J. B. Kramer, L. L. Cook, R. S. Fulton, D. L. Johnson, P. J. Minx, S. W. Clifton, T. Hawkins, E. Branscomb, P. Predki, P. Richardson, S. Wenning, T. Slezak, N. Doggett, J. F. Cheng, A. Olsen, S. Lucas, C. Elkin, E. Uberbacher, M. Frazier, R. A. Gibbs, D. M. Muzny, S. E. Scherer, J. B. Bouck, E. J. Sodergren, K. C. Worley, C. M. Rives, J. H. Gorrell, M. L. Metzker, S. L. Naylor, R. S. Kucherlapati, D. L. Nelson, G. M. Weinstock, Y. Sakaki, A. Fujiyama, M. Hattori, T. Yada, A. Toyoda, T. Itoh, C. Kawagoe, H. Watanabe, Y. Totoki, T. Taylor, J. Weissenbach, R. Heilig, W. Saurin, F. Artiguenave, P. Brottier, T. Bruls, E. Pelletier, C. Robert, P. Wincker, D. R. Smith, L. Doucette-Stamm, M. Rubenfield, K. Weinstock, H. M. Lee, J. Dubois, A. Rosenthal, M. Platzer, G. Nyakatura, S. Taudien, A. Rump, H. Yang, J. Yu, J. Wang, G. Huang, J. Gu, L. Hood, L. Rowen, A. Madan, S. Qin, R. W. Davis, N. A. Federspiel, A. P. Abola, M. J. Proctor, R. M. Myers, J. Schmutz, M. Dickson, J. Grimwood, D. R. Cox, M. V. Olson, R. Kaul, C. Raymond, N. Shimizu, K. Kawasaki, S. Minoshima, G. A. Evans, M. Athanasiou, R. Schultz, B. A. Roe, F. Chen, H. Pan, J. Ramser, H. Lehrach, R. Reinhardt, W. R. McCombie, M. de la Bastide, N. Dedhia, H. Blocker, K. Hornischer, G. Nordsiek, R. Agarwala, L. Aravind, J. A. Bailey, A. Bateman, S. Batzoglou, E. Birney, P. Bork, D. G. Brown, C. B. Burge, L. Cerutti, H. C. Chen, D. Church, M. Clamp, R. R. Copley, T. Doerks, S. R. Eddy, E. E. Eichler, T. S. Furey, J. Galagan, J. G. Gilbert, C. Harmon, Y. Hayashizaki, D. Haussler, H. Hermjakob, K. Hokamp, W. Jang, L. S. Johnson, T. A. Jones, S. Kasif, A. Kaspryzk, S. Kennedy, W. J. Kent, P. Kitts, E. V. Koonin, I. Korf, D. Kulp, D. Lancet, T. M. Lowe, A. McLysaght, T. Mikkelsen, J. V. Moran, N. Mulder, V. J. Pollara, C. P. Ponting, G. Schuler, J. Schultz, G. Slater, A. F. Smit, E. Stupka, J. Szustakowki, D. Thierry-Mieg, J. Thierry-Mieg, L. Wagner, J. Wallis, R. Wheeler, A. Williams, Y. I. Wolf, K. H. Wolfe, S. P. Yang, R. F. Yeh, F. Collins, M. S. Guyer, J. Peterson, A. Felsenfeld, K. A. Wetterstrand, A. Patrinos, M. J. Morgan, P. de Jong, J. J. Catanese, K. Osoegawa, H. Shizuya, S. Choi, Y. J. Chen, J. Szustakowki, and Consortium International Human Genome Sequencing. 2001. Initial sequencing and analysis of the human genome. *Nature* 409 (6822):860–921. doi: 10.1038/35057062.

Langmead, B., and S. L. Salzberg. 2012. Fast gapped-read alignment with Bowtie 2. *Nat Methods* 9 (4):357–359. doi: 10.1038/nmeth.1923.

Laslett, D., and B. Canback. 2004. ARAGORN, a program to detect tRNA genes and tmRNA genes in nucleotide sequences. *Nucleic Acids Res* 32 (1):11–16. doi: 10.1093/nar/gkh152.

Li, D., C. M. Liu, R. Luo, K. Sadakane, and T. W. Lam. 2015. MEGAHIT: an ultra-fast single-node solution for large and complex metagenomics assembly via succinct de Bruijn graph. *Bioinformatics* 31 (10):1674–1676. doi: 10.1093/bioinformatics/btv033.

Lomans, Bart P., Renato de Paula, Brett Geissler, Cor A. T. Kuijvenhoven, and Nicolas Tsesmetzis. 2016. *Proposal of improved biomonitoring standard for purpose of microbiologically influenced corrosion risk assessment. SPE International Oilfield Corrosion Conference and Exhibition*, Aberdeen, Scotland, UK, 2016/5/9/.

Love, M. I., W. Huber, and S. Anders. 2014. Moderated estimation of fold change and dispersion for RNA-seq data with DESeq2. *Genome Biol* 15 (12):550. doi: 10.1186/s13059-014-0550-8.

Lozupone, C., M. E. Lladser, D. Knights, J. Stombaugh, and R. Knight. 2011. UniFrac: an effective distance metric for microbial community comparison. *ISME J* 5 (2):169–172. doi: 10.1038/ismej.2010.133.

Machel, H. G. 2001. Bacterial and thermochemical sulfate reduction in diagenetic settings — old and new insights. *Sediment Geol* 140 (1):143–175. doi: 10.1016/S0037-0738(00)00176-7.

Mandal, S., W. Van Treuren, R. A. White, M. Eggesbo, R. Knight, and S. D. Peddada. 2015. Analysis of composition of microbiomes: a novel method for studying microbial composition. *Microb Ecol Health Dis* 26:27663. doi: 10.3402/mehd.v26.27663.

Manor, O., and E. Borenstein. 2015. MUSiCC: a marker genes based framework for metagenomic normalization and accurate profiling of gene abundances in the microbiome. *Genome Biol* 16:53. doi: 10.1186/s13059-015-0610-8.

Manuel, Alex. 2020. QUOREM: Query, unify, and organize research on the ecology of microorganisms. Accessed 30 July 2020. https://github.com/alexmanuele/QUOREM.

Marks, P. C., M. Bigler, E. B. Alsop, A. Vigneron, B. P. Lomans, R. De Paula, B. Geissler, and N. Tsesmetzis. 2018. MetaHCR: a web-enabled metagenome data management system for hydrocarbon resources. *Database (Oxford)* 2018:1–10. doi: 10.1093/database/bay087.

Maus, I., D. E. Koeck, K. G. Cibis, S. Hahnke, Y. S. Kim, T. Langer, J. Kreubel, M. Erhard, A. Bremges, S. Off, Y. Stolze, S. Jaenicke, A. Goesmann, A. Sczyrba, P. Scherer, H. Konig, W. H. Schwarz, V. V. Zverlov, W. Liebl, A. Puhler, A. Schluter, and M. Klocke. 2016. Unraveling the microbiome of a thermophilic biogas plant by metagenome and metatranscriptome analysis complemented by characterization of bacterial and archaeal isolates. *Biotechnol Biofuels* 9:171. doi: 10.1186/s13068-016-0581-3.

McIntyre, A. B. R., R. Ounit, E. Afshinnekoo, R. J. Prill, E. Henaff, N. Alexander, S. S. Minot, D. Danko, J. Foox, S. Ahsanuddin, S. Tighe, N. A. Hasan, P. Subramanian, K. Moffat, S. Levy, S. Lonardi, N. Greenfield, R. R. Colwell, G. L. Rosen, and C. E. Mason. 2017. Comprehensive benchmarking and ensemble approaches for metagenomic classifiers. *Genome Biol* 18 (1):182. doi: 10.1186/s13059-017-1299-7.

McMurdie, P. J., and S. Holmes. 2013. phyloseq: an R package for reproducible interactive analysis and graphics of microbiome census data. *PLoS One* 8 (4):e61217. doi: 10.1371/journal.pone.0061217.

McMurdie, P. J., and S. Holmes. 2014. Waste not, want not: why rarefying microbiome data is inadmissible. *PLoS Comput Biol* 10 (4):e1003531. doi: 10.1371/journal.pcbi.1003531.

Menke, S., M. A. Gillingham, K. Wilhelm, and S. Sommer. 2017. Home-made cost effective preservation buffer is a better alternative to commercial preservation methods for microbiome research. *Front Microbiol* 8:102. doi: 10.3389/fmicb.2017.00102.

Menzel, P., K. L. Ng, and A. Krogh. 2016. Fast and sensitive taxonomic classification for metagenomics with Kaiju. *Nat Commun* 7:11257. doi: 10.1038/ncomms11257.

Meyer, F., D. Paarmann, M. D'Souza, R. Olson, E. M. Glass, M. Kubal, T. Paczian, A. Rodriguez, R. Stevens, A. Wilke, J. Wilkening, and R. A. Edwards. 2008. The metagenomics RAST server - a public resource for the automatic phylogenetic and functional analysis of metagenomes. *BMC Bioinf* 9:386. doi: 10.1186/1471-2105-9-386.

Minich, J. J., J. G. Sanders, A. Amir, G. Humphrey, J. A. Gilbert, and R. Knight. 2019. Quantifying and understanding well-to-well contamination in microbiome research. *mSystems* 4 (4). doi: 10.1128/mSystems.00186-19.

Miranda, J., J. Seoane, A. Esteban, and E. Espi. 2019. Microbial exploration techniques: An offshore case study. In *Oilfield Microbiology*. ed. T. Skovhus, and C. Whitby, 271–297. Boca Raton: CRC Press, doi: 10.1201/9781315164700 271-297.

Mitraka, E. 2019. *MICON: The microbiologically influenced corrosion ontology*. Paper presented at the *Annual Meeting of the ISMOS*, Halifax.

Morton, J. T., J. Sanders, R. A. Quinn, D. McDonald, A. Gonzalez, Y. Vazquez-Baeza, J. A. Navas-Molina, S. J. Song, J. L. Metcalf, E. R. Hyde, M. Lladser, P. C. Dorrestein, and R. Knight. 2017. Balance trees reveal microbial niche differentiation. *mSystems* 2 (1). doi: 10.1128/mSystems.00162-16.

NACE TM0212-2018-SG, Detection, Testing, and evaluation of microbiologically influenced corrosion on internal surfaces of pipelines. 2018.

NACE TM0106-2016, Detection, testing, and evaluation of microbiologically influenced corrosion (MIC) on external surfaces of buried pipelines. 2016.

Narayanasamy, S., Y. Jarosz, E. E. Muller, A. Heintz-Buschart, M. Herold, A. Kaysen, C. C. Laczny, N. Pinel, P. May, and P. Wilmes. 2016. IMP: a pipeline for reproducible reference-independent integrated metagenomic and metatranscriptomic analyses. *Genome Biol* 17 (1):260. doi: 10.1186/s13059-016-1116-8.

Nissen, J. N., C. K. Sønderby, J. J. A. Armenteros, C. H. Grønbech, H. B. Nielsen, T. N. Petersen, O. Winther, and S. Rasmussen. 2018. Binning microbial genomes using deep learning. bioRxiv:490078. doi: 10.1101/490078.

NSERC IRC Petroleum Microbiology. 2020. https://ucalgary.ca/research/scholars/voordouw-gerrit. Accessed 30 July 2020. https://ucalgary.ca/research/scholars/voordouw-gerrit.

Nurk, S., D. Meleshko, A. Korobeynikov, and P. A. Pevzner. 2017. metaSPAdes: a new versatile metagenomic assembler. *Genome Res* 27 (5):824–834. doi: 10.1101/gr.213959.116.

Pannekens, M., L. Kroll, H. Muller, F. T. Mbow, and R. U. Meckenstock. 2019. Oil reservoirs, an exceptional habitat for microorganisms. *N Biotechnol* 49:1–9. doi: 10.1016/j.nbt.2018.11.006.

Parks, D. H., M. Imelfort, C. T. Skennerton, P. Hugenholtz, and G. W. Tyson. 2015. CheckM: assessing the quality of microbial genomes recovered from isolates, single cells, and metagenomes. *Genome Res* 25 (7):1043–1055. doi: 10.1101/gr.186072.114.

Paulson, J. N., O. C. Stine, H. C. Bravo, and M. Pop. 2013. Differential abundance analysis for microbial marker-gene surveys. *Nat Methods* 10 (12):1200–1202. doi: 10.1038/nmeth.2658.

Pearson, K. 1895. Note on Regression and inheritance in the case of two parents. *Proceedings of the Royal Society of London* 58:240–242.

Percak-Dennett, E., J. Liu, H. Shojaei, U. Luke, and I. Thomas. 2019. *High resolution dynamic drainage height estimations using subsurface DNA diagnostics. SPE Western Regional Meeting*, San Jose, California, USA, 2019/4/22/.

Pereira, M. B., M. Wallroth, V. Jonsson, and E. Kristiansson. 2018. Comparison of normalization methods for the analysis of metagenomic gene abundance data. *BMC Genomics* 19 (1):274. doi: 10.1186/s12864-018-4637-6.

Piro, V. C., T. H. Dadi, E. Seiler, K. Reinert, and B. Y. Renard. 2020. ganon: precise metagenomics classification against large and up-to-date sets of reference sequences. *Bioinformatics* 36 (Supplement_1):i12–i20. doi: 10.1093/bioinformatics/btaa458.

Quast, C., E. Pruesse, P. Yilmaz, J. Gerken, T. Schweer, P. Yarza, J. Peplies, and F. O. Glockner. 2013. The SILVA ribosomal RNA gene database project: improved data processing and web-based tools. *Nucleic Acids Res* 41 (Database issue):D590–D596. doi: 10.1093/nar/gks1219.

Rachel, N., M. Gieg, Preserving microbial community integrity in oilfield produced water. *Front Microbiol.* 2020 Oct 19;11:581387. doi: 10.3389/fmicb.2020.581387.

Reis, A. L. M., I. W. Deveson, T. Wong, B. S. Madala, C. Barker, J. Blackburn, E. Marcellin, and T. R. Mercer. 2020. A universal and independent synthetic DNA ladder for the quantitative measurement of genomic features. *Nat Commun* 11 (1):3609. doi: 10.1038/s41467-020-17445-5.

Robinson, M. D., and A. Oshlack. 2010. A scaling normalization method for differential expression analysis of RNA-seq data. *Genome Biol* 11 (3):R25. doi: 10.1186/gb-2010-11-3-r25.

Salter, S. J., M. J. Cox, E. M. Turek, S. T. Calus, W. O. Cookson, M. F. Moffatt, P. Turner, J. Parkhill, N. J. Loman, and A. W. Walker. 2014. Reagent and laboratory contamination can critically impact sequence-based microbiome analyses. *BMC Biol* 12:87. doi: 10.1186/s12915-014-0087-z.

Schloss, P. D., and J. Handelsman. 2005. Metagenomics for studying unculturable microorganisms: cutting the Gordian knot. *Genome Biol* 6 (8):229. doi: 10.1186/gb-2005-6-8-229.

Schloss, P. D., S. L. Westcott, T. Ryabin, J. R. Hall, M. Hartmann, E. B. Hollister, R. A. Lesniewski, B. B. Oakley, D. H. Parks, C. J. Robinson, J. W. Sahl, B. Stres, G. G. Thallinger, D. J. Van Horn, and C. F. Weber. 2009. Introducing mothur: open-source, platform-independent, community-supported software for describing and comparing microbial communities. *Appl Environ Microbiol* 75 (23):7537–7541. doi: 10.1128/AEM.01541-09.

Schoch, C. L., K. A. Seifert, S. Huhndorf, V. Robert, J. L. Spouge, C. A. Levesque, W. Chen, Consortium Fungal Barcoding, and List Fungal Barcoding Consortium Author. 2012. Nuclear ribosomal internal transcribed spacer (ITS) region as a universal DNA barcode marker for Fungi. *Proc Natl Acad Sci U S A* 109 (16):6241–6246. doi: 10.1073/pnas.1117018109.

Schumacher, Dietmar. 1996. Hydrocarbon-induced alteration of soils and sediments. *AAPG Memoir* 66.

Seemann, T. 2013. Barrnap: BAsic Rapid Ribosomal RNA Predictor.

Seemann, T. 2014. Prokka: rapid prokaryotic genome annotation. *Bioinformatics* 30 (14):2068–2069. doi: 10.1093/bioinformatics/btu153.

Shakya, M., C. C. Lo, and P. S. G. Chain. 2019. Advances and challenges in metatranscriptomic analysis. *Front Genet* 10:904. doi: 10.3389/fgene.2019.00904.

Shannon, C. E. 1948. A mathematical theory of communication. *Bell System Technical Journal* 27 (4):623–656. doi: 10.1002/j.1538-7305.1948.tb00917.x.

Sharma, M., H. Liub, N. Tsesmetzis, J. Handy, A. Kapronczai, T. Place, and L. Gieg. 2020. Diagnosing microbiologically influenced corrosion at a crude oil pipeline facility leak site – a holistic approach. under review.

Sharma, M., T. Place, N. Tsesmetzis, and L. Gieg. 2020b. Failure analysis for internal corrosion of crude oil transporting pipelines. In *Failure Analysis of Microbiologically Influenced Corrosion*, edited by R. B. Eckert and T. L. Skovhus. Boca Raton: CRC Press.

Sierra-Garcia, I. N., D. R. B. Belgini, A. Torres-Ballesteros, D. Paez-Espino, R. Capilla, E. V. Santos Neto, N. Gray, and V. M. de Oliveira. 2020. In depth metagenomic analysis in contrasting oil wells reveals syntrophic bacterial and archaeal associations for oil biodegradation in petroleum reservoirs. *Sci Total Environ* 715:136646. doi: 10.1016/j.scitotenv.2020.136646.

Sinha, R., G. Abu-Ali, E. Vogtmann, A. A. Fodor, B. Ren, A. Amir, E. Schwager, J. Crabtree, S. Ma, Consortium Microbiome Quality Control Project, C. C. Abnet, R. Knight,

O. White, and C. Huttenhower. 2017. Assessment of variation in microbial community amplicon sequencing by the Microbiome Quality Control (MBQC) project consortium. *Nat Biotechnol* 35 (11):1077–1086. doi: 10.1038/nbt.3981.

Skinner, K. A., and T. D. Leathers. 2004. Bacterial contaminants of fuel ethanol production. *J Ind Microbiol Biotechnol* 31 (9):401–408. doi: 10.1007/s10295-004-0159-0.

Skovhus, T. L., D. Enning, and J. S. Lee. (ed) 2017. *Microbiologically Influenced Corrosion in the Upstream Oil and Gas Industry*, 1st ed. edited by Torben L. Skovhus, Dennis Enning and Jason S. Lee. Boca Raton: CRC Press.

Song, S. J., A. Amir, J. L. Metcalf, K. R. Amato, Z. Z. Xu, G. Humphrey, and R. Knight. 2016. Preservation methods differ in fecal microbiome stability, affecting suitability for field studies. *mSystems* 1 (3). doi: 10.1128/mSystems.00021-16.

Spearman, C. 1904. The proof and measurement of association between two things. *The American Journal of Psychology* 15 (1):72–101. doi: 10.2307/1412159.

Sui, H. Y., A. A. Weil, E. Nuwagira, F. Qadri, E. T. Ryan, M. P. Mezzari, W. Phipatanakul, and P. S. Lai. 2020. Impact of DNA extraction method on variation in human and built environment microbial community and functional profiles assessed by shotgun metagenomics sequencing. *Front Microbiol* 11:953. doi: 10.3389/fmicb.2020.00953.

Sun, S., R. B. Jones, and A. A. Fodor. 2020. Inference-based accuracy of metagenome prediction tools varies across sample types and functional categories. *Microbiome* 8 (1):46. doi: 10.1186/s40168-020-00815-y.

Tamames, J., and F. Puente-Sanchez. 2018. SqueezeMeta, A highly portable, fully automatic metagenomic analysis pipeline. *Front Microbiol* 9:3349. doi: 10.3389/fmicb.2018.03349.

Tanizawa, Y., T. Fujisawa, and Y. Nakamura. 2018. DFAST: a flexible prokaryotic genome annotation pipeline for faster genome publication. *Bioinformatics* 34 (6):1037–1039. doi: 10.1093/bioinformatics/btx713.

Te Stroet, Chris, Jonathan Zwaan, Gerard de Jager, Roy Montijn, and Frank Schuren. 2017. *Predicting sweet spots in shale plays by DNA fingerprinting and machine learning. SPE/AAPG/SEG Unconventional Resources Technology Conference*, Austin, Texas, USA, 2017/7/24/.

Tourlousse, D. M., S. Yoshiike, A. Ohashi, S. Matsukura, N. Noda, and Y. Sekiguchi. 2017. Synthetic spike-in standards for high-throughput 16S rRNA gene amplicon sequencing. *Nucleic Acids Res* 45 (4):e23. doi: 10.1093/nar/gkw984.

Truong, D. T., E. A. Franzosa, T. L. Tickle, M. Scholz, G. Weingart, E. Pasolli, A. Tett, C. Huttenhower, and N. Segata. 2015. MetaPhlAn2 for enhanced metagenomic taxonomic profiling. *Nat Methods* 12 (10):902–903. doi: 10.1038/nmeth.3589.

Tsesmetzis, N., E. B. Alsop, A. Vigneron, F. Marcelis, I. M. Head, and B. P. Lomans. 2018. Microbial community analysis of three hydrocarbon reservoir cores provides valuable insights for the assessment of reservoir souring potential. *Int Biodeterior Biodegrad* 126:177–188. doi: 10.1016/j.ibiod.2016.09.002.

Tsesmetzis, N., M. J. Maguire, I. M. Head, and B. P. Lomans. 2017. Protocols for investigating the microbial communities of oil and gas reservoirs. In *Hydrocarbon and Lipid Microbiology Protocols: Field Studies*, edited by Terry J. McGenity, Kenneth N. Timmis and Balbina Nogales, 65–109. Berlin, Heidelberg: Springer Berlin Heidelberg.

Tsesmetzis, N., P. Yilmaz, P. C. Marks, N. C. Kyrpides, I. M. Head, and B. P. Lomans. 2016. MIxS-HCR: a MIxS extension defining a minimal information standard for sequence data from environments pertaining to hydrocarbon resources. *Stand Genomic Sci* 11:78. doi: 10.1186/s40793-016-0203-5.

Utro, F., N. Haiminen, E. Siragusa, L. J. Gardiner, E. Seabolt, R. Krishna, J. H. Kaufman, and L. Parida. 2020. Hierarchically labeled database indexing allows scalable characterization of microbiomes. *iScience* 23 (4):100988. doi: 10.1016/j.isci.2020.100988.

Vigneron, A., E. B. Alsop, B. Chambers, B. P. Lomans, I. M. Head, and N. Tsesmetzis. 2016. Complementary microorganisms in highly corrosive biofilms from an offshore oil production facility. *Appl Environ Microbiol* 82 (8):2545–2554. doi: 10.1128/AEM.03842-15.

Vigneron, A., E. B. Alsop, P. Cruaud, G. Philibert, B. King, L. Baksmaty, D. Lavallee, P. Lomans, N. C. Kyrpides, I. M. Head, and N. Tsesmetzis. 2017. Comparative metagenomics of hydrocarbon and methane seeps of the Gulf of Mexico. *Sci Rep* 7 (1):16015. doi: 10.1038/s41598-017-16375-5.

Vigneron, A., E. B. Alsop, P. Cruaud, G. Philibert, B. King, L. Baksmaty, D. Lavallee, B. P. Lomans, E. Eloe-Fadrosh, N. C. Kyrpides, I. M. Head, and N. Tsesmetzis. 2019. Contrasting pathways for anaerobic methane oxidation in gulf of Mexico cold seep sediments. *mSystems* 4 (1). doi: 10.1128/mSystems.00091-18.

Vigneron, A., E. B. Alsop, B. P. Lomans, N. C. Kyrpides, I. M. Head, and N. Tsesmetzis. 2017b. Succession in the petroleum reservoir microbiome through an oil field production lifecycle. *ISME J* 11 (9):2141–2154. doi: 10.1038/ismej.2017.78.

Vigneron, A., A. Bishop, E. B. Alsop, K. Hull, I. Rhodes, R. Hendricks, I. M. Head, and N. Tsesmetzis. 2017c. Microbial and isotopic evidence for methane cycling in hydrocarbon-containing groundwater from the Pennsylvania region. *Front Microbiol* 8:593. doi: 10.3389/fmicb.2017.00593.

Vigneron, A., P. Cruaud, E. Alsop, J. R. de Rezende, I. M. Head, and N. Tsesmetzis. 2018. Beyond the tip of the iceberg; a new view of the diversity of sulfite- and sulfate-reducing microorganisms. *ISME J* 12 (8):2096–2099. doi: 10.1038/s41396-018-0155-4.

Vigneron, A., I. M. Head, and N. Tsesmetzis. 2018b. Damage to offshore production facilities by corrosive microbial biofilms. *Appl Microbiol Biotechnol* 102 (6):2525–2533. doi: 10.1007/s00253-018-8808-9.

Voordouw, Gerrit, Brenton Buziak, Shiping Lin, Alexander Grigoriyan, Krista M. Kaster, Gary Edward Jenneman, and Joseph John Arensdorf. 2007. *Use of nitrate or nitrite for the management of the sulfur cycle in oil and gas fields. International Symposium on Oilfield Chemistry*, Houston, Texas, U.S.A., 2007/1/1.

Voordouw, G., A. A. Grigoryan, A. Lambo, S. Lin, H. S. Park, T. R. Jack, D. Coombe, B. Clay, F. Zhang, R. Ertmoed, K. Miner, and J. J. Arensdorf. 2009. Sulfide remediation by pulsed injection of nitrate into a low temperature Canadian heavy oil reservoir. *Environ Sci Technol* 43 (24):9512–9518. doi: 10.1021/es902211j.

Walker, A. W., J. C. Martin, P. Scott, J. Parkhill, H. J. Flint, and K. P. Scott. 2015. 16S rRNA gene-based profiling of the human infant gut microbiota is strongly influenced by sample processing and PCR primer choice. *Microbiome* 3:26. doi: 10.1186/s40168-015-0087-4.

Walters, W., E. R. Hyde, D. Berg-Lyons, G. Ackermann, G. Humphrey, A. Parada, J. A. Gilbert, J. K. Jansson, J. G. Caporaso, J. A. Fuhrman, A. Apprill, and R. Knight. 2016. Improved bacterial 16S rRNA gene (V4 and V4-5) and fungal internal transcribed spacer marker gene primers for microbial community surveys. *mSystems* 1 (1). doi: 10.1128/mSystems.00009-15.

Walters, W. A., J. G. Caporaso, C. L. Lauber, D. Berg-Lyons, N. Fierer, and R. Knight. 2011. PrimerProspector: de novo design and taxonomic analysis of barcoded polymerase chain reaction primers. *Bioinformatics* 27 (8):1159–1161. doi: 10.1093/bioinformatics/btr087.

Wang, J., M. Liu, H. Xiao, W. Wu, M. Xie, M. Sun, C. Zhu, and P. Li. 2013. Bacterial community structure in cooling water and biofilm in an industrial recirculating cooling water system. *Water Sci Technol* 68 (4):940–947. doi: 10.2166/wst.2013.334.

Wang, Y., K. Wang, Y. Y. Lu, and F. Sun. 2017. Improving contig binning of metagenomic data using [Formula: see text] oligonucleotide frequency dissimilarity. *BMC Bioinformatics* 18 (1):425. doi: 10.1186/s12859-017-1835-1.

Wedemeyer, A., L. Kliemann, A. Srivastav, C. Schielke, T. B. Reusch, and P. Rosenstiel. 2017. An improved filtering algorithm for big read datasets and its application to single-cell assembly. *BMC Bioinformatics* 18 (1):324. doi: 10.1186/s12859-017-1724-7.

Wemheuer, F., J. A. Taylor, R. Daniel, E. Johnston, P. Meinicke, T. Thomas, and B. Wemheuer. 2020. Tax4Fun2: prediction of habitat-specific functional profiles and functional redundancy based on 16S rRNA gene sequences. *Environmental Microbiome* 15 (1):11. doi: 10.1186/s40793-020-00358-7.

Wetterstrand, K.A. 2019. The cost of sequencing a human genome. Accessed 30 July 2020. www.genome.gov/about-genomics/fact-sheets/Sequencing-Human-Genome-cost.

Wheeler, D. A., M. Srinivasan, M. Egholm, Y. Shen, L. Chen, A. McGuire, W. He, Y. J. Chen, V. Makhijani, G. T. Roth, X. Gomes, K. Tartaro, F. Niazi, C. L. Turcotte, G. P. Irzyk, J. R. Lupski, C. Chinault, X. Z. Song, Y. Liu, Y. Yuan, L. Nazareth, X. Qin, D. M. Muzny, M. Margulies, G. M. Weinstock, R. A. Gibbs, and J. M. Rothberg. 2008. The complete genome of an individual by massively parallel DNA sequencing. *Nature* 452 (7189):872–876. doi: 10.1038/nature06884.

Wilkinson, T. J., S. A. Huws, J. E. Edwards, A. H. Kingston-Smith, K. Siu-Ting, M. Hughes, F. Rubino, M. Friedersdorff, and C. J. Creevey. 2018. CowPI: A rumen microbiome focussed version of the PICRUSt functional inference software. *Front Microbiol* 9:1095. doi: 10.3389/fmicb.2018.01095.

Wood, D. E., J. Lu, and B. Langmead. 2019. Improved metagenomic analysis with Kraken 2. *Genome Biol* 20 (1):257. doi: 10.1186/s13059-019-1891-0.

Wu, Y. W., B. A. Simmons, and S. W. Singer. 2016. MaxBin 2.0: an automated binning algorithm to recover genomes from multiple metagenomic datasets. *Bioinformatics* 32 (4):605–607. doi: 10.1093/bioinformatics/btv638.

Zhao, R., Z. M. Summers, G. D. Christman, K. M. Yoshimura, and J. F. Biddle. 2020. Metagenomic views of microbial dynamics influenced by hydrocarbon seepage in sediments of the Gulf of Mexico. *Sci Rep* 10 (1):5772. doi: 10.1038/s41598-020-62840-z.

3 A Case for Molecular Biology

How This Information Can Help in Optimizing Petroleum Top Side Facility Operations

Geert M. van der Kraan and Nora Eibergen

CONTENTS

3.1 Introduction ... 42
3.2 The Oilfield as an Interlinked Set of Artificial Ecosystems ... 44
3.2.1 Limits of Life, Seen in the Light of Oilfields ... 44
3.2.2 The Oilfield and Its Topside Facilities Described as Industrial Ecosystems ... 45
3.2.2.1 Seawater Intake ... 45
3.2.2.2 The Oil Reservoir ... 46
3.2.2.3 Production Wells ... 46
3.2.2.4 Gathering and Trunk Lines ... 46
3.2.2.5 Oil Water Separators ... 47
3.2.2.6 Water Processing ... 47
3.2.2.7 Injectors ... 47
3.2.2.8 Open Ponds ... 47
3.2.3 Microbial Life Classified according to Redox Reactions, Providing Energy ... 47
3.2.4 Different Classification Systems of Microorganisms ... 48
3.3 Monitoring Programs ... 52
3.3.1 The Molecular Biology Toolbox & Diversity Studies ... 53
3.3.2 Quantitative Polymerase Chain Reaction (qPCR) ... 54
3.3.3 What Does Such a Monitoring Program Help Answer & What Can Be Done with This Information? ... 54
3.4 Case Studies ... 56
3.4.1 The Effect of Adding Oxygen Scavenger to Produced Water That Contains Low Sulfate ... 56

3.4.2 The Effect of Nitrate Addition When Deaerated Seawater Is Mixed with Sulfide Containing Produced Water at Injectors (FPSO-System)....58
3.4.3 How Molecular Biology Can Determine Where in a System Issues Occur....61
3.4.4 How Effective Biocide Treatment Can Significantly Reduce Biogenically Formed Suspended FeS (Oilfield in Latin America).....64
3.5 Conclusions....66
References....66

3.1 INTRODUCTION

Microbial activity in petroleum environments and waters has been known for over a century. Despite this fact, in today's petroleum industry, the concept of microbiology is still viewed as unfamiliar. This is somewhat understandable, as the petroleum industry is a macro industry in which engineers typically deal with processes that occur on a large scale. Examples thereof are the large water volumes deployed during seawater flooding of oil reservoirs, the vast amounts of produced oil and multiphase fluids, and large manmade constructions like offshore drilling platforms. In contrast, microbiological processes and their associated biogenic chemical conversions play a role at the microscale. A typical microbial cell has the size of one cubic micrometer ($1\ \mu m^3$) and is therefore orders of magnitude smaller than the dimensions at which the petroleum industry operates.

What is critical to understand, however, is that the biogenic processes taking place at the microscale level (microns) do have a significant impact on the macro-scale operations of this industry. For example, one only has to consider the technical and economic impact biogenic reservoir souring has. Reservoirs are soured when H_2S is produced by sulfate-reducing microorganisms that metabolize sulfate introduced through secondary oil recovery methods of sea waterflooding. This process can significantly decrease the value of hydrocarbon crudes [1]. Another significant and persistent problem initiated on the micro scale is Microbially-Influenced Corrosion (MIC). This is not a special form of corrosion but merely a biogenically accelerated form of common galvanic corrosion. The acceleration arises from various metabolic processes that microbes perform in petroleum environments when in close proximity to carbon steel, one of the main materials found in oilfield tanks, pipelines, and other topside facilities. The OPEX/CAPEX expenses associated with a pipeline failure arising from MIC have been estimated to be $10 to 20 million. This estimate does not include the required environmental cleanup from an oil spill that might result from such a failure [2].

Other examples are 1) the downtime a plugged injector can cause for the oilfield, and 2) The biogenic formation of FeS & FeS_2 scale and Schmoo (a thick wax like substance of biomass, oil components, and iron-sulfur scale). These byproducts of souring and MIC can plug various oilfield components, causing downtime and resulting in OPEX economic impacts to oil production. Together, these three issues, which can be directly linked to microbial processes occurring on the microscale, represent the majority of macroscale issues in the petroleum industry with a biogenic origin. While microbial life and its associated microscale processes have significant negative consequences to the macroscale operations of oil & gas operations, microbial contamination can sometimes be overlooked in this industry. To achieve a more complete understanding of how microbial processes affect macroscale processes in oilfield operations and to subsequently contain these, the following items should be implemented for every upstream or midstream petroleum-related industrial system that carries water or a multiphase flow: 1) a monitoring program to detect the presence and activities of microbes and 2) a microbial control program tailored towards controlling the population and activities of microbes in oilfield topside facilities. When these monitoring and treatment practices are effectively implemented in parallel, the unwanted effects of microorganisms dwelling in oilfield systems can be contained.

Ideally, monitoring practices are implemented in such a way that the following questions can be answered: 1) Where in my system does microbial contamination occur and how heavy is this contamination? 2) What is the best way to contain the contamination? (where and how to treat), and 3) Is my treatment working? Today's molecular biology toolbox offers a set of technologies which are perfectly suited to this task. Since the mid-1990s, the use of molecular biology techniques has taken flight and has also quietly entered the petroleum industry. Since the mid-2000s, molecular biology has slowly gained in popularity as a monitoring tool for oilfield topside facilities. However, despite their recent rapid progression and vast potential, these techniques are still not yet *routinely* used in the petroleum industry. This raises the following question: why are these sophisticated molecular biology technologies used so infrequently to monitor oilfield assets and drive chemical treatment decisions? Part of the answer is likely due to an information gap that lies between the format in which results are presented (i.e. a list of microbial genera and their relative abundances or cell counts of specific microbial groups of interest) and (chemical) engineering solutions and actions that can be deployed in the oilfield to diminish the effects of problematic microorganisms (i.e. biogenic H_2S, MIC, and FeS accumulation). In this chapter, an attempt is made to bridge this gap by covering three interlinked angles of view. First an oilfield and its topside facility will be described as a system of different artificially created ecosystems, which are connected via various liquid streams; along with a short overview of groups of microorganisms that can thrive in these different ecosystems. Second, a short overview is given regarding the most frequently used molecular biology tools for monitoring programs and the results they generate. Finally, a set of case studies will be presented in which the deployment of tools and interpretation of these results has led to 1) valuable insights into the microbial status of the oilfield and topside operations, 2) actions that were deployed in the field to mitigate issues, 3) changes regarding chemical treatment and chemical engineering to obtain better control of the system, and 4) learnings about the efficacy of other deployed functional chemicals. In this way, the chapter aims to provide more understanding about how molecular tools can and should be used to enhance their impact in improving day-to-day oilfield operations, reducing overall chemical spend, and resolving issues caused by microbial contamination.

3.2 THE OILFIELD AS AN INTERLINKED SET OF ARTIFICIAL ECOSYSTEMS

3.2.1 Limits of Life, Seen in the Light of Oilfields

Oilfields and their topside facilities can be viewed as industrially created ecosystems. In order to understand the types of microorganisms, which can inhabit such systems, we must first view them in the context of the physical limits of life (Table 3.1).

Let's first consider the physical property of temperature. On the low end of the spectrum Archaea, single celled microorganisms that differ from Bacteria, have been found to survive in arctic ice at a temperature of −20°C [3]. Microorganisms that can survive the other end of the temperature spectrum can be found around so-called "black smokers" on the ocean floor [4]. At such locations, microorganisms have been discovered, which are able to survive at temperatures up to 121°C. On the other hand, it has been shown in oil reservoirs, which hold additional stringent conditions, that propagation of microbial life becomes challenging and has empirically not been observed above 80°C [5].

Microbial life has also been found to survive across a wide spectrum of salinities that range from zero salt to salt saturation (250+ grams/liter) [6]. The salinity, however, does have a large impact on the type of microorganisms that can thrive. Halotolerant and halophilic microorganisms have developed strategies to deal with elevated salinity levels and e.g. have a salt-in strategy which helps them to deal with the osmotic pressure caused by their environment [7]. In addition, salt levels can also have an impact in which redox reactions are favorable from a biogenic energy perspective. Oilfield waters can range from saline (~24 g/L) to highly saline (>75 g/L) [8]. Salinity has an impact on the microbial diversity but is not an influencer regarding the presence or absence of life.

Microbial life has also proven itself to be versatile across almost the entire pH range from pH 0 to pH 11 (the full pH range however, runs from 0 to 14). While some microorganisms have been able to survive in acidic lakes around volcanic areas [9], other types of microbes can survive at high alkalinities. For example, soda lakes, which have a pH between 9 and 12, have been shown to be teeming with life [10]. Oilfields and their topside facilities typically hold a pH between 5.5 and 7. Seawater injection systems however are often of a basic pH since seawater has a pH between

TABLE 3.1
Physical Limits of Life

Parameter	Lower Limit	Upper Limit	Notes
Temperature (°C)	−20	121	~80°C for oilfields
Ph	0	11[a]	
Salinity (gram/liter)	0	Saturation	
Water activity (A_w)	0.65	1[b]	

[a] The common pH range is from 0 to 14.
[b] Pure water.

7.5 and 8.6. While the pH of an environment will have a significant impact on the type of microorganisms that can thrive, pH, unlike temperature, is not a qualifier to exclude if microorganisms can survive or not.

The fourth physical constraint to microbial life is the availability of liquid water, often reported as the water activity A_w. Microbial life cannot exist when there is no liquid water available. In oilfield systems, it should not be a surprise that there is plenty of water available, and this physical constraint is therefore not a constraint of relevance. In conclusion, the only physical property that has some power of exclusion for microbial life is temperature. All the other oilfield environmental parameters are well within the limits of microbial life and only impact the type of microorganisms that can survive.

3.2.2 The Oilfield and Its Topside Facilities Described as Industrial Ecosystems

From a microbiological perspective, an oilfield and its topside service facility is a connected network of different ecosystems. This network extends from the reservoir itself, to the production wells, to all the topside industrial elements (pipes, tanks, filters, separators, and pumps), and ends at the injectors. Each of these elements hold their own environment with distinct temperatures, pHs, salinities, and hydraulic retention times of liquid streams. Chemically some of these environments also hold heavily alternating conditions. For example, dissolved oxygen levels can vary widely among the various elements of an oilfield. In addition, different elements may be exposed to different chemical treatments of corrosion inhibitors, de-emulsifiers, biocides, oxygen scavengers, etc. These variations in chemical environment can dictate which reduction/oxidation couples an environment holds. To illustrate how widely these conditions can differ throughout an oilfield system, the most commonly encountered industry elements will be discussed. A basic description about the encountered ecosystems in an oilfield is given. The focus is on the differences encountered in each industrial element [11].

3.2.2.1 Seawater Intake

Seawater intake often is several meters below the air/water interface. Depending where the system is built, the temperature of the seawater can range from 4°C to 28°C [12]. To prepare it for injection, the seawater is either electro chlorinated or treated with an oxidizing biocide to reduce the microbial load. Then it is passed over coarse filters to remove most of the large organic particles in the water. As an unwanted consequence, some of the dissolved organic molecules will get oxidized as well, albeit in cool seawater, the organic load is in general low. The water is then often sent over an additional water clarifier and passed through a fine filter (2 micrometer) before it is subsequently passed over a deaeration tower. Here, the water is trickled down in a large vessel, where the surface area is enlarged by the inclusion of packing material, (viz. porous cylinders or O-rings). The deaeration may proceed via either vacuum stripping or wet combustion. In vacuum stripping, O_2 is stripped from the water by replacement with N_2, and the resulting N_2/O_2 gas mixture is reacted with methanol, regenerating the N_2 for gas stripping (minox process). As an unwanted consequence, methanol might enter the water system and provide an additional organic food source for microbes to utilize. A mechanical deaeration tower can bring down the O_2 content of the water from 8–10 ppm to 30 to 50 ppb (or 100 ppb in case of older deaerators). The trickling of water over a large surface area creates an environment that is prone to biofilm formation. Therefore, deaerator towers are often incubators for microbial life. In most cases, chemical scavenging of oxygen is performed after the water has passed via a sump in the bottom of the tower. This is often done via the addition of ammonium bisulfite or sodium bisulfite. Sulfite (SO_3^{2-}) ions can be considered an "activated" form of sulfate ions (SO_4^{2-}) and can therefore serve as an alternate, attractive electron acceptor for microbes, thereby inducing (additional) H_2S formation.

After the deaeration, a nitrate addition point is sometimes put in place. Nitrate (NO_3^-) provides a more potent electron acceptor than sulfate or sulfite and the ruling theory is that nitrate-reducing microbes can outcompete sulfate/sulfite reducing cells via substrate exclusion (more on this topic later) [13]. It should be noted that nitrate introduces yet another electron acceptor into the water and that when dosed inappropriately can yield an opposite effect [14,15]. The location where nitrate may be added is typically also the place where organic biocides are added, and other functional chemicals are added like corrosion inhibitors and oxygen scavengers. Cross-reactivity of chemicals added should therefore always be checked. The deaerated seawater is pumped via pipelines to injectors, and here, the water stream can be co-mingled with the separated produced water before re-injection to the reservoir. Produced water may contain residual H_2S, which is now in combination with seawater that could contain traces of oxygen and/or nitrate. In a proper functioning oxygen management system, there should however be no oxygen left anymore.

3.2.2.2 The Oil Reservoir

Produced multiphase fluids start their journey in the reservoir. Here, where oil & (formation or injected) water are contained in porous rock (reservoir stone), the fluids often see elevated temperatures, pressures, and salinities. The temperatures can range from 35 to 150°C but often are in the range of 50 to 90°C. Pressures can go up to a range of 150 bars.

3.2.2.3 Production Wells

The multiphase fluids are produced at production wells and are pushed out by the internal field pressure or via secondary recovery processes usually involving (sea) water injection. The fluids come into contact with metal from the well casing and are pumped to the surface. It is here, where the multiphase fluid sees often its first chemical treatment, mostly injected down well. When produced, the multiphase fluid passes a decompression tank and is brought to a lower pressure, e.g. 4 bar. At the wells, chemicals are frequently added to the multiphase fluids, e.g. H_2S scavengers, biocides, or corrosion inhibitors.

3.2.2.4 Gathering and Trunk Lines

The wells are almost always connected to local storage tanks and are transported there via trunk lines. The temperature then has decreased by several degrees °C vs. the reservoir. Pipelines can vary in their internal diameter and flowrate, so conditions differ in these pipelines as well.

3.2.2.5 Oil Water Separators

From local storage tanks, the multiphase fluid is transported via pipelines to oil water separators, here a de-emulsifier is often added, and the oil and water are separated from each other. Separators in principle come in two forms, horizontal and vertical, and differ slightly in mode of operation. The separated oil is then sent to a local storage tank and the separated water is then transported to a water processing site in order to further clean it and remove carried-over traces of oil remaining.

3.2.2.6 Water Processing

Residual oil is removed using filtration units and or hydrocyclones, and the remaining oil is sent back to the separator tank. It is important to note that this returned water and oil traces create a feedback loop and thus some of the water here is recycled and pumped to an earlier part of the topside facility. Often oxygen scavenger is added together with biocides and corrosion inhibitors to the water treatment plant.

3.2.2.7 Injectors

The separated water is then sent via pipelines toward injector or disposal wells, where either alone or mixed with deaerated seawater, it is re-injected into the oil reservoir. By the time it is re-injected, the temperature of the water has often dropped below 45°C toward so-called mesophilic conditions. The separated oil often is kept locally at the site when onshore, where water further separates from the oil phase, this water is often re-added to the oilfields water cycle at the location of the separator. It should be noted that also here, a water feedback loop is created. Feedback loops create a situation where microbial cells are retained and therefore create an unwanted "bioreactor" setup (recycling of biomass).

3.2.2.8 Open Ponds

Open ponds function to balance water in the system are often fully aerobic. Due to the possible ingress of this water into the system, traces of O_2 might enter the system unnoticed.

An important conclusion to draw here is that the water changes in ionic composition, temperature, pH, redox state, and salinity all the time. Which in turn means that different sets of microorganisms are best suited to thrive in the different environments provided at the various locations within the system. Thus, the local environmental conditions of a particular component (tank, pipe, separator, well, etc.) will determine the rise of the most dominant metabolic activity in that component and whether the activity is detrimental to the component's function and integrity, In the next session of this chapter, the types of microorganisms that thrive under the different environments will be described.

3.2.3 Microbial Life Classified according to Redox Reactions, Providing Energy

Life on this planet has diversified into every niche it could find. From a chemical point of view, life harvests the energy that is generated from reduction/oxidation reactions (or redox reactions). In these reactions, one molecule donates electrons to another molecule, generating energy to fuel biochemical reactions and thus microbial life itself. In the absence of oxygen, however, microbial cells can deploy various other reducer/oxidizer combinations to generate chemical energy. This is important to consider for oxygen-depleted or oxygen-diminished oilfield components where many such electron acceptors can be found. The oil phase holds a variety of different electron donors in the form of the petroleum molecules themselves [16], but the injection and formation water can also hold usable components, such as short organic acids. Through seawater flooding, significant amounts of sulfate, an electron acceptor, may enter petroleum systems. This can be partially counterbalanced by the implementation of sulfate stripping membranes. Since the mid-2000s, nitrate is sometimes added to injection water to stimulate growth of nitrate reducers at the expense of H_2S-producing sulfate-reducers via competitive inhibition, since both groups compete for the same substrates. This practice adds yet another type of electron acceptor to the system.

In addition, all the metals of which wells and topside facilities are mostly built can potentially serve as a source of electrons to fuel biogenic life [17]. On the reverse end, various metal ions can be used as electron shuttles (either donor or acceptor) to facilitate other redox reactions. Then there is the process of fermentation which can be simplified as one part of the same molecule gets oxidized and one part reduced. Lastly in the deep anoxic zones of oilfield system, special classes of microbes can utilize CO_2 as an electron acceptor (often present as HCO_3^- (bicarbonate)). Even the organic functional chemicals applied may be used by microbes as electron donors, and this will be covered later in the chapter. The relevant and frequently found groups of microbes in oilfield systems are mostly classified according to which electron acceptor they utilize. A simplified overview of the electron acceptors found in the various components of an oilfield is given in Table 3.2. As mentioned before, the amount of electron donors is plenty in the form of functional chemicals, the petroleum molecules itself, and various organic acids in the water, which will be shortly described below.

3.2.4 Different Classification Systems of Microorganisms

As indicated by Table 3.2, fluids and multiphase fluid streams in petroleum systems provide a variety of environmental conditions and different chemical compositions. The note "everything is everywhere, the environment selects" [18] refers to the fact that microorganisms are everywhere, but that the conditions of a given environment will select which ones will become dominant in that environment. Microorganisms can be classified according to several different types of criteria. Some examples of these criteria include: 1) the environmental conditions under which a microorganism thrives, 2) the electron acceptors a microorganism uses, 3) the genetic diversity of a microorganism, and 4) the state in which a microorganism grows in their environment. These criteria are all interconnected. For example, when a microorganism is adapted to thrive at high temperatures, the genes that enable them to cope with these high temperatures will be reflected in their genetic material (DNA & RNA). Life can be divided into three major so-called kingdoms. The Bacteria, Archaea, and Eukarya. The first two only hold unicellular organisms and thus are relevant for oilfields, the focus of this paragraph is on subgroups of those two kingdoms. Archaea are less familiar than Bacteria, Archaea are often called life extremists and have diversified to

TABLE 3.2
Environments, Their Conditions, and Potential Redox Couples[a]

	T in °C	Pres[b]. (Bar)	Salinity (%)	pH	Donor	Acceptor[c]
Reservoir	>60	50–150	0.5-satur.	5.5–7	Oilphase VFA[d]	**$CO_2(HCO_3^-)$** **SO_4^2** **$S_2O_3^{2-}$** S^0 Fe^{2+}
Wells	>55	50–150	0.5-satur.	6.5–7	Oilphase VFA Functional-chemicals	$CO_2(HCO_3^-)$ SO_4^{2-} $S_2O_3^2$ S^0 Fe^{2+}
Multiph[e]. tank	20–70	4	0.5-satur.	6.5–7	Oilphase VFA Functional-chemicals	$CO_2(HCO_3^-)^2$ **SO_4^{2-}** $S_2O_3^{2-}$ S^0 Fe^0/Fe^{2+}
Water tank	10–25	4	0.5-satur.	6.5–7	Functional-chemicals VFA Oil trace	$CO_2(HCO_3^-)$ SO_3^2 **Fe^0** /[h]Fe^{x+} O_2 trace
Pipeline	4–50	4	0.5-satur.	6.5–7	Functional-chemicals VFA Oil trace	$CO_2(HCO_3^-)$ SO_3^2 Fe^0 /Fe^{x+} O_2 trace
Oil/ Water-Separator	35–55	4	0.5-satur.	6.5–7	Oil phase VFA Functional-chemicals	$CO_2(HCO_3^-)$ SO_4^{2-} SO_3^{2-} Fe^0 /Fe^{x+} O_2 trace[g]
Open pond	ambient	1	variable	7	Oil phase	**O_2**
Filter	ambient	4	0.5-satur.	6.5–7	Trace oil VFA Functional-chemicals	$CO_2(HCO_3^-)$ SO_4^{2-} SO_3^{2-} Fe^0 /Fe^{x+} O_2 trace
Injector	ambient	4	0.5-satur.	6.5–7	Trace oil VFA Functional-chemicals	$CO_2(HCO_3^-)$ SO_4^2 SO_3^{2-} Fe^0 /Fe^{x+} O_2 trace

(*Continued*)

TABLE 3.2 (Continued)
Environments, Their Conditions, and Potential Redox Couples

	T in °C	Pres[b]. (Bar)	Salinity (%)	pH	Donor	Acceptor[c]
Injector	10–40	Above Reservoir Pressure	0.5-satur.	6.5–7	VFA	$CO_2(HCO_3^-)$
					Functional-chemicals	SO_4^{2}
						SO_3^{2} Fe^0/Fe^{x+} O_2 trace NO_3^-
Seawater intake	4–25	1	3.1–3.8	7.5–8.6	VFA organics	$\mathbf{O_2}$
Chlorination	4–25	1	3.1–3.8	7.5–8.6	VFA	$\mathbf{O_2}$ $HClO_3^-$
					Oxid[f]. VFA Functional-Chemicals Organics	O_2
Deaeration tower	10–25	1	3.1–3.8	7	VFA	$\mathbf{O_2}$
					Oxid. VFA Functional-Chemicals (methanol)	NO_3^-(post) SO_4^{2-}
						SO_4^{2-}

[a] Units depicted in this table are *average numbers*, based on various oilfields, exceptions can always be found.
[b] Pressure.
[c] In bold, most likely the most dominant electron acceptors in the environment.
[d] Volatile fatty acids.
[e] Multiphase.
[f] Oxidized.
[g] Traces of oxygen should not be in the system under superb field operation, various practical examples have however shown that traces of oxygen do occur in systems.
[h] Iron ions under oxidizing conditions can be 2+ or 3+, when sufficient O_2 is present 3+.

occupy hostile ecosystems, hence they can be found often in oilfield environments. Also, the metabolic pathway of methanogenesis, meaning the utilization of CO_2 as a final electron acceptor forming methane (CH_4), is in the sole hands of the archaeal domain of life. Methanogenesis kicks generally when no other electron acceptors are available but CO_2 (HCO_3^-). When temperatures, e.g. rise above 70°C, often the waters become dominated by Archaeal species. In petroleum environments often, groups are indicated via their prime dissimilatory metabolism. This prime dissimilatory metabolism depends on the most potent and prevalent electron acceptor present at that time in the system and is partially dependent on concentration gradients. Both largely dictate which organism type will dominate the population. A summary of groups is

TABLE 3.3
Classification of Microorganisms Based on a Prevailing Environmental Parameter

Environmental Parameter	Classification of Microorganism	Where Found (Example)
Cold temperatures −20°C to 10°C	Psychrophilic	Cold seawater intake
Cold seawater intake 10°C to 45°C	Mesophilic	Cool zone, pond, injector
High temperatures 41°C to 70°C	Thermophilic	Reservoir, well, after heat exchange
Extreme temperatures 65°C to 121°C	Hyper thermophilic	Reservoir
Saline 7%–15%	Halophilic/Halotolerant[a]	Halophilic/Halotolerant[a]
Extremely saline >15%	Extremely halophilic	Hypersaline formation water
High pH 8.5 to 11	Alkaliphilic	Alkaliphilic
Medium pH 5.5 to 8.5	Neutrophilic	Various oilfield topside related waters
Low pH 1 to 5.5	Acidophilic	Formation water/soured waters

[a] Halotolerant indicates species that can tolerate high salt levels, halophilic species require high salt for growth.

provided in Table 3.3. Here, the focus is on genetic diversity as a tool for microbial classification. The genetic diversity of a microorganism is determined by the sequence of its 16S rRNA gene, a chosen marker gene [19,20], which allows determination of microbial diversity; this is covered in detail in other parts of this book.

A simple differentiation can also be made based on how microorganisms appear in their environment, either in suspension, free floating in the water phase (planktonic), or in the form of a biofilm attached to a surface (like a pipeline or tank wall), often referred to as 'sessile.' The vast majority of microorganisms appear in biofilms attached to a surface. Microorganisms can also be classified regarding the environment in which they thrive best. Table 3.4 lists the most common examples found in petroleum-related environments.

TABLE 3.4
Relevant Groups of Microbes in Oilfield Systems

Classification	Metabolic Activity	Issues Caused	Examples of Genera
Sulfate reducers [21]	Convert sulfate to H_2S	Biogenic H_2S causing souring and exacerbate chemical corrosion via chemical reaction with metals. Quality reduction of the oil phase	*Desulfovibrio* *Desulfomicrobium* *Desulfohalobium* *Nitrospirae* *Thermodesulfovibrio* *Thermodesulfobacteria* *Archaeoglobus*
Sulfur and thiosulfate reducers [22]	Convert elemental sulfur (S^0) or thiosulfate ($S_2O_3^{2-}$) to H_2S.	Associated with heavy MIC and the same issues as sulfate reducers.	*Anaerobaculum* *Halanerobium* *Fervidobacterium* *Thermotoga* *Desulfuromonas* *Proteus*

(Continued)

TABLE 3.4 (Continued)
Relevant Groups of Microbes in Oilfield Systems

Classification	Metabolic Activity	Issues Caused	Examples of Genera
Nitrate reducers [23]	Convert NO_3^- to either NO_2^- or dinitrogen gas (N_2) depending on temperature	Associated with biocorrosion, pit formation due to biofilms and potential production of ammonia (NH_4^+) or nitrite. Induction of MIC in some cases.	Scattered across the entire phylogenetic tree of life. Many *Pseudomomas* members are nitrate reducing
Methanogenic Archaea [24]	Convert CO_2 (HCO_3^-) to Methane	Utilize H_2 or small organic molecules for their metabolism, found in deep anoxic environments. Can induce MIC	*Methanosarcinales* *Methanocalculus* *Methanobacteriales*
Sulfide and sulfur oxidizers (SOPs) [25]	Oxidize H_2S back to elemental sulfur (or sulfate). Oxidation of Sulfur to Sulfate under a reduced oxygen environment or when nitrate is present	Heavily associated with MIC. This type of microbes will create elemental sulfur which is detrimental to metals. Can also occur as a consequence of oxygen scavenger dosing (bisulfite)	*Sulfurospirillum* *Thiomicrospira* *Sulfurimonas* *Sulfurovum* *Arcobacter*
'Acid producers'	Cells which deploy fermentation processes and thereby excrete small organic acids	MIC via acidification and indirectly create substrates for sulfate reducers.	Very diverse. *Clostridia*

3.3 MONITORING PROGRAMS

Monitoring programs traditionally use culture-dependent methods to provide a rough estimation of the severity of the microbial contamination at different locations in a system. A microbial monitoring program should be an integral part of the management of an oilfield and its topside facility. Typically, the monitoring program should go hand in hand with a microbial control mitigation program to avoid and reduce the severity of the issues caused by microbial life in petroleum systems, namely MIC, souring, biogenic scale, and plugging. When done correctly, these programs can not only help in reducing these issues but can also help optimize the day-to-day operation of an oilfield topside facility. Since this Chapter focusses on newer molecular biology technologies, the authors direct readers who are seeking information and a discussion of the traditionally used culture-based monitoring methods, to the reference documents below:

- For the evaluation of pipelines, NACE Standard test method TM0212-2018, which details the detection, testing and evaluation of MIC on internal surfaces. This includes also molecular biology techniques;

- NACE Standard test method TM0194-2014, provides a detailed overview of the traditional serial extinction dilution (MPN) method for SRB, acid producers and "total viable bacteria counts."

The rest of this chapter segment will focus on molecular biology techniques.

3.3.1 The Molecular Biology Toolbox & Diversity Studies

Microorganisms thrive in industrial ecosystems and can cause a variety of issues. Since the turn of the millennium, there has been a rise of various types of molecular biology analyses into the petroleum industry. As many microorganisms are unculturable, the use of molecular biology has proven to be effective at shedding more light on the microbial inhabitants of environmental samples. The application of these techniques is predominantly targeted at maintaining asset integrity (viz. top site facilities). The main driver for this has been the limitations that are intrinsically bound to the deployment of microbiological culture techniques (only 0.1% to 10% of microorganisms are culturable [26], long duration times of tests).

One of the important molecular biology techniques that has been developed and extensively reported in scientific publications are the Next Generation Sequencing (NGS) technologies (e.g. illumina and the decommissioned 454 pyrosequencing platform [27,28]). The deployment of these types of technologies has dramatically increased the available knowledge on microbial diversity. Most ecological studies that are performed are targeted at the 16S rRNA gene, which has been used since the late 1980s as a phylogenetic classifier for species determination. Due to its function in the cellular process, it is a highly conserved DNA sequence. The 16S rRNA gene sequences isolated from environmental samples are often used to determine which microorganism types thrive in these environments. The 16S rRNA genes isolated from environmental samples of unknown microbial composition, such as those collected from oil or gas associated ecosystems, can be sequenced using NGS technologies. These sequences can then be matched to databases of sequences from known microorganisms (e.g. via the RDP, SILVA, or GreenGenes database).

The 16S rRNA gene sequence alone cannot be used a classifier for metabolism or activities. However, oilfield-associated environments are special ecosystems. As has been extensively explained in the earlier parts of this chapter, the metabolic activities of a set of microorganisms in the reservoir, wells, and other topside facilities of an oilfield are determined by the most potent and prevailing electron acceptor and the environmental conditions. Thus, a genus and, in some cases, a species can be linked to information about the environmental conditions and electron acceptors present at the site of sampling to make accurate predictions about active metabolic processes in a given environment. Since metabolism in oilfields is electron acceptor dependent, 16S rRNA gene phylogenomics still provides some information about basic metabolism. But for this reason, when NGS technologies are deployed, an effort should be made as well to describe the ecosystem from which the samples were collected. At minimum, the temperature, salinity, and pH of the water/multiphase fluid should be recorded. Preferably, the type of oil (paraffinic or aromatic), which functional chemicals are in the system, and whether the environment is oxic or anoxic should also be noted. Finally, an ionic composition of the water will be helpful for analysis. From this combination of data, ecosystem description, diversity study, and quantification of the number of prokaryotic cells, the microbial status of a system becomes an emergent property. A detailed overview of all relevant parameters which could have an impact on the environment and thus the microbial population is provided by [29]. As will be shown in the case studies described later in this chapter, this type of analysis can provide information about the presence of a detrimental metabolic process and the efficacy of added functional chemicals. This is based on the general idea that the microbes found in an environment, are a reflection of their environment and thus reveal something about it. While NGS is a powerful monitoring technology when combined with the reporting of a chemical and physical description of the ecosystem sampled, the outcome of NGS is always relative. It does not provide an absolute cell quantification, but rather a relative abundance estimate based on the number of 16S rRNA genes sequenced. To determine the absolute cell levels, present in a given sample volume, a different set of technologies has to be utilized. One of these technologies, quantitative PCR (qPCR), uses molecular biology to quantify the number of cells in a given sample. This technique will be briefly discussed in the next paragraphs, since it is covered in detail in other parts of the book.

3.3.2 Quantitative Polymerase Chain Reaction (qPCR)

Quantitative PCR (qPCR) is a DNA-based method that specifically detects and quantifies a given marker gene. These marker genes are usually chosen to identify cells with potential of certain metabolic processes (for example, hydrogen sulfide production). This allows an absolute quantification of total cells and subgroups of cells with particular characteristics. One drawback of the molecular methods is the fact that these are difficult to perform in the field without sending samples off-site for analysis and that the timelines to obtain results are relatively long (days to weeks, but still shorter than e.g. the 28 days of e.g. SRP bug bottles). Another notification should be made, which is that qPCR for some time will also detect cells that are no longer alive. Therefore, these should be combined with a quick and field deployed technology. ATP analysis, while not discussed in this chapter, is a good tool for this task.

3.3.3 What Does Such a Monitoring Program Help Answer & What Can Be Done with This Information?

In the previous paragraphs, the outcomes of a preferred set of recommended techniques for a robust monitoring program has been provided, and justification as to why these technologies are preferred has been described. But what set of relevant questions does such a monitoring program help answer? When deploying combinations of techniques, one can obtain answers to the following questions:

1) How many cells are there in the fluid of each of my sampled locations? (bug bottles, qPCR)

2) How active are the microorganisms in that particular environment? (bug bottles, ATP)
3) Which microbe types are present? – (mostly up to the genus level) (NGS – diversity)
4) Are the microbial numbers (microbial contamination) increasing or not? (requires sampling longitudinal in time), (NGS diversity + ATP measured overtime, sometimes qPCR)
5) Is the population performing a detrimental process and what does this entail? – this requires an overlay with the geochemical description of each environment. (NGS diversity + geochemical description of the environment)
6) Is the population shifting toward a more unwanted one (for example when something has changed in the system, different chemical treatment regime, changes in the overall chemical engineering setup of the system, etc.) (all the recommended technologies)
7) What is the effectiveness of my current microbial control program? (Bug bottles, ATP analysis, and qPCR)
8) Is the population responding well to biocide treatment and to suggested changes to this program after the first analysis of the system? – Where do I actually need to treat?
9) Are other functional chemicals performing or at risk? (NGS diversity study + metabolic interpretation)
10) What are actionable items one can derive from this information to better the situation in case of microbial issues? – this includes changes in both a chemical regime and engineering changes (all the discussed technologies).

A recommended combination of methods is summarized in Table 3.5. In the next part of the chapter, several case studies will be discussed that describe actionable items that can be derived from such monitoring programs. This will also include practical examples that will resonate with current system operation.

TABLE 3.5
Summarizing Table Regarding Advised Time Windows to Deploy Certain Techniques When Monitoring

Monitoring Method	Type of Information	How Often
Most probable Number (MPN) (bug bottles)	Quantitative information About sulfate reducers and acid producers, trends are more important than the absolute values. High levels of detection indicate an issue, low levels do not indicate the absence of an issue.	Weekly
ATP analysis	Quantitative information is provided about a combination of activity and level of cells. Like bug bottles, trends are more important than absolute values.	Biweekly
Geochemical analysis of liquid streams	Determination of temperature, pH, salinity, ion composition of the water, type of oil, presence of H_2S, chemical treatments, etc.	Monthly

(Continued)

TABLE 3.5 (Continued)
Summarizing Table Regarding Advised Time Windows to Deploy Certain Techniques When Monitoring

Monitoring Method	Type of Information	How Often
qPCR analysis	Quantitative analysis of microorganisms of certain types and those with genes that impart certain functions that are important in petroleum certain functions that are important in petroleum in absolute cells/ml. (or cells/cm^2 in case of biofilms)	Quarterly
16S rRNA gene-based diversity study (Phylogenomics) Utilizing next Generation Sequencing (NGS)	Provides information on which genera (and potentially) Provides information on which genera (and potentially) population changes over time in response to an environmental change or change in chemical treatment. NGS-derived diversity data should always be analyzed along with data from the geochemical analysis, to obtain a holistic view of the system.	Quarterly

3.4 CASE STUDIES

3.4.1 The Effect of Adding Oxygen Scavenger to Produced Water That Contains Low Sulfate

In this case study, which is published and extensively discussed here [30], the microbial diversity of an onshore production and water separation system was investigated since some unaccounted for H_2S was detected. Since this study was performed in 2007, the microbial diversity was investigated using an older technology; clone library construction. The system in this case study involves a mature oilfield in shallow marine deposit. The assets relevant to this study include a primary separator, a secondary separator, and an oxygen scavenger tank, and the focus of this case study will be on the chemical and environmental changes between the primary oil-water separator and the third oxygen scrubber tank that are imposed enforced by system operation.

In this system, the multiphase fluids produced via several production wells was around 55°C, and the surface water had a pH of around 6.4 and a salinity of 73 grams/liter. Very little, if any, sulfate was present in these fluids. The multiphase fluids were pumped to an oil separation site where the oil is separated from the water via several flotation tanks. Fluids arriving at the primary separator were about 35°C. In the secondary separator, the fluids were passed over a heat exchanger to aid the separation, boosting the temperature to slightly over 50°C. Oxygen removal via chemical addition of 15 ppm ammonium bisulfite (NH_4HSO_3) was performed in a third tank. The majority of O_2 in this tank is introduced via tap water, which is added to lower the salinity of the water to around 58 grams/liter. These conditions are summarized in Table 3.6 and the microbial diversity in each environment (taken and simplified from [30]) is shown in Figure 3.1.

The most important observation from the diversity study was that fundamentally different diversities were observed when the chemical composition of the water was altered. This alteration was induced by the field operation in the form of adding

TABLE 3.6
Main Differences between the Primary Separator and the Oxygen Scrubber

Parameter	Primary Separator	Oxygen Scrubber Tank
Temperature	35°C	~53°C
Electron acceptor	HCO_3^-/metal ions/ferm	Metal ions/SO_3^{2-}
Salinity	73 grams/liter	58 grams/liter

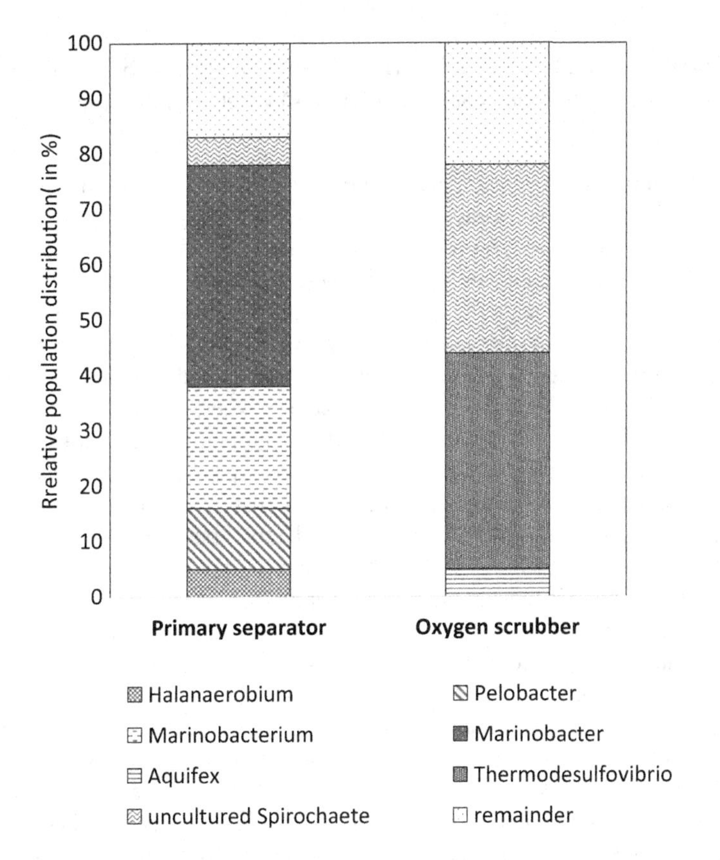

FIGURE 3.1 Microbial diversity in the primary separator vs. the oxygen scrubber scavenger.

ammonium (nitrogen source), increasing the temperature from 35°C to above 50°C (mesophilic to thermophilic), adding tap water to lower salinity, and most importantly, adding sulfite (an electron acceptor) to scavenge oxygen. These environmental and chemical changes led to the emergence of *Thermodesulfovibrio* in the oxygen scrubber tank. The presence of these microorganisms likely explains why the oxygen scrubber tank gave off H_2S at times. Another lineage found in the scrubber tank was *Aquifex*, members of this genus are able to utilize a variety of compounds, like thiosulfate, elemental sulfur, and molecular hydrogen. Hydrogen sulfide was not detected in the other tanks and was therefore likely induced by the addition of the ammonium bisulfite to the oxygen scrubber tank. The recommendations for this field were to minimize the amount of oxygen scavenger added so that excess electron acceptor is not present after reacting with oxygen. While the results from this case study might seem obvious, they do demonstrate how simple changes in an environment can greatly impact the microbial community and metabolic processes present in that environment. The use of technologies from the molecular biology toolbox is an excellent way to detect the impact of seemingly simple operational changes and to investigate the "microbial status" of a system.

3.4.2 The Effect of Nitrate Addition When Deaerated Seawater Is Mixed with Sulfide Containing Produced Water at Injectors (FPSO-System)

The focus of this case study is on the combination of diversity and cell levels throughout a Floating Production Storage and Offloading (FPSO) system. This FPSO combines a seawater intake with a production water stream, and because these two water sources have totally different compositions, the effect of their mixing on the microbial population can be significant. The full details of this study are described in [31], and an abbreviated description of the system will be provided in this chapter. The investigated FPSO system was designed to take in seawater, filter it, and chlorinate it via electrochlorination. The water then passed over a deaerator where it was stripped of oxygen. The water was then amended with nitrate (NO_3^-) and pumped to an injector where it was co-mingled with separated produced water. This separated produced water was the aqueous product from separating produced multiphase fluid in several three phase separators. This produced water stream contained >100 ppm H_2S. Figure 3.2 shows the qPCR levels of the total bacteria. A qPCR analysis determined absolute cell level amounts in the different topside facilities and indicated where microbial proliferation was highest. Figure 3.2 lists the population distributions determined through NGS analysis, which was done to identify potential harmful populations in the system.

As shown in Figure 3.2, qPCR analysis revealed a significant increase in the number of cells (after passing through the deaerator. Indeed, the cell counts of samples taken downstream of the deaerator were 4–5 orders of magnitude greater than those taken upstream of the deaerator. This is an indication that a deaerator is a place in which the microbial population is enriched. Deaeration towers are filled with water trickling over a large surface area, which is created by filling the tower up with cylinders or other packings which magnify the surface area over which the water trickles. Microbes tend to grow on surfaces where there is liquid flow present. This topside industrial element is a prime environment for microbial growth. What can be observed from the cell levels is that nowhere in this system is the cell count so high as in the effluent of the deaeration tower but that deeper in the system, more in the direction of the injectors still remain high, albeit there being a totally different microbial population. The cell levels indicate that they are highest in the actual deaeration tower and remain high after this. The microbial populations however are totally different. The microbial diversity analyses for several water samples from this system are shown in Figure 3.3. The diversity data is shown for water samples downstream of the

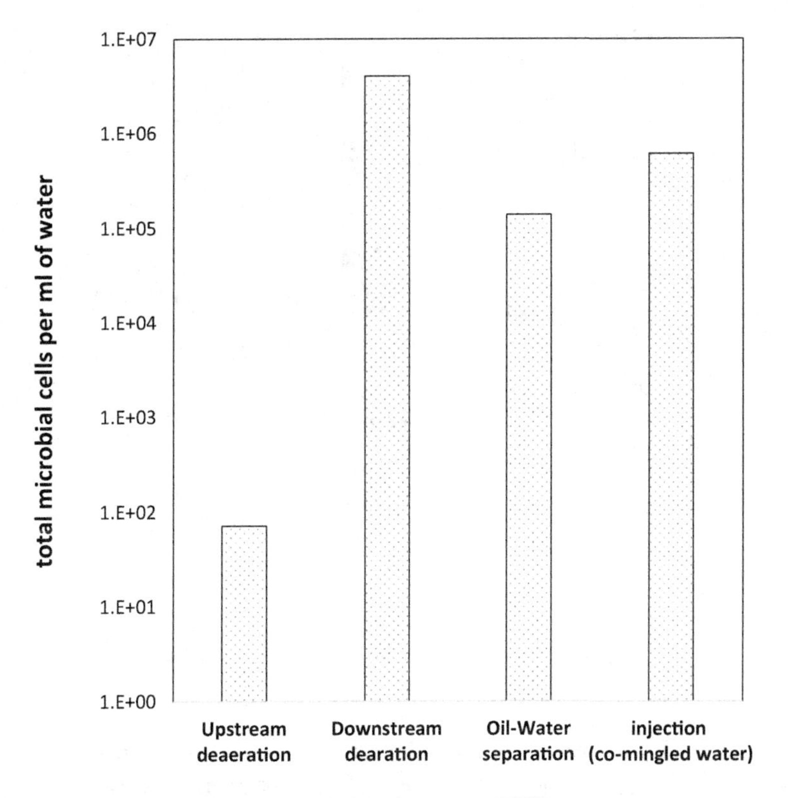

FIGURE 3.2 qPCR result of the sampled environments FPSO.

deaeration tower, production water samples from the three phase separators, and comingled water samples at the injection pumps. High cell counts were measured for all these samples (Figure 3.3); however, the diversity studies show that the microbial populations present in these samples are very different and well suited to the artificially created ecosystems from which they were collected.

Several observations could be made from these population distributions. First, the water flowing out of the deaeration tower shows a significant presence of members of the genus *Methylophaga* [32]. *Methylophaga* are methylotrophs and many species belonging to this genus are specialized in utilizing C1-components (methanol or methane). A speculation could be that this deaeration tower was operated using N_2 and methanol service and that perhaps traces of methanol were accidentally entering the process water. This provided a plausible explanation for the presence of *Methylophaga*. The produced water taken from one of the separators could be summarized as a combination community. About three quarters of this community belonged to the genus *Desulfocaldus*; the other quarter belonged to a genus consisting mostly of fermenting members, called *Caminicella* [33]. This combination of fermenting species with sulfate reducers is consistent with the microbial composition of most produced waters. Due to the dominance of sulfate-reducing *Desulfocaldus*, it should not come as a surprise that H_2S was detected at the separators. At the injectors, where these two water types were mixed, a very different microbial population

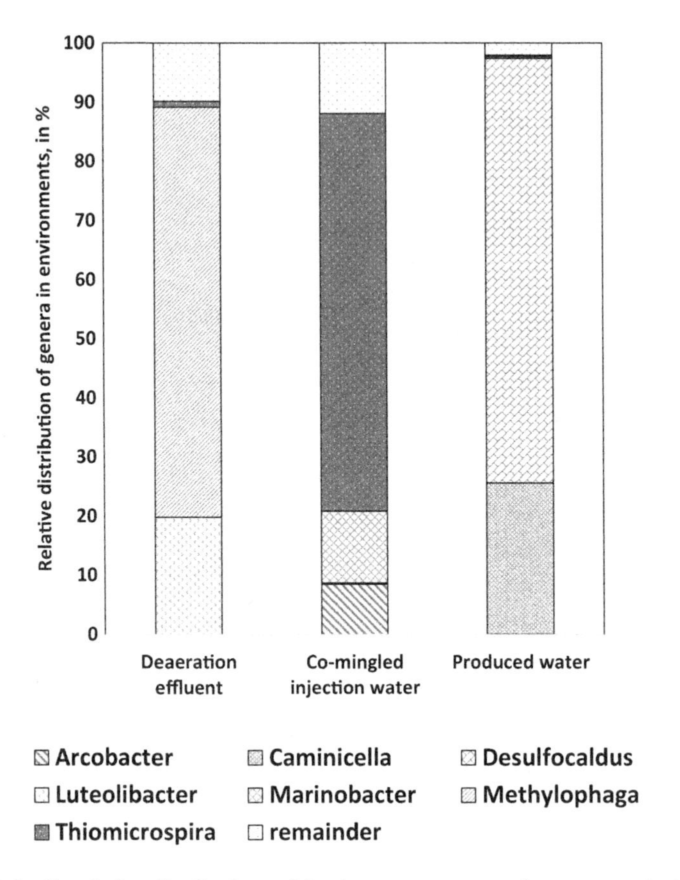

FIGURE 3.3 Population distributions of the three water sources (pyrosequencing).

emerged. Around 70% of the population consisted of a genus called *Thiomicrospira* [34]. This genus falls into the category of the epsilon proteobacteria and is specialized in sulfur/sulfide oxidation (discussed in the earlier parts of this chapter). Sulfur and sulfide oxidation can occur under oxygenic conditions or when nitrate is present. In this system, it is possible that both were present, albeit oxygen most likely under very limited conditions. The fact that these conditions existed in the same environment presented an issue for the injectors. Epsilon proteobacteria have the capacity to produce elemental sulfur and polysulfur, which is highly corrosive to metal.

In summary, the conditions at this FPSO were far from optimal from a microbial control perspective, and the system needed to be placed back under microbial control. The cell levels after the deaerator and in the rest of the system were high, the separators were dominated by a sulfate reducing community, and the injectors were populated by a large community of epsilon proteobacteria with the ability to create a scale or MIC hazard. This case study demonstrates the strength of the combining qPCR with microbial diversity studies. Using these molecular biology techniques together, one can identify and locate issues in a particular system and can estimate the risk that the issue presents based on the type of microbial contamination detected. In order to make proper recommendations, more samples must be taken longitudinally over time, during which an effective treatment program should be implemented.

This being said, one clear learning from this study was that the nitrate addition point needed to be moved. Nitrate should not be added directly after the deaerator, but rather later in the system to avoid the enrichment of epsilon proteobacteria and unwanted utilization of this nitrate topside. This is a very good example of how one can use diversity analyses to observe how added functional chemicals are performing and what unintended effects they may be having in the system. What this case study shows is that molecular biology techniques can indeed be used as an alternate way of monitoring the system and can provide valuable information on system performance.

3.4.3 How Molecular Biology Can Determine Where in a System Issues Occur

We will next summarize the results of a case study in which a molecular biology-based monitoring program was deployed in an oilfield topside facility in Latin America. Full details on this study were previously published [35]. In this facility, multiphase fluids flowed through a complex network of wells and local storage tanks before they were pumped to an oil/water processing site. A simplified overview of this system can be found in Figure 3.4.

The monitoring program for this system used qPCR and NGS analysis to assess the microbial status of the system. The qPCR analysis was performed to quantify Bacteria, Archaea, and sulfate reducers (Figure 3.5). From this analysis, three important observations were made. First, samples collected on the production side (upstream of the entrance to the water treatment system) contained low levels of microbial cells in the water phase. Most samples contained 10^2 to 10^3 cells/mL of

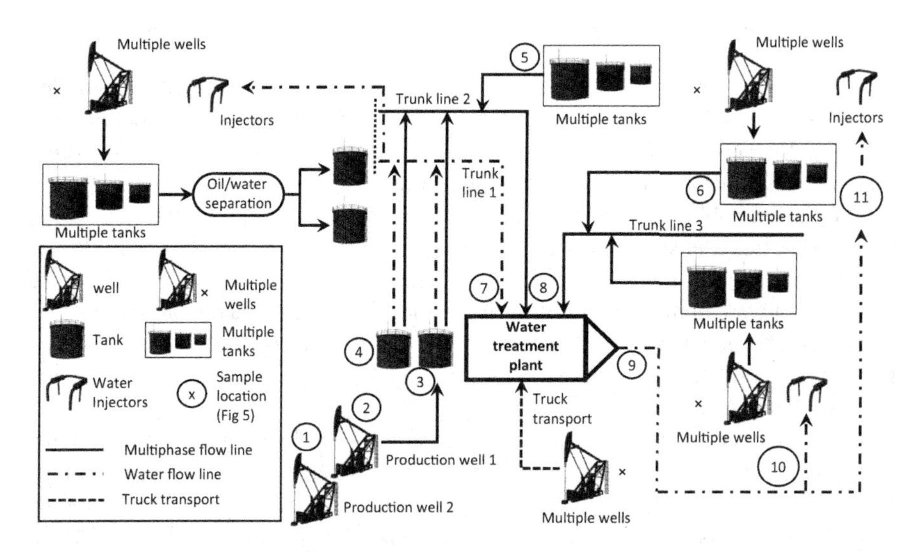

FIGURE 3.4 A simplified overview of the oilfield topside facility under investigation.

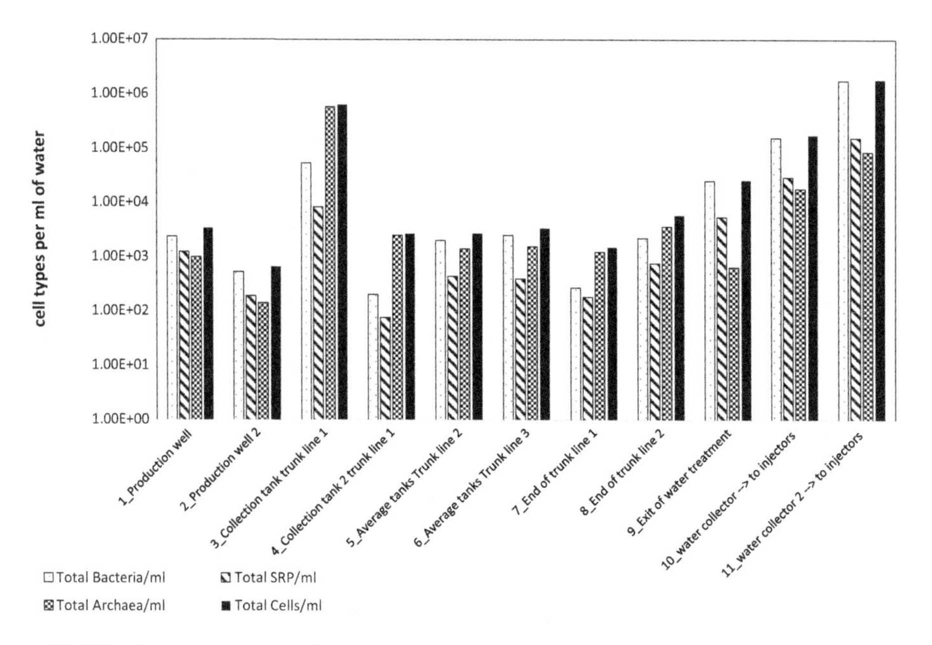

FIGURE 3.5 qPCR results of the sampled topside facilities.

each of the quantified classes of microorganisms. The exception to this, however, was a local storage tank in the production system (indicated in Figure 3.5, *collection tank trunk line 1*). In this tank, cell levels spiked to close to 10^6 cells/ml and most of these cells were Archaea.

This finding illustrates how molecular biology techniques can help define particular microbial control issues in a system such that it can be appropriately addressed. How the presence of Archaea could change a biocide dosing strategy has yet to be determined; it is however beyond the scope of this chapter.

The second observation was that a cell level increase was measured for samples that were collected closest to the water treatment system and post separator, with the highest cell levels found at the water treatment site exit. In these samples, the presence of Archaea dropped, and the population was dominated by Bacteria.

The third notable observation from this data was a shift in microbial population for samples taken upstream and downstream of the water treatment system. While samples upstream of the water treatment system were dominated by Archaea, those downstream were dominated by Bacteria. In summary, microbial contamination was not an issue for samples collected from the production side of the system. However, for samples collected from the water injection system, cell levels were high, which may be detrimental to the injectors.

To get a better indication of how detrimental the microbial populations might be, the microbial diversity of the samples needs to be evaluated and analyzed in the context of the cell count data. For example, the sample collected from Local Storage Tank 2 was one for which a dense microbial population was detected by qPCR. In particular, the sample was found to be dominated by Archaea (Figure 3.6). Microbial

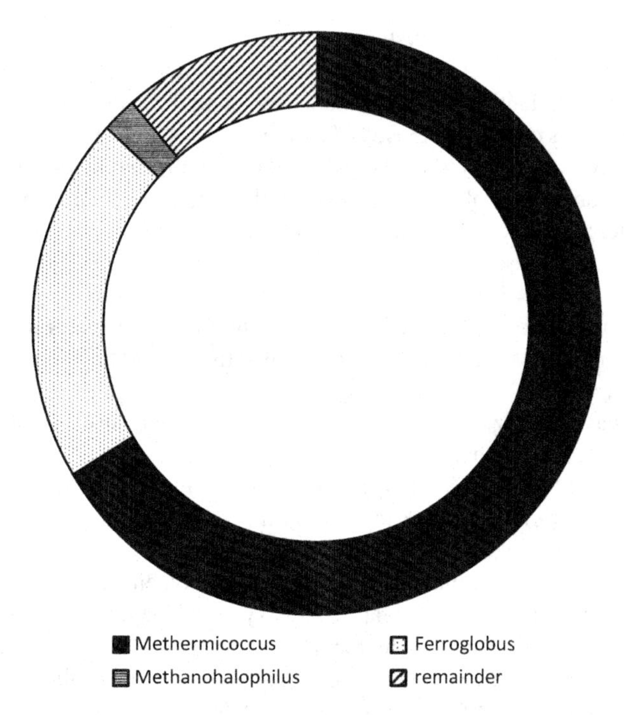

FIGURE 3.6 Archaeal microbial population distribution in the local storage tank (with elevated cell level).

diversity studies on this sample revealed that the sample was dominated by two archaeal genera (Figure 3.6). The most dominant genus, making up about 70% of the archaeal population, was *Methermicoccus* [36]. Species belonging to this genus are thermophilic and use simple organic components, such as methylamine, for methanogenesis. The second most-abundant archaeal genus in this sample was *Ferroglobus* [37], which made up about 20% of the population. Members of this genus are thermophilic and are known for their ability to oxidize Fe^{2+} to Fe^{3+} [38]. The significant presence of these members in the community indicates an MIC risk.

A comparison of the microbial diversities of multiple samples revealed significant fluctuations in microbial populations throughout the topside facility. At the production wells, there was a large presence of Actinobacteria. The local storage tanks were shown to have the most diverse population consisting of Clostridia, Deltaproteobacteria, Epsilonproteobacteria, Synergistia, Thermotogaea, and Gamma proteobacteria. At the main gathering line, an increased abundance of Delta-, Epsilon-, and Gamma proteobacteria was detected. By the time the water exited the water treatment site, over 90% of the population consisted of Gamma proteobacteria and was particularly rich in *Pseudomonas*. This comparative analysis of the microbial diversity data revealed some interesting trends. Most importantly, the community changes indicated an increase in oxygen intrusion and reflected the decrease in through the system. The large presence of Deltaproteobacteria suggests that H_2S is being produced. Additionally, Epsilonproteobacteria appear, which pinpoints subsequent biogenic

H_2S oxidation, possibly due to systemic O_2 intrusion or nitrate, leading to a diminished oxygenic environment. The presence of *Pseudomonas* in the injection system is likely due to sufficiently low temperatures and enough O_2 intrusion to support the growth of this genus comprised mostly of aerobic species. This, taken in combination with the observed cell level increase closer to the injectors, led to the conclusion that the major microbial control problems were within the water treatment site and the injection system. In conclusion, this field audit of the topside facility yielded the following four conclusions:

1) On the production site, there were little to no microbial issues with microbes. The recommendation was made to monitor these sampling sites occasionally for changes;
2) There was a localized issue in one of the local storage tanks. Treatment of this tank was recommended;
3) There was a severe microbial control issue in the water treatment site. The recommendation was made to develop a proper microbial control treatment program for this site;
4) The diversity studies indicated an increased oxygen intrusion closer to the injectors, which indicated a gap in the oxygen management program used in the field's topside facility. The recommendation was to investigate if the current treatment was sufficiently eliminating oxygen from the system and/or identify where the oxygen intrusion was occurring.

3.4.4 How Effective Biocide Treatment Can Significantly Reduce Biogenically Formed Suspended FeS

In the previous case study of this chapter, the results of a field audit were described, and the recommended solutions and monitoring plans presented. In this case study, solution implementation and monitoring effectiveness of this solution is included in the study. The oilfield in question suffered from a topside suspended biogenic FeS issue, viz. blackwater. The temperature of the producer wells was 70°C but the liquid temperature dropped to 39°C at the injection wells. The salinity of the field was around 7% total salts and had a complex matrix of bivalent ions. Some of the wells reported trace amounts of elemental sulfur. Suspended FeS scale passed through the entire system as part of the total suspended solids. This blackwater issue was particularly apparent at the injector wells where several acid jobs [39] had been performed in an effort to remediate the problem. It was clear that the current microbial control program was not sufficient and that an effective solution for the topside facility was required.

NGS analysis of the microbial communities present in this system identified a diverse set of genera, including those known to contribute to topside souring and MIC issues. Examples of these genera, all of which contain sulfate reducing species, included *Thermodesulforhabdus*, *Desulfacinum*, *Desulfotomaculum*, and *Desulfitobacter* [21]. Sulfur-reducing genera, such as *Thermotoga*, *Kosmotoga*, and *Thermosipho* [40], were also found. In addition, *Halanerobium* [41], a genus often associated with MIC, was detected.

qPCR analysis revealed that cell levels varied greatly throughout the system. High cell levels (10^6 total cells/ml) were found at the injectors. The local storage tanks of the production system showed relatively high cell levels (10^5 cells/ml) and 10^4 cells/ml were detected for samples collected from the wells. In the water processing site, several filters had been deployed to clean up the water. In this case, filtration had an effect. The water directly downstream of a filter lacked any black precipitate, and cell levels in this water and at the water storage tanks downstream were reduced to 10^2 cells/ml.

In summary, the issue for this field, topside FeS formation, was most problematic at the injector wells. Cell levels were high in several locations within the system and problematic genera known to form H_2S from either sulfate or sulfur were present. Geochemical analysis of the water showed elevated levels of iron ions, which under these circumstances, combined with the formed H_2S to form FeS.

To identify the best solution for this problem, an enrichment culture was prepared from various water samples collected from the system. In a laboratory study that has been published previously [42], several biocidal solutions were tested for performance against this enrichment culture. The best performing biocidal solution, both from a technical performance and cost perspective, was a combination of Glutaraldehyde, a Quaternary ammonium salt, and Tris hydroxy nitromethane (THNM). This solution was implemented at the oilfield site, and the efficacy of the solution was monitored extensively for a year. Figure 3.7 shows the reduction in the number of interventions after the optimized biocidal solution was implemented.

After the implementation of the optimized biocidal solution, the amount of required interventions reduced by an average of 40%. The reduction in interventions was accompanied by a significant reduction in OPEX costs for the field. Here the direct benefit of a strong monitoring program containing qPCR and microbial diversity studies, determining where the microbial contamination is the highest, determining where biocide treatment is most needed. qPCR determined where in the

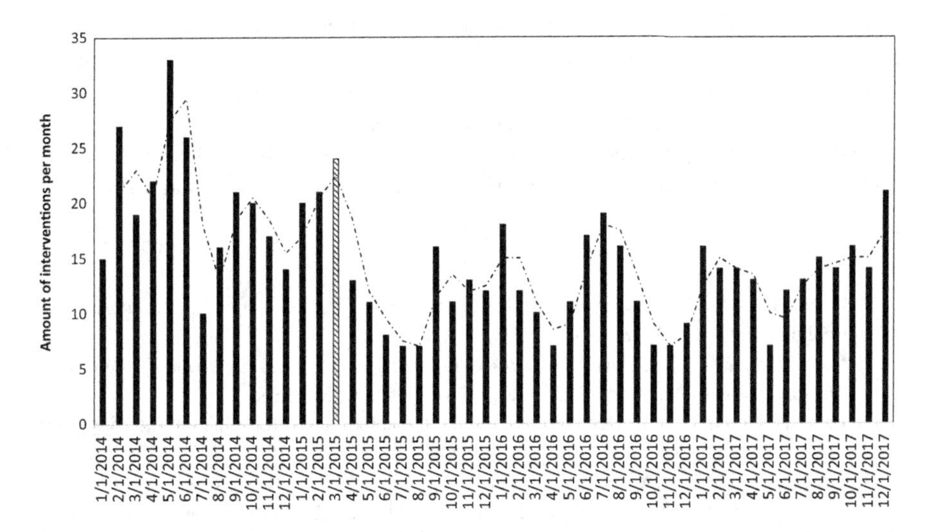

FIGURE 3.7 Amount of interventions required per month before and after implementation.

system the microbial contamination is highest and required a higher treatment level, the NGS data determined which populations were detrimental and of concern, it allowed for a holistic layout of the topside facility, indicating in a combination of both techniques, where the highest concern areas were. It also allowed for a carefully selected enriched culture, on which efficacy tests could be performed, tailored specifically to this top side facility. In this case study, it is clearly shown that such a carefully designed program can significantly benefit oilfield operations and is therefore worthwhile considering from an economic perspective.

3.5 CONCLUSIONS

The primary aim of this chapter was to describe how a microbiological monitoring program utilizing molecular biology technologies can be deployed to improve the overall operation of an oilfield and its topside facility. An ecological view of an oilfield has been presented that described an oilfield system as a set of interconnected ecosystems with their own differentiated environmental parameters. The great importance that electron acceptors have in determining the microbial and metabolic composition of a given environment was discussed. Finally, various classes of microorganisms that can thrive under the combined presence of electron acceptors and different environmental conditions were described.

Several case studies were presented that aimed to illustrate how molecular biology techniques can be deployed to monitor oilfield topside facilities and how, most importantly, the data from these techniques can be used to guide changes to enhance oilfield operations. Molecular biology techniques can help in identifying if system contamination is severe via absolute determination of cell levels and can help identify what types of microorganisms are present such that the risk of microbial contamination to specific assets can be estimated.

What has been shown is that molecular biology technologies can be deployed within the development and execution of a larger and solid microbial monitoring program for topside facilities. It has been demonstrated how the outcome of such analyses can lead to actionable engineering related items that can be executed to optimize oilfield topside facility operations. This includes how the dosing (both concentration, rate, and location) of functional chemicals, including but not limited to biocides, can be optimized. This, as was demonstrated in the final case study, can significantly optimize an oilfield operation to reduce downtime and operational costs.

REFERENCES

1. Robert Bacon, S.T., Crude oil price differentials and differences in oil qualities: A statistical analysis, in *ESMAP Technical Paper*. 2005.
2. Crolet, J.L., *Microbial Corrosion in the Oil Industry: A Corrosionist's View*, in *Petroleum Microbiology*, M.M. Bernard Ollivier, Editor. 2005, ASM Press.
3. Eric Collins, R., Jody W. Deming. Persistence of bacterial and archaeal communities in sea ice through an Arctic winter. *Environmental Microbiology*, 2010, 12(7): p. 1828–1841.
4. Gregory, J. Dick, K.A., Brett J. Baker, Meng Li, Daniel C. Reed, Cody S. Sheik. The microbiology of deep-sea hydrothermal vent plumes: Ecological and biogeographic linkages to seafloor and water column habitats. *Fronteers in Microbiology*, 2013, 4:124.

5. Magot, M., Indigenous microbial communities in oil fields, in *Petroleum Microbiology*, M.M. Bernard Ollivier, Editor. 2005, ASM Press.
6. Oren, A., Life at high salt concentrations, in *The Prokaryotes*, E.F.D. Eugene Rosenberg, Stephen Lory Erko Stackebrandt, Fabiano Thompson, Editor. 2013, Springer.
7. Nina Gunde-Cimerman, A.P., Aharon Oren, Strategies of adaptation of microorganisms of the three domains of life to high salt concentrations. *FEMS Microbiology Reviews*, 2018, 42(3): p. 353–375.
8. Collins, A.G., Editor *Classification of Oilfield Waters*, in *Developments in Petroleum Science*. 1975, Elsevier.
9. Brian, M. Hynek, K.L.R., Monique Antunovich, Geoffroy Avard, and Guillermo E. Alvarado, Lack of microbial diversity in an extreme mars analog setting: Poás Volcano, Costa Rica. *Astrobiology*, 2018, 18(7): 923–933.
10. Dimitry, Y Sorokin, H.L.B. Gerard Muyzer, Functional microbiology of soda lakes. *Current Opinion in Microbiology*, 2015, 25: p. 88–96.
11. *Guidelines for the Management of Microbiologically Influenced Corrosion in Oil and Gas Production*, 1st ed., Vol. 1., 2015, Energy Institute, London.
12. Lord, J.P., Impact of seawater temperature on growth and recruitment of invasive fouling species at the global scale. *Marine Ecology*, 2016, 38(2): 1–10.
13. Eilen Arctander Vik, A.O.J., F. Kristi Garshol, Liv Bruas Henninge, Stian Engebretsen, Cornelis Kuijvenhoven, David Oilphant, Willem Pieter Hendriks, *Nitrate Based Souring Mitigation of Produced Water - Side Effects and Challenges from the Draugen Produced Water re-Injection Pilot*, in *International Symposium on Oilfield Chemistry*, 2007, Society of Petroleum Engineers: Houston Texas.
14. Gunhild Bødtker, T.T., Bente-Lise P.L., Bente E. Thorbjørnsen, Rikke Helen Ulvøen, Egil Sunde, Terje Torsvik, The effect of long-term nitrate treatment on SRB activity, corrosion rate and bacterial community composition in offshore water injection systems. *Journal of Industrial Microbiology & Biotechnology*, 2008, 35: p. 1625–1636.
15. Rizk, T.Y., Nitrate treatment — effect on corrosion and implementation guidelines. *Saudi Aramco Journal of Technology*, 2015, 55: 36–40.
16. Head, I.M., D.M. Jones, S.R. Larter, Biological activity in the deep subsurface and the origin of heavy oil. *Nature*, 2003, 426(6964): p. 344–352.
17. Dennis Enning, J.G., Corrosion of iron by sulfate-reducing bacteria: New views of an old problem. *Applied and Environmental Microbiology*, 2014, 80(4): p. 1226–1236.
18. Baas Becking, L.G.M., *Geobiologie of inleiding tot de milieukunde*. 1934, W.P. Van Stockum & Zoon.
19. Woese, C.R., et al., Conservation of primary structure in 16S ribosomal RNA. *Nature*, 1975, 254(5495): p. 83–86.
20. Woese, C.R., Bacterial evolution. *Microbiology Review*, 1987, 51(2): p. 221–271.
21. Muyzer, G., A.J.M. Stams, The ecology and biotechnology of sulphate-reducing bacteria. *Nature Reviews Microbiology*, 2008, 6(6): p. 441–454.
22. Ravot, G., et al., Thiosulfate reduction, an important physiological feature shared by members of the order thermotogales. *Applied and Environmental Microbiology*, 1995, 61(5): p. 2053–2055.
23. Moreno-Vivián, C., et al., Prokaryotic nitrate reduction: Molecular properties and functional distinction among bacterial nitrate reductases. *Journal Bacteriology*, 1999, 181(21): p. 6573–6584.
24. Thauer, R.K., et al., Methanogenic archaea: Ecologically relevant differences in energy conservation. *Nature Reviews Microbiology*, 2008, 6(8): p. 579–591.
25. Friedrich, C.G., Physiology and genetics of sulfur-oxidizing bacteria. *Advances in Microbial Physiology*, 1998, 39: p. 235–289.

26. Amann, R.I., W. Ludwig, K.H. Schleifer, Phylogenetic identification and in situ detection of individual microbial cells without cultivation. *Microbiology Review*, 1995, 59(1): p. 143–169.
27. Sogin, M.L., et al., Microbial diversity in the deep sea and the underexplored "rare biosphere". *Proceedings of the National Academy of Sciences*, 2006, 103(32): p. 12115–12120.
28. Ambardar, S., et al., High throughput sequencing: An overview of sequencing chemistry. *Indian Journal of Microbiology*, 2016, 56(4): p. 394–404.
29. Tsesmetzis, N., et al., MIxS-HCR: A MIxS extension defining a minimal information standard for sequence data from environments pertaining to hydrocarbon resources. *Standards in Genomic Sciences*, 2016, 11(1): p. 78.
30. Van Der Kraan, G.M., et al., Microbial diversity of an oil–water processing site and its associated oil field: The possible role of microorganisms as information carriers from oil-associated environments. *FEMS Microbiology Ecology*, 2010, 71(3): p. 428–443.
31. Van Der Kraan, Geert M., et al., *A microbiological audit of an FPSO*, in *Rio Oil&Gas* 2014: Rio de Janeiro.
32. Boden, R., Emended description of the genus Methylophaga Janvier et al. 1985. *International Journal of Systematic and Evolutionary Microbiology*, 2012, 62(Pt_7): p. 1644–1646.
33. Alain, K., et al., Caminicella sporogenes gen. nov., sp. nov., a novel thermophilic spore-forming bacterium isolated from an East-Pacific Rise hydrothermal vent. *International Journal of Systematic and Evolutionary Microbiology*, 2002, 52(Pt 5): p. 1621–1628.
34. Kuenen, J.G., H. Veldkamp, Thiomicrospira pelophila, gen. n., sp. n., a new obligately chemolithotrophic colourless sulfur bacterium. *Antonie van Leeuwenhoek*, 1972, 38(1): p. 241–256.
35. van der Kraan, G.M., et al., *Using integrated microbial auditing for an argentinian oil-field, enabled an improved complete chemical treatment regime developme NT*. in *Rio Oil&Gas*. 2018, Rio de Janeiro.
36. Cheng, L., et al., Methermicoccus shengliensis gen. nov., sp. nov., a thermophilic, methylotrophic methanogen isolated from oil-production water, and proposal of Methermicoccaceae fam. nov. *International Journal of Systematic and Evolutionary Microbiology*, 2007, 57(Pt 12): p. 2964–2969.
37. Hafenbradl, D., et al., Ferroglobus placidus gen. nov., sp. nov., a novel hyperthermophilic archaeum that oxidizes Fe2+ at neutral pH under anoxic conditions. *Archives of Microbiology*, 1996, 166(5): p. 308–314.
38. Smith, J.A., et al., Mechanisms Involved in Fe(III) Respiration by the Hyperthermophilic Archaeon Ferroglobus placidus. *Applied and Environmental Microbiology*, 2015, 81(8): p. 2735–2744.
39. Cavallaro, A., et al., *Improving water injectivity in barrancas mature field with produced water reinjection: A team approach, 8th European Formation Damage Conference*, 2009, Scheveningen, The Netherlands.
40. Frock, A.D., J.S. Notey, R.M. Kelly, The genus Thermotoga: Recent developments. *Environmental Technology*, 2010, 31(10): p. 1169–1181.
41. Liang, R., et al., Metabolic capability of a predominant halanaerobium sp. in hydraulically fractured gas wells and its implication in pipeline corrosion. *Frontiers Microbiology*, 2016, 7: p. 988.
42. Curci, A.E., et al., *Un enfoque Diferente Para Optimizar Los Tratamientos Antimicrobianos Ensayo En El Yacimiento Barrancas*, in *INGEPET EXPL-IP-EC-3-E: Un enfoque diferente para optimizar los tratamientos antimicrobianos. Ensayo en el yacimiento Barrancas. Eduardo Curci, YPF, Argentina y Verónica Silva*, 2018, Dow Química Argentina, Argentina.

4 Using a Holistic Monitoring Approach to Set Effective Control Strategies

How Data Can Lead to Information and Subsequently to "Wisdom"

Heike Hoffmann

CONTENTS

4.1 Introduction 69
4.2 Common Practice – Monitoring and Setting a Strategy 71
4.3 Case Studies 73
 4.3.1 Case Study 1: Seawater Injection System in the North Sea 74
 4.3.2 Case Study 2: Produced Water Re-Injection System in the North Sea 76
4.4 Conclusion – A Holistic Monitoring Approach 78
Acknowledgments 79
References 79

4.1 INTRODUCTION

Shortcomings in effectively controlling microbiological activity within oil and gas installations can have detrimental effects on process systems both operationally and financially, due to, for example, microbiologically influenced corrosion (MIC) leading to failures of pipelines or the onset of microbial reservoir souring leading to necessary process system amendments. Historically, these issues are attributed to the activity of sulfate-reducing prokaryotes (SRP) in seawater processing equipment and within waterflooded reservoirs (Maxwell 1986, Maxwell and Spark 2004, Sanders and Latifi 1994). Wherever there is water, there is a high probability of an environment conducive for microorganisms. Water from numerous different sources (e.g. seawater, aquifer) is used in many oil production processes, such as water injection into reservoirs to aid secondary oil recovery. The pipeline material (e.g. steel) in combination with water will supply abundant electron acceptors such as iron, sulfate, and carbon substrates as electron donors. Conditions in the pipeline often become anaerobic, creating an environment conducive for certain microbiological growth, especially SRP. Traditionally, it was postulated that one of the main MIC mechanisms was SRP growth and its biogenic H_2S production in combination with the development of iron sulfides. However, now there is evidence of a direct mechanism as described by various investigators (Enning et al. 2012, Venzlaff et al. 2013) whereby some SRP strains are using metallic iron as the sole electron donor and contribute to corrosion by direct displacement of electrons. This discovery resulted in introduction of the terms electrical microbiological influenced corrosion (EMIC) and chemically microbiological influenced corrosion (CMIC). A review by Enning (Enning and Garrelfs 2014) highlights the complexity of the various and often interlinking processes involved.

Most studies and monitoring strategies concentrate on the effects of SRP and, to a lesser extent, organic acid-producing bacteria (APB) which contribute to corrosion by causing an acidic pH from carbon dioxide and organic acid production. Conclusions drawn from the research work conducted by implementing molecular methods, such as metagenomics, have linked other physiological groups to iron corrosion, such as thiosulfate-reducing bacteria (Magot et al. 1997), nitrate-reducing bacteria (Hubert et al. 2005), iron-reducing bacteria (Valencia-Cantero and Peña-Cabriales 2014), and methanogenic archaea (Dinh et al. 2004). Furthermore, through the identification of microbiological taxa and functional genes in biofilm communities, the potential for several different but complementary metabolic processes involved in MIC have been proposed by Vigneron (Vigneron et al. 2016). For example, they suggest that in the presence of hydrogen producing fermenters, *Desulfovibrio* may have the ability to catalyze sulfate and iron oxide reduction concurrently, which would remove any iron oxide coating exposing the steel surface to corrosive products like H_2S produced at the same time. Other bacteria such as *Pelobacter* might grow by fermentation of various hydrocarbon-derived substrates, generating hydrogen that might be utilized syntrophically by Methanogens or *Acetobacterium* (Schink and Pfennig 1982); however, this mechanism is dependent on specific environmental conditions. Indeed, organic acids produced from fermentation, such as lactate, will be available to *Desulfovibrio* species as an electron donor, resulting in sulfide production and corrosion, while the organic acids itself might also dissolve iron and therefore can enhance corrosion rates.

While complex the above theories elegantly describe mechanisms which could operate in an "ideal" system. However, there are numerous uncontrollable variables in an operating system that add increased complexity, such as intermittent addition of biocides and chemicals; periods of high flow and low flow; excursions in oxygen control; etc. These physicochemical variations may act to inhibit or stimulate microbiological activity and corrosion rates, either due to altering the biology of the system or entirely independently as an electrochemical process. In this chapter, we explore how the knowledge gained through advanced technology, such as next generation sequencing (NGS) and data analyses, has aided a better understanding of the complex processes surrounding microbiological growth within an oil and gas system. Including how this can lead to a holistic monitoring approach and in turn to a design of mitigation strategies, followed by implementing controls and measuring performance.

4.2 COMMON PRACTICE – MONITORING AND SETTING A STRATEGY

In an oilfield system, where the activity of SRP is considered the main corrosion mechanism, the monitoring of SRP numbers as well as ensuring that internal surfaces of the system are free from SRP biofilm formation is considered a key element in MIC management. While it is generally easier to keep a "clean" system clean, once microbiological growth is established, it becomes more difficult to set effective control strategies. A direct means of maintaining low numbers of microorganisms in the system is the application of biocides.

While this approach appears straightforward, efficient mitigation of MIC depends on a wide range of biotic and abiotic factors, many of which are specific to pipeline operations and design, as well as the nature of the fluids or gases transported within the lines. It is important to note that when using the generic terms SRP and APB, which are common to all systems, the extant genera present will likely be very different in a cold seawater pipeline to those present in the main oil pipeline or a hot three-phase pipeline (gas-oil-water). The microbial structure may also vary in response to system changes such as biocide treatment or any other operational changes. Operational adjustments in different oilfield systems can lead to variations in pH, temperature, or nutrient availability which can affect the growth of microorganisms within the systems. Additionally, monitoring and control within these systems can become more complicated when a corrosion event or the risk of MIC cannot be explained by the "classic" corrosion mechanisms. While the main focus of microbiological growth is often related to MIC, the presence of specific microorganisms can cause a variety of different issues, such as blockages due to biofilm build-up, oil-water separation issue due to microemulsion formation, and H_2S or other gas formation. A program of routine monitoring and mitigation must be conducted to prevent issues related to microbiological growth. The program should be simple enough to be performed in the field without constant expert personnel involvement but simultaneously needs to be sufficiently detailed to cover all the parameters of importance. Thus, MIC risk assessment and monitoring, and biocide treatment optimization should not be considered a straightforward operation.

However, there is a difference between monitoring and attempting to control microorganisms and actively mitigating MIC. Predictive models are well established and can aid in successful mitigation strategies. Even with the uncertainty of the MIC mechanism, it is generally accepted that the rate of MIC is not necessarily linked to planktonic bacteria numbers within the system. Though this data may provide a relative qualitative measure of biocide performance, it often may not be representative for the impact of chemicals on the sessile population in the same operating system and does not provide sufficient corrosion mitigation data. Other parameters are required to facilitate an all-inclusive approach in controlling microbiological growth, such as:

- Microbiological activity in biofilms on the pipe walls/vessel internal surfaces
- Water content and water availability
- Water chemistry

- Flow velocity
- Deposition of solids and scale
- Organic carbon source
- Concentration of dissolved gases (CO_2 & O_2)
- Temperature, pH, and Eh

For this reason, the MIC model presented by Pots (Pots et al. 2002) places emphasis on deposit formation and removal, oxygen ingress, flow velocity, and physicochemical conditions. Pots model was modified by Maxwell (Maxwell 2006) to include modules for biofilm development activity and its control. Since 2006, a number of prediction models have been developed to include data obtained by molecular methods, and thus further improving the accuracy of outcome (Skovhus et al. 2017).

A combination of prediction, modeling, and review of data is suggested to be the most efficient approach in tackling and managing the MIC challenges and risk. An ongoing prediction can be calibrated to the system performance by applying the appropriate metrics. This approach allows personnel to design, implement, and monitor mitigation strategies which are practical and cost-effective and which identify the critical controls and key performance indicators (KPIs) for specific system sensitivities; i.e. removal of solids, killing of SRP, dissolved oxygen control, etc.

Traditionally, monitoring of microbiological numbers in the Oil and Gas industry was achieved with culture-dependent methods such as most probable numbers (MPN) counts. During last decade, molecular methods have become widely available and opened up ample opportunities in monitoring microbiological groups and numbers, which might be underestimated by MPN counts. Molecular methods such as fluorescence *in situ* hybridization (FISH), quantitative polymerase chain reaction (qPCR), and next generation sequencing (NGS) do not only help gain a better insight into microbial communities and their impact within the oil and gas industry, they can also be utilized for targeted monitoring programs (Caffrey 2009, Hoffmann et al. 2008, Mand et al. 2016). Next generation sequencing techniques generate vast amounts of data, which without the availability of high-performance computers and bioinformatics techniques would quickly become unmanageable. Over the last decades, hundreds of microbial genomes have been sequenced and archived for public research and allowed for more accurate identification of bacteria strains. Commonly, bioinformatic approaches are used to analyze sequencing data and are primarily run using command-line programs utilizing pipelines built from open-source software. However, bioinformatics datasets must be translated into standardized outputs to provide meaningful information. This requires the capability to filter the data and allocate statistical significance to changes between different microbiological communities from various locations in order to make sense of the data obtained. This in turn will allow for a better judgement and understanding of processes at play within an oil and gas installation aiding in setting up targeted monitoring and mitigation strategies. Several bioinformatic tools have been developed to help analyzing huge sequence data such as web servers like MG-RAST (Meyer et al. 2008) or pipelines such as MEGAN (Huson and Weber 2013), to name a couple.

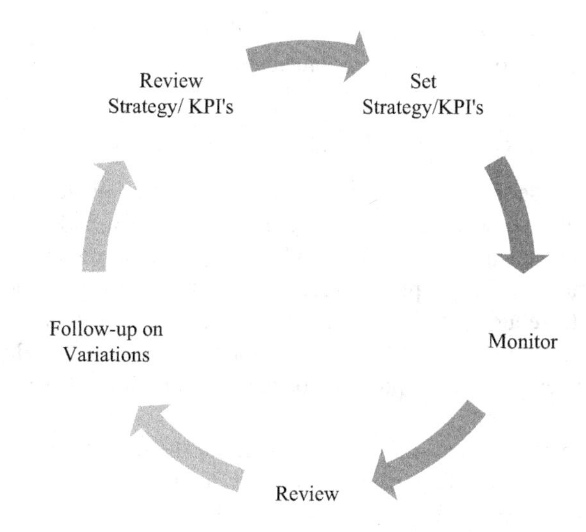

FIGURE 4.1 Feedback loop of monitoring and mitigation strategy.

Having all these tools available, it is vital to set a monitoring and control strategy applicable to the system of concern. Whatever the strategy is, the critical element in maintaining the integrity of systems is using a multi-disciplinary and more importantly a holistic management approach, which creates a feedback loop between strategy and monitoring. This involves creating a scheduled monitoring and mitigation plan, regularly reviewing of obtained data, following up variations in results and reviewing the strategy and KPI's accordingly, followed by implementing any required changes to the plan (see Figure 4.1).

4.3 CASE STUDIES

A sound model system is the seawater injection system as there is a plethora of information and guidance available, and the set-up, chemicals used, and fluids present are relatively simple compared to for example production systems. Nevertheless, if not maintained sufficiently, seawater injection systems can develop microbiological problems and difficulties which can have a broader impact on other areas of the oil and gas installation, such as groove corrosion in seawater transport pipelines and microbiological contamination of deaerators. The design of the deaerator is to maximize the oxygen removal through increased surface areas, inadvertently making it ideal for the promotion of anaerobic bacterial growth within the tank. Once biological matter has established within the tank, it will reduce the oxygen removal efficiency as the required surface area for removal has been decreased. Additionally, the oxygen scavenger itself being an intermediate of the reduction pathway from sulfate to sulfide may be utilized as a terminal electron acceptor. This is energetically more favorable to the SRP as it does not require the energy consuming activation step of sulfate (SO_4) to adenosine-phospho-sulfate (APS) and thus adding to the loss of oxygen removal efficiency. Ultimately, the introduction of unwanted microorganisms to the reservoir, via downhole seawater injection, will lead to issues, as they could cause undesirable reservoir pore plugging or contribute to reservoir souring.

4.3.1 Case Study 1: Seawater Injection System in the North Sea

Seawater is often used as injection water to aid secondary oil recovery. The main challenges related to seawater injection are the introduction of SRP, organic material, and solids in the water injection system and consequently to the injection reservoir. Routinely, measures are put in place to combat the most common issues, such as filtration systems to reduce the entry of solids; primary and secondary biocide treatment upstream and downstream of the deaerator (DA) system to reduce biological content; and oxygen removal to prevent corrosion. A monitoring strategy and KPI's are set to maintain control.

Various parameters can be monitored throughout the topside system to maintain system integrity and control souring potential of downstream systems, for example residual chlorine (upstream DA), microbiological numbers, total suspended solids (TSS) (upstream DA), dissolved oxygen levels (downstream DA). Maintaining these parameters within set guideline limits will aid in continued operations and overall integrity of vessels and pipework. For example, ensuring optimal filtration performance will help controlling microbiological growth, and solid and scale build-up within the seawater injection system, and in turn also minimizes the introduction of these into the reservoir, preventing well plugging by solids and control of biogenic sulfide generation within the reservoir. Moreover, the control and measurement of oxygen levels supplies a measure in controlling the risk from oxygen corrosion. Thus, maintaining and controlling microbiological growth within the seawater injection system will ultimately support mitigation of reservoir souring.

This case study describes a site following the above guidelines. Though data was collected on a regular basis, the data review was missed, and the loop could not be closed and eventually increased microbiological numbers were detected downstream of the DA along with warnings from nondestructive testing (NDT) due to thinning of some pipe sections. A review of the data was conducted, including reviewing the set strategies versus their application manual used by the site. A few inconsistencies were observed, mainly concerning the biocide treatment strategies. However, when reviewing the data, a clear trend was observed, suggesting that if data sets were reviewed on a regular basis, the risk could have been highlighted at an earlier stage and allow for timely action. Though the site was aware of intermittent chlorine dosage which did not achieve recommended residual chlorine levels, the link was not made to the slow increase in overall microbiological numbers upstream of the DA (Figure 4.2). Additionally, microbiological numbers exiting the DA also increased over time (Figure 4.3) with a slight delay compared to the increased numbers entering the system. The further review revealed that the secondary biocide treatment (not chlorine) was introduced downstream of the DA, leaving the DA tower unprotected. The biofilm was allowed to build up, resulting in the requirement for a mechanical cleaning operation of DA tower and temporary water injection system shutdown leading to loss of production.

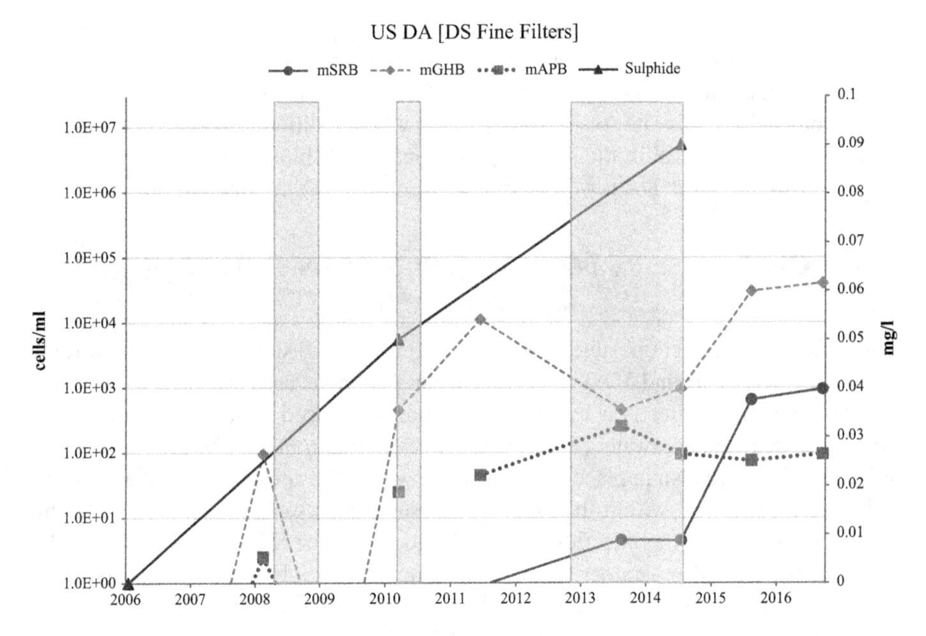

FIGURE 4.2 Case Study 1: Data trend US of DA overtime showing an increase in microbiological numbers. Shaded areas reflect out of spec time for residual chlorine.

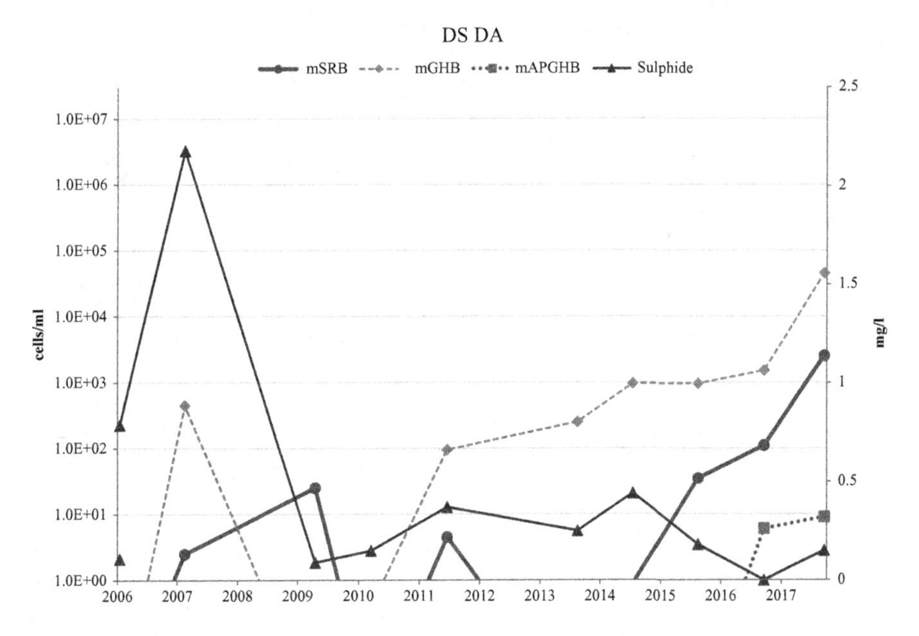

FIGURE 4.3 Case Study 1: Data trend DS of DA overtime showing increase in microbiological numbers.

This case study demonstrates an example of how important it is not only to review data regularly, and follow up on observed changes in numbers, but also highlights the importance of communication between the different departments. The ongoing review of data has helped this site to set trigger levels when additional biocide treatment is required to keep control in the system. Additionally, the biocide injection point was moved upstream of the DA tower, helping to minimize DA contamination.

4.3.2 Case Study 2: Produced Water Re-Injection System in the North Sea

At this floating production, storage, and offloading (FPSO) site, produced water reinjection (PWRI) of around 55000 barrels of water per day is used to enhance oil recovery. The injection water is a mixture of offshore produced water (PW) containing sulfide and deaerated seawater (SW) (Figure 4.4). Similar routine measures as outlined in Section 4.3.1 are also in place on this site. However, the site experienced a build-up of biological material within the medium pressure (MP) separator as well as in the flowline upstream of the injection pumps. Historically, the SRP numbers in the PW (downstream of MP separator) were low but increased slightly over the years up to 10^4 cells per ml. In comparison, relatively low SRP numbers (10^1 cells per ml) were

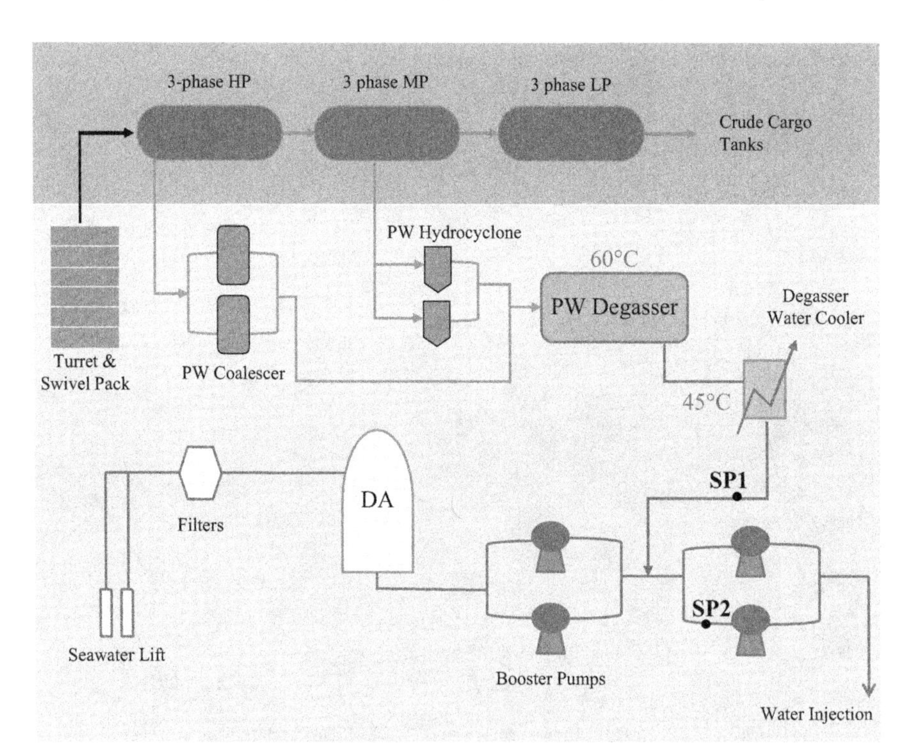

FIGURE 4.4 Case Study 2 – Operational diagram of FPSO indicating sampling points SP1 (produced water before PWRI unit) and SP2 (PWRI water), a three-stage separation process (high pressure (HP), medium pressure (MP) and low pressure (LP) separators) is utilized. Modified from Mand, J. 2014.

entering the offshore system upstream of DA. Also, the downstream DA sample (Figure 4.4; SP2), a mix of PW and SW, enumerated higher SRP levels of 10^5 cells per ml. The initial conclusion was that the influx of SRP in the PWRI system originated from the production system. In order to understand the mechanism and possible root of the contamination, samples were taken from the production system and PWRI system (Figure 4.4; SP1 and SP2) and additional testing utilizing metagenomics was conducted (Mand et al. 2014). DNA was isolated, and following a two-step amplification procedure, PCR products were sent for pyrosequencing. The resulting pyrosequencing data were analyzed using the Phoenix software package (Soh et al. 2013). The results revealed that PW community was dominated by *Desulfuromonas* (47%), the APB or oil-degrading bacterium *Caminicella* (24%), the SRB *Desulfovibrio* (18%), and the methanogen *Methanothermococcus* (5%). In comparison, the PWRI microbial community of PWRI water consisted of a different population containing a high fraction of the sulfide-oxidizing bacterium (SOB) *Arcobacter* (25%), *Bacteroidetes* (25%), the aerobic or nitrate-reducing hydocarbon-degrading *Defluvibacter* (10%), the SRB *Desulfovibrio* (5%), and the APB or oil-degrading *Firmicutes* (6%) and *Fusibacter* (4%) (Figure 4.5).

As outlined above, MIC has historically been attributed to SRP and APB activity, which routine MPN monitoring did confirm. In this case, the metagenomic analysis showed that the SRP present within both samples might belong to the genus *Desulfovibrio*, reducing sulfate by oxidizing lactate (commonly used as an electron donor in growth media for culture-based enumeration). Another sulfate reducer, *Desulfobacter*, which can use acetate as an electron donor for sulfate reduction, was detected in the PWRI sample. Acetate may also serve as a substrate in the PW sample for sulfur-reducing bacteria (SuRB) of the genus *Desulfurimonas* (47%) by reducing sulfur to sulfide using acetate or other organics as an electron donor.

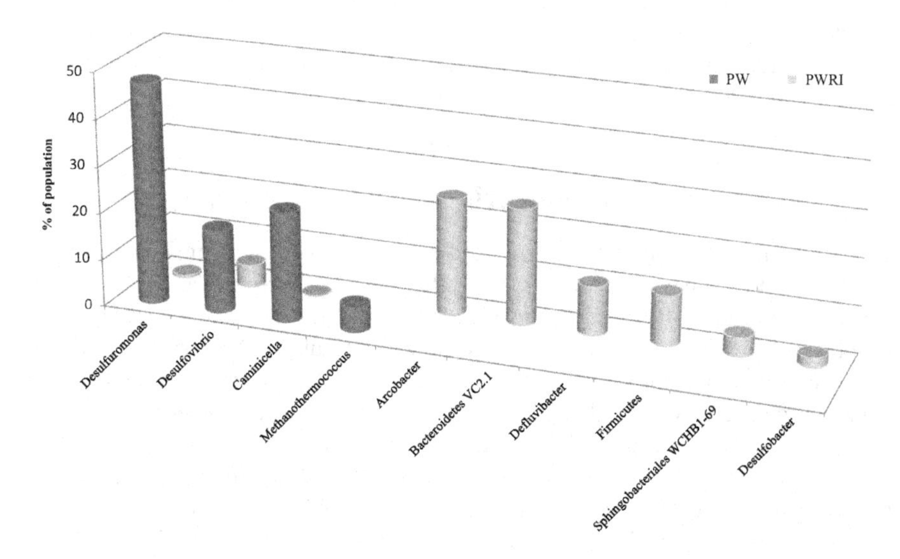

FIGURE 4.5 Comparison of phylogenetic classification for the PW and PWRI samples. Displayed according to their average fraction present within the two samples.

Of interest was the very high fraction of the sulfide-oxidizing bacterium *Acrobacter* in the PWRI water sample (25%), as well as that of the related *Sulfurimonas* (1.0% in PWRI water) and *Sulfurospirillum* (0.5% in PWRI water), but not in the PW sample. These taxa oxidize sulfide while reducing nitrate or oxygen. However, nitrate treatment was not used at this site, and a DA tower is used to deaerate the seawater, seemingly eliminating the obvious sources for nitrate or oxygen. Though the seawater is also treated with an oxygen scavenger, the possibility of residual oxygen concentration was investigated. Records showed that during a maintenance shut down, fluids were left untreated in the DA as well as in associated pipework and could potentially have added to the increase of microbiological numbers. Additionally, oxygen readings had been above the platform specification (<5 ppb) occasionally, which in itself poses a corrosion issue. If this partially oxygenated seawater is combined with sulfide-containing produced water, ideal conditions are created for the growth of *Arcobacte*r and other sulfide-oxidizing bacteria (SOB). These can produce sulfur, which can, in turn, be reduced to sulfide by SuRB of the order *Desulfuromonadales*. Hence, the data indicates cycling of sulfur in the PWRI water, but possibly also in the PW. Because sulfur is highly corrosive toward carbon steel, the microbial community compositions found in these samples suggests an environment which is more corrosive than one in which only SRP are present. Sulfur as a significant intermediate increases the probability of corrosion damage.

This case study has demonstrated that along with regular monitoring and review of data, additional analysis using advanced technology in the domain of life science including the application of bioinformatics helps in combating challenges and mitigating risks caused by microbiological proliferation. It also shows the importance for incorporating testing and reviewing of operational and chemical parameters as part of the monitoring matrix. Gathering data (microbiological quantities as well as types; chemical data) has led to a better understanding of the processes within the system and has led to a better monitoring regimen. This regimen included adding a biocide injection point upstream of the PW and SW comingling point, as well as regular biocide treatment of the MP separator. Close monitoring and review of microbiological numbers in combination with residual oxygen levels has helped the site to gain microbial control.

4.4 CONCLUSION – A HOLISTIC MONITORING APPROACH

The importance of a sound monitoring and control strategy is undeniable; however, setting a sound strategy is not always straightforward. As described earlier, a monitoring program needs to be easy, practical, and efficient, but one strategy will not fit all systems, and the selected strategy has to be aligned carefully with the system in focus.

It is essential to build-up knowledge to set a meaningful program, which will need to include data from various disciplines. For example, the data obtained in case study 2 revealed the requirement of the key indicators to be included in the routine, such as closer control of residual oxygen and microbiological numbers. In this case, advanced DNA sequencing methods identified the presence of microbes that utilize elemental sulfur. Active cycling of elemental sulfur may point to aggressive corrosion scenarios. In pipelines, SOB numbers can be warning signs, like canaries in coal mines, as a high fraction of these microorganisms makes aggressive sulfur corrosion likely. Thus, the detection of SOB should be a strong incentive to take remedial measures (i.e. use oxygen scavengers to altogether remove oxygen from the seawater before mixing with produced water). Whereas the data from case study 1 demonstrates the importance of reviewing the data regularly as the increase of microbiological contamination could have been foreseen earlier, and additional measures should have been put in place.

When utilizing NGS analyses, a vast flood of sequence information is created and it is crucial to have the right technology in place in order to analyze and filter the data in order to obtain meaningful information. The use of mathematical modeling techniques and statistical analysis can assist in identifying common patterns and key microorganism within a data set. By identifying key microorganism, such as the SOB in case study 2, and aligning these with other meta data, such as residual oxygen levels, should trigger action. By building up data sets and understanding the interaction between different microbiological communities and operational conditions, a risk-based management system can be set up. Additionally, once these data sets are built up and incorporated in the monitoring programs, they have the potential to identify and monitor key areas prone to suffer from the introduction of microbiological matter and associated issues.

The set-up and chosen parameters for a successful strategy will have to be assessed for different systems. A means to do this requires data accumulation and trending, which will need to be linked to any operational events. In doing this, indicators, like the SOB numbers in case study 2, can be used in order to create required action plans, such as increasing biocide rates or applying shock biocide treatment. Ultimately, a successful monitoring program should be based on both traditional microbiological techniques and contemporary high-throughput molecular approaches and developed from knowledge gained on a case-by-case basis, appropriate for specific systems.

ACKNOWLEDGMENTS

Part of the work in Section 4.3.2 was conducted in cooperation with University of Calgary through a Natural Sciences and Engineering Research Council (NSERC) Industrial Chair Award to Gerrit Voordouw, which was also supported by Baker Hughes, BP, Computer Modelling Group Limited, ConocoPhillips Company, Intertek, Dow Microbial Control, Enbridge, Enerplus Corporation, Oil Search Limited, Shell Global Solutions International BV, Suncor Energy Inc. and Yara Norge AS, as well as by Alberta Innovates – Energy and Environment Solutions (AIEES). Special thanks to Jaspreet Mand and Gerrit Voordouw for the analytical support.

REFERENCES

Caffrey SM. 2009. Application of molecular microbiological methods to the oil industry to analyze DNA, RNA and proteins. In *Applied Microbiology and Molecular Biology in Oilfield Systems*. ed. C. Whitby and T.L. Skovhus, 19–26. Springer Science & Business Media.

Dinh HT, Kuever J, Mußmann M, Hassel AW, Stratmann M, Widdel F. 2004. Iron corrosion by novel anaerobic microorganisms. *Nature* 427:829–832.

Enning D, Garrelfs J. 2014. Corrosion of Iron by Sulfate-Reducing Bacteria: New Views of an Old Problem. *App. Environ. Microbiol.* 80 (4): 1226–1236.

Enning D, Venzlaff H, Garrelfs J, Dinh, H T, Meyer V, Mayrhofer K, Hassel A W, Stratmann M, Widdle F. 2012. Marine sulfate-reducing bacteria cause serious corrosion of iron under electroconductive biogenic mineral crust. *Environ. Microbiol.* 14 (7). 1772–1787.

Hoffmann H, Devine C, Maxwell S. 2008. Application of DGGE and FISH analyses in seawater injection systems. *NACE Corrosion 2008*. Paper 08513.

Hubert C, Nemati M, Jenneman G, Voordouw G. 2005. Corrosion risk associated with microbial souring control using nitrate or nitrite. *Appl. Microbiol. Biotechnol.* 68:272–282.

Huson DH, Weber N. 2013. Microbial community analysis using MEGAN. *Methods Enzymol.* 531:465–485.

Magot M, Ravot G, Campaignolle X, Ollivier B, Patel BK, Fardeau ML, Thomas P, Crolet JL, Garcia JL. 1997. Dethiosulfovibrio peptidovorans gen. nov., sp. nov., a new anaerobic slightly halophilic, thiosulfatereducing bacterium from corroding offshore oil wells. *Int. J. Syst. Bacteriol.* 47:818–824.

Mand J, Park HS, Jack TR, Voordouw G, Hoffmann H. 2014. Use of molecular methods (pyrosequencing) for evaluating mic potential in water systems for oil production in the North Sea. SPE-169638-MS.

Mand J, Park HS, Voordouw G, Hoffmann H. 2016. Linking sulfur cycling and MIC in offshore water transporting pipelines. *NACE Corrosion 2016 Paper 7578.*

Maxwell S. 1986. The assessment of sulphide corrosion risks in offshore systems by biological monitoring. *SPE Prod. Eng.* 1(5): 363–368.

Maxwell S. 2006. *SPE 100519. Predicting microbially influenced corrosion (MIC) in seawater injection systems. SPE Corrosion Conference*, Aberdeen.

Maxwell S, Spark I. 2004. *Souring of reservoirs by bacterial activity during seawater flooding. SPE 93231, SPE Oilfield Chemicals Symposium*, Houston, Texas, USA.

Meyer F, Paarmann D, D'Souza M. et al. 2008. The metagenomics RAST server—a public resource for the automatic phylogenetic and functional analysis of metagenomes. *BMC Bioinf.*. 9:386.

Pots BFM, John RC, Rippom IJ, Thomas MJ, Kapusta SD, 2002. Improvements on De Waard-Milliams corrosion prediction and Applications to corrosion management. *Corrosion 2002, Paper 02235.*

Sanders PF, Latifi L. 1994. *On-site evaluation of organic biocides for cost effective control of sessile bacteria and microbially influenced corrosion. Second International Conference on Chemistry in Industry*. Manama, Bahrain, 242–257.

Schink B, Pfennig N. 1982. Fermentation of trihydroxybenzenes by Pelobacter acidigallici gen. nov. sp. nov., a new strictly anaerobic, nonsporeforming bacterium. *Arch. Microbiol.* 133:195–201.

Skovhus TL, Eckert RB, Rodrigues E. 2017. Management and control of microbiologically influenced corrosion (MIC) in the oil and gas industry—Overview and a North Sea case study. *J. Biotechnol.* 256:31–45.

Soh J, Dong X, Caffrey SM, Voordouw G, Sensen CW. 2013. Phoenix 2: A locally installable large-scale 16S. rRNA gene sequence analysis pipeline with web interface. *J. Biotechnol.* 167:393–403.

Valencia-Cantero E, Peña-Cabriales JJ. 2014. Effects of iron-reducing bacteria on carbon steel corrosion induced by thermophilic sulfate-reducing consortia. *J. Microbiol. Biotechnol.* 24(2):280–286.

Venzlaff H, Enning D, Srinivasan J, Mayrhofer K, Hassel AW, Widdel F, Stratmann M. 2013. Accelerated cathodic reaction in microbial corrosion of iron due to direct electron uptake by sulfate-reducing bacteria. *Corros. Sci.* 66:88–96.

Vigneron A, Alsop EB, Chambers, B., Lomans BP, Head IM, Tsesmetzis N. 2016. Complementary microorganisms in highly corrosive biofilms from an offshore oil production facility. *Appl. Environ. Microbiol.* 82(8):2545–2554.

5 Influence of Chemical Treatments and Topside Processes on the Dominant Microbial Communities at Conventional Oilfields

Matheus Paschoalino

CONTENTS

5.1 Introduction 81
5.2 Selected Oilfields and Considerations on Main Topside Zones 82
 5.2.1 Production Water (PW) 84
 5.2.2 Inlet of Water Treatment Plant (WTPin) 85
 5.2.3 Outlet of Water Treatment Plant (WTPout) 85
 5.2.4 Injection Wells or Injection Water (IW) 85
5.3 Dominant Microbial Communities 86
 5.3.1 Bacterial Contamination 86
  5.3.1.1 Variation of Topside Bacterial Profiles 86
 5.3.2 Archaeal Contamination 92
5.4 Conclusion 94
Acknowledgments 94
References 95

5.1 INTRODUCTION

In a conventional oilfield, water flooding is an economically viable technique for recovery of additional oil from mature fields (Civan 2016). It consists of injecting water into an oil reservoir to maintain formation pressure. This practice leads to an overall rise in water cuts in produced fluids and typically uses seawater or recycled produced water for re-injection in secondary oil recovery operations (Van der Kraan 2018). During this process, oil and water flow from the production wells to be treated in a series of different stages topside. Each water treatment step results in changes in physicochemical conditions that exert selective pressure on the microbial communities in each one of these unique environments (Salgar-Chaparro et al. 2020a).

Microorganisms in oilfields can cause many problems such as H_2S generation, microbiologically influenced corrosion (MIC), hydrocarbon degradation, biofilm formation, clogging, a decrease in injectivity/productivity, and an increase in energy and maintenance costs (Enning and Garrelfs 2014). Adequate monitoring of these organisms is crucial to guide effective microbial control strategies to mitigate risks to assets, employees, and the reservoir.

Unfortunately, microbial dynamics in these systems is often misunderstood and, coupled with a lack of knowledge of oilfield chemistry, has resulted in undesirable consequences for oilfield operators over the last decade. Understanding that microorganisms are the root cause of several oilfield problems could prevent the overdose or overuse of corrosion/scaling inhibitors, H_2S scavengers, dispersants, and different types of interventions. Another important but unexplored point is the influence of some chemical components other than the active ingredients, such as surfactants, glycols, and solvents, on shifting of microbial communities and the negative effects they can have over the microbial control strategy (Tidwell et al. 2017).

The selection of effective microbial control approaches remains an important challenge to this industry. While most chemical additives are rigorously tested under model field conditions, biocides are still regularly selected based simply on cost without a deep understanding of interactions with other chemicals or how established microbial communities respond to treatment.

The advent of bioinformatics has started a paradigm shift in microbial control in the oil and gas industry enabling to identify microbial root causes of oilfield problems, and also preventing future issues by the study of past cases under similar conditions. This is possible by linking environmental changes on topside to the eventual microbial community shifts (Van der Kraan et al. 2010).

The goal of this chapter is to review the microbial ecology found during bioaudits carried out from 2014 to 2019 at 12 South American oilfields and illustrate with specific examples the effect of process or chemical alterations that shifted microbial communities and resulted in unintended consequences on some of these fields.

5.2 SELECTED OILFIELDS AND CONSIDERATIONS ON MAIN TOPSIDE ZONES

Twelve conventional oilfields using water flooding for secondary recovery were selected for discussion in this work. Oilfields included in this survey are shown in Figure 5.1 and enumerated A to L. The basic fluid characteristics, functional chemicals, and main reported problems for each oilfield are presented in Table 5.1.

Among the 12 oilfields, only G and H are located offshore and operate using seawater injection. All other fields are located onshore and recycle treated produced water.

FIGURE 5.1 Approximate locations of the selected oilfields (A-L).

TABLE 5.1
Main Characteristics of Selected Oilfields

Field	Max. Salinity (mg/L)	Max. Temp. (°C)	Biocide Actives			Oxygen Scavenger	Main Concern
			PW	WTP	IW		
A	2500	60	-	THPS/QAC	-	ABS	H_2S
B	93000	55	-	Glut/QAC	-	SBS	H_2S
C	2000	65	-	THPS	THPS/QAC	SBS	MIC
D	85000	65	-	-	-	SBS	MIC
E	100000	50	-	-	-	SBS	H_2S
F	60000	45	-	THPS	QAC	SBS	MIC
G	30000	35	-	Cl_2/DBNPA	THPS/QAC	SBS	Biofouling
H	30000	30	-	Cl_2/DBNPA	-	SBS	MIC
I	30000	70	THPS	THPS	THPS	SBS	MIC
J	50000	70	THPS	THPS	THPS/QAC	ABS	MIC

(Continued)

TABLE 5.1 (Continued)
Main Characteristics of Selected Oilfields

Field	Max. Salinity (mg/L)	Max. Temp. (°C)	Biocide Actives PW	Biocide Actives WTP	Biocide Actives IW	Oxygen Scavenger	Main Concern
K	50000	65	THPS	THPS	THPS/QAC	SBS	Injectivity
L	5000	70	THPS	THPS	THPS/QAC	SBS	MIC

THPS - Tetrakis(hydroxymethyl)phosphonium sulfate; QAC – Quaternary Ammonium Compound; Glut – Glutaraldehyde; ABS - Ammonium Bisulfite; SBS – Sodium Bisulfite; H_2S – hydrogen sulfide gas; DBNPA - 2,2-dibromo-3-nitrilopropionamide; MIC – Microbiologically Influenced Corrosion.

The topside operations of each oilfield were divided into three main zones to compare microbial changes between regions: Production (PW), Water Treatment Plant (WTP), and Injection (IW) areas. These zones and audit sample sites selected in this study are shown in Figure 5.2.

5.2.1 Production Water (PW)

Microbial communities can vary considerably even in production wells of the same oilfield (Sianawati et al. 2016). To obtain an average overall microbial composition, samples in this zone were collected from production batteries or gathering stations prior to chemical treatment. In general, this sample point represents an environment of lower oxygen and highest salinity and temperature, compared to the other areas. Production water is a key sample point since it is indicative of microbiota thriving in the reservoir and near wellbore and how that microbiota is impacted by topside chemical treatment (Vigneron et al. 2017).

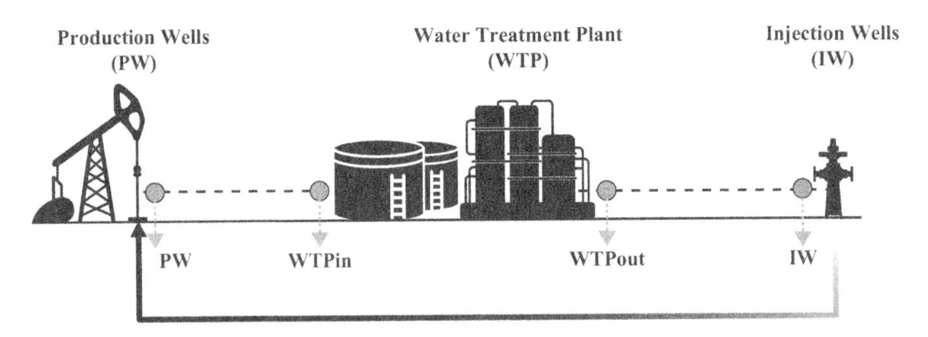

FIGURE 5.2 Simplified layout of the main topside zones in a conventional onshore waterflooding system. Arrows indicate the sampling point regions used to compare microbial shift.

5.2.2 Inlet of Water Treatment Plant (WTPin)

Samples were collected at the end of the main production pipelines before entering the separators on the WTP. Usually, WTPin samples were dosed with demulsifiers and could also contain biocides if the operator has selected to treat production systems (Fields I, J, K, L). The water treatment inlet is a point of high turnover, and the planktonic bacterial community may reflect the dominant attached microbial community located on internal surfaces of transmission pipelines (Marshall 2013). Temperatures in the WTPin tend to be 5°C to 30°C lower than in the production zone, depending on distance, flow-rate, pipeline material, and local weather. Oxygen concentration is generally higher than in production fluid, depending on the sampling point.

5.2.3 Outlet of Water Treatment Plant (WTPout)

This collection point represents the end of a series of processes that vary from plant to plant, depending on the number of stages used to separate oil from water and to treat the water intended for re-injection into the reservoir. The WTPout sample is very significant because it represents a system with reduced fluid velocities or stagnated flows since most WTPs include filters, membranes, flotators, deaerators, and storage tanks with long residence times. A variety of chemicals can impact the resident microbial community such as scale inhibitors, coagulants, corrosion inhibitors, H_2S scavengers, oxidizers, oxygen scavengers, and biocides. Temperature can also vary depending on the use of heat treatment to improve oil separation, which often results in regrowth of thermophilic organisms. Salinity levels at the WTPout tend to be lower, especially when fresh water is introduced to supplement production water to be injected. Generally, this step results in changes in microbiota related to oxygen levels, depending on existing chemical treatment and the sampling point. Cell counts tend to be lower (10^2–10^5 cells/mL) than the WTPin depending on the biocide dosage, active ingredients, type of application (batch or continuous), and distance from dosage point to the sampling point. Typical locations of contamination hot-spots on the WTPs are flotation systems, filters, and deaerators.

5.2.4 Injection Wells or Injection Water (IW)

These points represent the end of injection lines, typically on injection wellheads or manifolds. It is an area of high turnover similar to WTPin samples, and the planktonic community reflects, typically, dominant attached microbiota located in biofilms in transmission pipelines. This location is especially impacted by the continuous injection of oxygen scavenger, creating an anoxic environment ideal for anaerobes but under lower temperature (30°C to 45°C) than the ones found originally in the reservoir. In most fields, this is a highly contaminated area (10^4 to 10^7 cells/mL), mainly because traditional biocides like THPS are deactivated by iron sulfide scaling on the transmission pipelines (Paschoalino et al. 2019a). Therefore, the injection water is a critical monitoring point for the introduction of organisms or nutrients to the formation, since untreated or undertreated injection water could change the diversity of indigenous organisms and result in significant changes in production fluids (Vigneron et al. 2017).

5.3 DOMINANT MICROBIAL COMMUNITIES

The data presented in this chapter was selected from bioaudits using 16S phylogenomics carried out on 12 conventional onshore/offshore oilfields from 2014 to 2019. Microbial DNA was extracted from selected water samples and analyzed using quantitative polymerase chain reaction (qPCR) and next generation sequencing (NGS) using optimized methods as previously reported (Van der Kraan et al. 2011, 2014).

Although 16S rRNA gene sequences do not provide direct information on the metabolic properties of an organism, their basic potential metabolisms were discussed based on existing information in the literature.

5.3.1 Bacterial Contamination

Eleven classes of bacteria were most common with relative abundances higher than 10% in at least one of the sampling zones. The results are shown in Figure 5.3 in descending order from the most frequent to least.

Clostridia was the most common class identified among all oilfields except for the offshore oilfield H, followed by Gammaproteobacteria, Alphaproteobacteria, and Epsilonproteobacteria found at least in seven oilfields. Actinobacteria, Deltaproteobacteria, Sphingobacteria, Thermotogae, Betaproteobacteria, Synergistia, and Deferribacteres were detected less frequently. Bacilli, Aquificae, Dictyoglomia, and Nitrospira were also detected in several fields (not shown in Figure 5.3).

The most diverse was class Gammaproteobacteria, where it was possible to identify 12 different genera. At least five genera were detected for the classes Deltaproteobacteria, Epsilonproteobacteria, and Thermotogae. Less than five genera were recognized for all other classes.

The most frequent genus was *Clostridium*, found in six oilfields, followed by *Marinobacter*, *Marinobacterium*, and *Arcobacter*, found in five oilfields. *Halanaerobium* was identified in four oilfields, while *Defulfotomaculum*, *Desulfovibrio*, and *Sediminibacterium* were detected in three oilfields.

The most diverse population was found in Oilfield D, with 15 genera from five classes and oilfield C, with 13 genera from eight classes, followed by E, F, J, and K with at least 10 genera each.

5.3.1.1 Variation of Topside Bacterial Profiles

Prokaryotes have a variety of mechanisms to attach to surfaces, but they also possess a range of strategies for detachment, like releasing daughter cells from the biofilm to return to a planktonic state (Marshall 2013). However, the planktonic microbial population profiles do not necessarily represent the same location sessile population profiles (Wrangham and Summer 2013). Therefore, instead of focusing on microbial variation at each stage individually, it is more relevant to assess potential risks of the entire top side system. Variations in the bacterial profile and total bacterial cells/mL from Production (PW) to Injection (IW) for each oilfield are shown in Figure 5.4.

Results demonstrate the dynamic changes of the microbial community from one stage to another, as well the unique population of each unit operation, even for oilfields in the same basin with similar water and process characteristics.

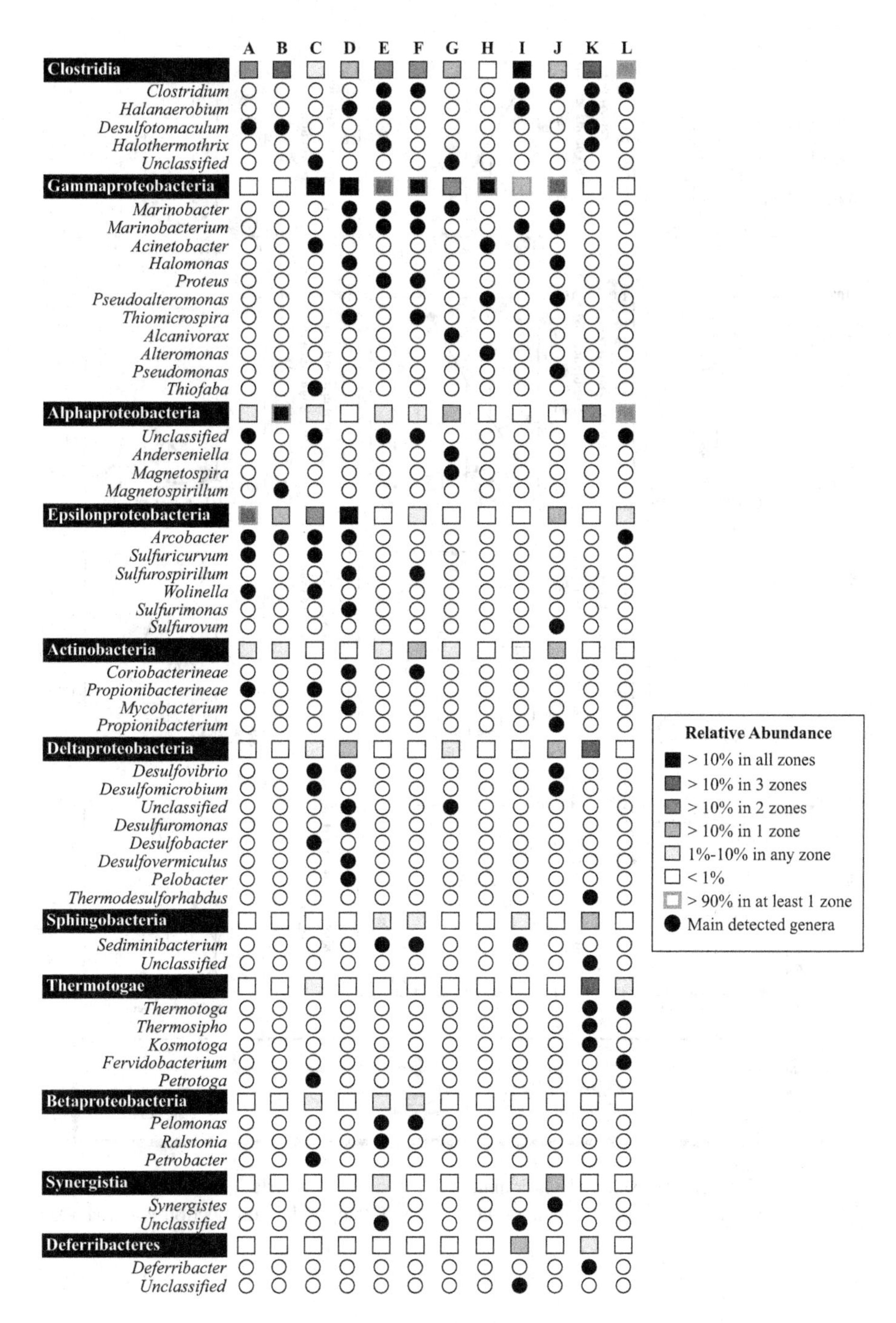

FIGURE 5.3 Heatmaps of taxonomic relative abundance of bacterial classes and main detected genera for the studied oilfields.

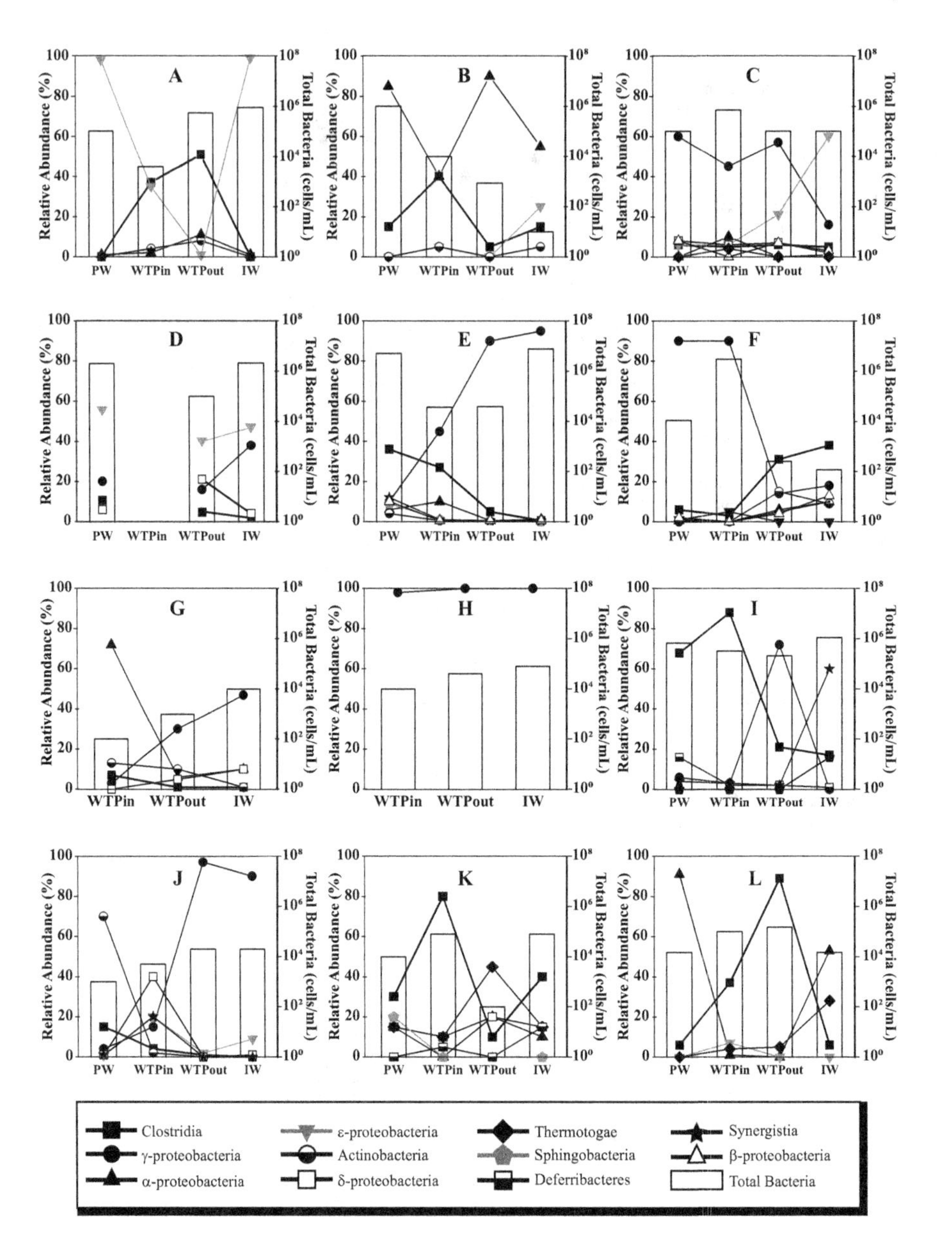

FIGURE 5.4 Relative bacterial class abundance and total bacteria levels from Production Water (PW) to Water Treatment Plant (WTPin and WTPout) and final Injection Water (IW).

The remainder of this section demonstrates that bioinformatics results from these audits allow correlation between the bacterial community and chemicals used in operations as well as examples of how underdosing or overdosing certain chemicals can be detrimental.

5.3.1.1.1 Antimicrobial Performance and Clostridia

As expected, abrupt changes in the microbial community occur downstream of biocide and oxygen scavenger dose points, reflected in the audit results for WTPout and IW. It is difficult to separate the influence of each one of these chemicals since both are dosed in nearby points, at various dose levels, but it is possible to observe that when biocide is not applied, like in oilfields D and E, the relative class abundances are more stable, or the trend observed from PW to WTPin is maintained to IW.

The presence of Clostridia in significant abundance on almost all oilfields studied in this work emphasizes the importance of a deeper understanding of their role in oilfields and the possible consequences when Clostridia grow out of control. Bacteria in this class are typically anaerobic acid-producing bacteria, and many members are spore-formers, such as *Clostridium*, *Desulfotomaculum*, and *Halothermothrix* (Hyun et al. 1983, Mavromatis et al. 2009, O'Sullivan et al. 2015, Monroy et al. 2019). Some members, like the sulfate-reducing bacteria (SRB) *Desulfotomaculum*, found in oilfields A, B, and K, have been previously described as relatively high aerotolerant (Bryukhanov et al. 2011). *Halanaerobium* (a halotolerant thiosulfate reducer) has been described as important in steel corrosion and reservoir souring (Daly et al. 2016). The genus *Clostridium* was the most frequent member, identified in half of the oilfields (E, F, I, J, K, L). Many species in this genus are acid-producing bacteria (APB) spore-formers. Recent studies indicate that they can be corrosive, more so than thiosulfate-reducing, or sulfate-reducing bacteria (Duncan et al. 2019; Monroy et al. 2019).

An effective antimicrobial treatment should control microorganisms on topside systems and support the frequent re-contamination that can occur during transport of the fluid to the injection wells (Kip & Van Veen 2015). Therefore, the selection of robust biocide treatment and dosage points must consider hot-spots as well as the possible interaction of biocides in the presence of other chemical products used in the WTPs. Depending upon water turnover rate and length of the water transport pipeline, the biocide may be required to remain active from hours to days. To further complicate matters, the waters transported from PW to WTP or from WTP to IW have different characteristics (i.e. sulfide content, salinity, pH, etc.) which can significantly impact the performance of a biocide treatment (Paschoalino et al., 2019b).

Based on the audit results, only oilfields B, treated by Glut/QAC (200 ppm a.i.) and F, treated by THPS (200 ppm a.i.) and QAC (50 ppm a.i.) appear to be under control. The high contamination ($> 10^6$ cells/mL) found on PW and IW for Oilfields D and E, both operating without any biocide application, confirms the importance of a microbial control program in place.

THPS is the most common biocide treatment, primarily because of its lower cost. Only one oilfield had adequate microbial control despite nine oilfields treated with THPS, clearly demonstrating that the use of this biocide needs to be reevaluated.

The lack of performance for any biocide is often related to underdosing (< 25 ppm a.i.), such as at oilfield I. This is a commonly ignored risk and could cause proliferation of tolerant organisms (Tidwell et al. 2017). Other treated fields received adequate dosages of THPS (150–250 ppm a.i.) in 2–3 h batches, typically twice a week yet still exhibited poor performance. This may be due to the fact all the fields (except the offshores G and H) are considered mature, with decades of water flooding operations, and in many of them, iron sulfide scaling is present in a significant amount. In these cases, THPS will work preferentially as an iron-sulfide scavenger before acting as a biocide, meaning that it will be deactivated and will not be available at an adequate concentration to provide optimal microbial control (Wang et al. 2015). Trivedi (2018) demonstrated that while 20 ppm a.i. of glutaraldehyde is slightly affected by the presence of 100 ppm of iron sulfide, THPS is severely limited and may require 50 ppm a.i. to maintain antimicrobial performance. This situation is illustrated in oilfield K, where THPS at 250 ppm a.i. performs very well close to the WTPout dosage point but loses effectiveness through the injection pipelines, where biofilm continuously inoculates the water. By the time the water reaches the IW, contamination levels have returned to levels prior to biocide dosage.

There are some occasions where knowledge of the dominant microbiota could help to narrow biocide options. The relevant abundance of Clostridia class for most of the studied oilfields is an example since many of these organisms are spore formers, which may give them advantages against harsh environmental conditions and biocidal chemistries (Christman et al. 2020). Sporicidal chemistries such as glutaraldehyde should be selected where known spore-formers Clostridia and Bacilli are dominant (Ascenzi 2019).

Improved performance of glutaraldehyde-based biocides against dominant Clostridia populations can be seen in oilfield B, where a 3-log reduction in bacteria is provided by glutaraldehyde at 200 ppm a.i., particularly in a community with about 40% *Desulfotomaculum*, a known spore-former (O'Sullivan et al. 2015). Similar results were obtained during field trials in oilfields A, C, K, and L, where THPS/QAC (200–250 ppm a.i.) was replaced by Glut (150–200 ppm a.i.) or Glut/QAC (200 ppm a.i.) treatments (Sianawati et al. 2016, Paschoalino et al. 2017, Paschoalino et al. 2018, Paschoalino et al. 2019a, 2019b).

5.3.1.1.2 Gammaproteobacteria Stimulus after Biocide Formulation Changes

Another example of microbiota changes due to interference of an expected non-functional chemical occurred in the offshore oilfield G. In this system, seawater is treated by electrochlorination and filters prior to nanofiltration membranes used to selectively remove sulfates from the water. Results in Figure 5.4 G demonstrate an abnormal increase in counts, even after high dosages of a 2,2-dibromo-3-nitrilopropionamide (DBNPA)-based biocide. Audit results indicated growth of Gammaproteobacteria.

Most organisms belonging to the class Gammaproteobacteria are aerobic heterotrophs and are often considered innocuous or less problematic compared to sulfate-reducing or iron-reducing bacteria. Typically defined as hydrocarbon degraders, some species have been described as important contributors to corrosion by providing ideal substrates for other organisms (Thornton 2017). For example, in oilfield H, *Pseudoalteromonas sp.* has been described as a pivotal contributor to carbon-steel or stainless-steel MIC in seawater injection systems (Moradi et al. 2014; Wu et al. 2016).

20% DBNPA is the standard biocide used to prevent biofouling on membranes and has been dosed in this system for more than five years. After a change on the biocide formulation, increases in membrane pressure indicated that fouling was occurring, to the point the system needed to be shut-down to avoid damaging the membranes. Chemical analysis of the new product indicated that the biocide active, DBNPA, was present at around 20%, but other components of the formulation were different from the previous supply. While the original product was formulated using the polyethylene glycol (PEG 300) and water, the new one was formulated using a lower molecular weight glycol, diethylene glycol (DEG). From a chemical standpoint, this glycol difference was not expected to cause any damage to the membranes; however, we decided to investigate the microbial community variation during and after a one-hour batch dosing of the new biocide. Results are presented in Figure 5.5.

NGS analysis of the permeate after a batch dose of the new 20% DBNPA formulation showed that the only bacterial genus increasing its relative distribution was *Alcanivorax*. Zadjelovic et al. (2020) demonstrated that members of this genus can degrade many types of polymers in the seawater environment, such as the ethylene glycol present in the novel DBNPA formulation.

The results suggest that the less complex glycol (Diethylene glycol, DEG) used in this new formulation could be stimulating microbial regrowth of the remaining population after the biocidal effect of the DBNPA active disappears. This hypothesis is corroborated by Haines and Alexander (1975) who showed a higher oxygen consumption by bacteria over a short time period (< 2 days) in the presence of DEG or ethylene glycol versus in the presence of polyethylene glycol (PEG).

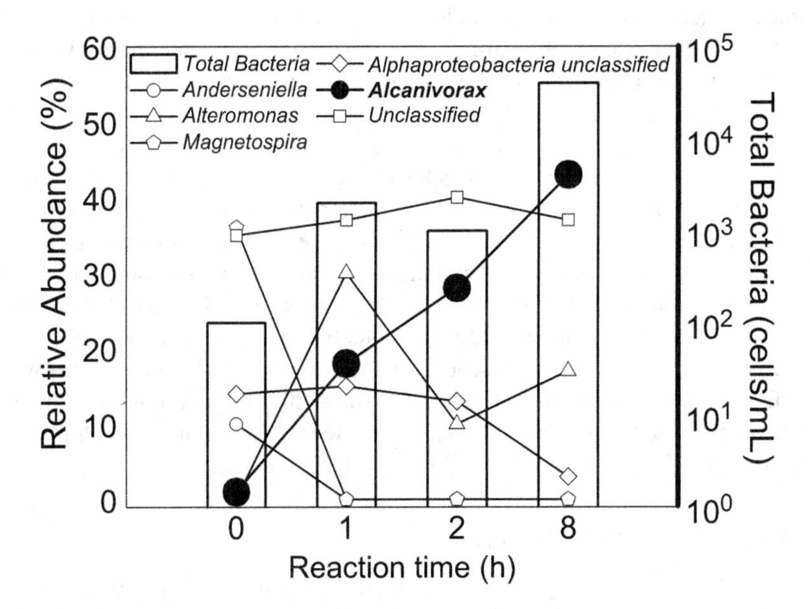

FIGURE 5.5 Variation in total bacterial cells and relative abundance of dominant genera after 20% DBNPA dosage.

All these facts suggest that it is preferable to utilize high molecular weight glycols, like PEG, in 20% DBNPA formulations since it can inhibit the metabolism of glycol residues by surviving bacteria after the quick-kill effect of DBNPA disappears. This example demonstrates the importance of not just selecting the appropriate active ingredient, but rigorously testing the formulation for compatibility in the system.

5.3.1.1.3 Solids Management, Oxygen Scavengers, and Sulfur-Oxidizing Epsilonproteobacteria

The presence of fine suspended particles of iron sulfide in oilfield fluids can cause numerous problems, such as efficiency reduction of oil-water separation in WTPs and loss of injectivity in injection wells (Nasr-El-Din and Al-Humaidan 2001). Historically, hydrogen peroxide was used on fields A, B, C, and D to control the suspended sulfide levels, by its oxidative conversion to sulfate. This process generates significant amounts of sulfate and oxygen as byproducts and requires a substantial increase in oxygen scavenger (bisulfite) dosing before IW. This practice was discontinued for the fields A, B, and D after almost one decade of use, while oilfield C continues to utilize hydrogen peroxide due to environmental regulations. In addition, oilfields C and D recently implemented the use of a tannin-based coagulant, typically synthesized via aminomethylation with the purpose of cationizing the tannin to improve the coagulation properties (Grenda et al. 2018).

Figure 5.4 (A, B, C, D) presents an analysis of dominant organisms in the oilfields included in the bio-audit. A significant shift of the microbiota from WTPout to IW was characterized by a substantial increase of Epsilonproteobacteria class (20% to 95% increase) in wells A, B, and C, while at D (no biocide), this class is the major dominant in all sampling points. These increases probably were likely influenced by chemical practices on these oilfields, since the similar phenomenon has been described when an excess of bisulfite was regularly dosed in pipelines, favoring the growth of Epsilonproteobacteria downstream to the oxygen scavenger dosage point (An et al. 2016).

The genera *Sulfuricurvum* and *Sulfurovum* are chemolithotrophs, described as Sulfur-Oxidizing Bacteria (SOB), which oxidize sulfide or sulfur with O_2 or reduce sulfur with H_2, consequently making this class a potentially powerful contributor to microbial corrosion (An et al. 2016). The presence of a nitrogenated nutrient, the coagulant, on C and D appears to result in high microbial diversity on these oilfields. The resultant microbial community is a concern in terms of MIC, since H_2S generation it is expected from most members of Deltaproteobacteria, such as *Desulfovibrio* (Wei et al. 2010), particularly in the presence of nutrients produced by fermenters (Widdel and Pfenning 1982, Alabbas et al. 2012, De Turris 2012). The produced sulfide can then be metabolized by the SOBs creating a cycle of interactions with a high risk of MIC (Mori and Suzuki 2008, Han and Perner 2015, An et al. 2016).

5.3.2 Archaeal Contamination

In addition to bacteria, a quantitative assessment using qPCR, from PW to IW points, was performed for Archaea only in oilfields C, D, E, F, I, J, K, and L. The comparative results (Archaea x Bacteria) are shown on Figure 5.6.

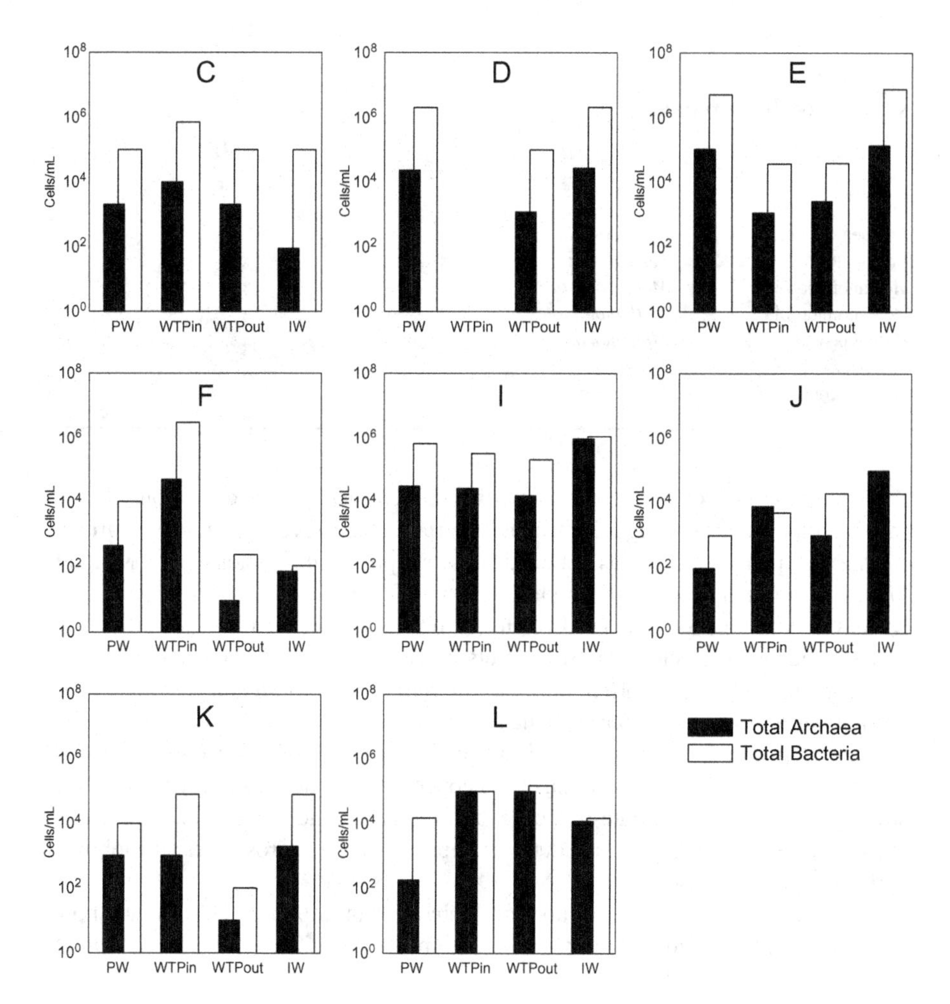

FIGURE 5.6 Total bacteria and Total Archaea levels from Production Water (PW) to Water Treatment Plant (WTPin and WTPout) and final Injection Water (IW).

It is possible to observe that for most of the sampling points (85%) Archaea were less abundant than Bacteria, generally representing just 1% to 9% of the total cells. In all fields, Archaeal contamination followed a similar trend, and the oilfields under best archaeal control were C, F, and K. The hot-spots on WTPin point for C and F were detected at heated tanks (> 65°C), favoring the proliferation of these thermophilic organisms.

The most significant Archaeal presence was found on oilfields with higher average temperatures (I, J, L), confirming recent findings that temperature is the crucial factor affecting Archaeal diversity and distribution in oilfields (Liang et al. 2018).

Underdosing of biocide on oilfield I appears to favor both Archaea and Bacteria, as I is the only field with high contamination levels (> 10^4 cells/mL) for all sampling points and for both types of microorganisms. Growth observed from the WTPout to

TABLE 5.2
Main Detected Archaea

		Oilfields				
Classes	**Genera**	**C**	**I**	**J**	**K**	**L**
Archaeglobi	*Archaeoglobus*					x
Archaeglobi	*Ferroglobus*	x		x	x	
Methanomicrobia	*Methanosaeta*	x		x	x	
Methanomicrobia	*Methanolobus*	x	x	x		
Methanococci	*Methanococcus*	x	x	x		
Thermococci	*Thermococcus*				x	
Thermoprotei	*Sulphobococcus*					x

IW in all oilfields, except C and L shows that the return anaerobic conditions after the addition of oxygen scavenger is also an important factor favoring Archaea regrowth. Archaeal sequencing data was obtained for select points in oilfields C, I, J, K, and L. The main dominant classes/genera are shown in Table 5.2.

As opposed to bacteria, the diversity of dominant Archaea was lower, and the variation between selected sampling points was less significant (data are not shown). Methanogenic organisms (*Methanosaeta*, *Methanolobus*, *Methanococcus*) were the most commonly detected contaminants.

Ferroglobus and *Archaeoglobus* were predominant on oilfields J and L, respectively. The presence of these organisms is concerning as members of *Archaeoglobus* are hyperthermophilic sulfate-reducing archaea, and some *Ferroglobus* species are known to be involved in iron reduction, representing a corrosion risk (Slobodkin 2005, Khelifi et al. 2010).

The lack of field methods to monitor Archaea is an important factor that emphasizes the need for a frequent microbial community evaluation using molecular methods for conventional oilfields.

5.4 CONCLUSION

The work described in this chapter gives a clear rationale for the importance of understanding microbial aspects in oilfield processes, even when the problem is not apparently related to microbial activity. Likewise, the examples presented illustrate that the use of bioinformatics tools could help prevent unintended consequences that would result from changes in microbial population dynamics, especially during the significant process and/or chemical changes in conventional water-flooded oilfields.

ACKNOWLEDGMENTS

All colleagues in DuPont Microbial Control that contributed to the bioaudits and discussions that enabled the publication of this data, especially Ethan Solomon, Emerentiana Sianawati, Fernando Mesquita, Geert van der Kraan, Kenneth Wunch, Sheila Tinetti, and Veronica Silva.

REFERENCES

Alabbas, F., Kakpovbia, A., Mishra, B. et al. 2012. *Effects of sulfate reducing bacteria on the corrosion of X-65 pipeline carbon steel.* In *NACE International Corrosion 2012 Conference & Expo.* Salt Lake City, C2012-0001140.

An D., Dong, X., An, A. et al. 2016. Metagenomic analysis indicates epsilonproteobacteria as a potential cause of microbial corrosion in pipelines injected with bisulfite. *Front. Microbiol.* 7:28.

Ascenzi, J. M. 2019. Glutaraldehyde-based disinfectants. In: *Handbook of Disinfectants and Antiseptics*, ed. J. M. Ascenzi, 111–127. Boca Raton: Taylor & Francis, CRC Press.

Bryukhanov, A. L., Korneeva, V. A., Kanapatskii, T. A. et al. 2011. Investigation of the sulfate-reducing bacterial community in the aerobic water and chemocline zone of the black sea by the FISH technique. *Microbiology* 80:108–116.

Christman, G. D., León-Zayas, R. I., Zhao, R., Summers, Z. M., & Biddle, J. F. 2020. Novel clostridial lineages recovered from metagenomes of a hot oil reservoir. *Sci. Rep.* 10: 8048.

Civan, F. 2016. *Reservoir Formation Damage: Fundamentals, Modeling, Assessment, and Mitigation.* New York: Gulf Professional Publishing.

Daly, R., Borton, M., Wilkins, M. et al. 2016. Microbial metabolisms in a 2.5-km-deep ecosystem created by hydraulic fracturing in shales. *Nat. Microbiol.* 1:16146.

De Turris, A., Haile, T., Gould, W. D. et al. 2012. *Correlation between the growth of a mixed culture of sulphate-reducing bacteria isolated from produced water and the corrosion of carbon steel.* In *NACE International Corrosion 2012 Conference & Expo.* Salt Lake City, C2012-0001126.

Duncan, K., DeGarmo, M., Tanner, R et al. 2019. *System-wide analysis of shale gas/liquids processing from an MIC point of view. Paper #77 presented at ISMOS 7 - International Symposium on Applied Microbiology and Molecular Biology in Oil Systems*, Halifax.

Enning, D. and Garrelfs, J. 2014. Corrosion of iron by sulfate-reducing bacteria: New views of an old problem. *Appl. Environ. Microbiol.* 80:1226–1236.

Grenda, K., Arnold, J., Hunkeler, D. et al. 2018. Tannin-based coagulants from laboratory to pilot plant scales for coloured wastewater treatment. *BioRes.*13:2727–2747.

Haines, J. R. and Alexander, M. (1975). Microbial degradation of polyethylene glycols. *Appl Microbiol.*, 29:621–625.

Han, Y. and Perner, M. 2015. The globally widespread genus *Sulfurimonas*: Versatile energy metabolisms and adaptations to redox clines. *Front. Microbiol.* 6:989.

Hyun, H. H., Zeikus, J. G., Longin, R. et al. 1983. Ultrastructure and extreme heat resistance of spores from thermophilic *Clostridium* species. *J. Bacteriol.* 156:1332–1337.

Khelifi, N., Grossi, V., Hamdi, M. 2010. Anaerobic oxidation of fatty acids and alkenes by the hyperthermophilic sulfate-reducing archaeon *Archaeoglobus fulgidus. Appl. Environ. Microbiol.* 76:3057–3060.

Kip, N. and van Veen, J. 2015. The dual role of microbes in corrosion. *ISME J.* 9:542–551.

Liang, B., Zhang, K., Wang L. Y. et al. 2018. Different diversity and distribution of archaeal community in the aqueous and oil phases of production fluid from high-temperature petroleum reservoirs. *Front Microbiol.* 9:841.

Marshall, K. C. 2013. Planktonic versus sessile life of prokaryotes. In: *The Prokaryotes*, ed. E. Rosenberg, E. F. DeLong, S. Lory, E. Stackebrandt and F. Thompson, 191–201. Berlin: Springer.

Mavromatis K., Ivanova N., Anderson I. et al. 2009. Genome analysis of the anaerobic thermohalophilic bacterium *Halothermothrix orenii. PLoS One.* 4:e4192.

Monroy, O. A. R., Ordaz, N. R., Gayosso, M. J. H. et al. 2019. The corrosion process caused by the activity of the anaerobic sporulated bacterium *Clostridium celerecrescens* on API XL 52 steel. *Environ. Sci. Pollut. Res.* 26:29991–30002.

Moradi, M., Song, Z., Yang, L., et.al. (2014). Effect of marine *Pseudoalteromonas sp.* on the microstructure and corrosion behavior of 2205 duplex stainless steel. *Corros. Sci.* 84:103–112.

Mori, K., Suzuki, K. 2008. *Thiofaba tepidiphila gen. nov., sp. nov.*, a novel obligately chemolithoautotrophic, sulfur-oxidizing bacterium of the Gammaproteobacteria isolated from a hot spring. *Int. J. Syst. Evol. Microbiol.* 58:1885–1891.

Nasr-El-Din, H. A., Al-Humaidan, A. I. 2001. *Iron sulfide scale: Formation, removal and prevention.* In *International Symposium on Oilfield Scale 2001*, Aberdeen, SPE-68315-MS.

O'Sullivan, L., Roussel, E., Weightman, A. et al. 2015. Survival of *Desulfotomaculum* spores from estuarine sediments after serial autoclaving and high-temperature exposure. *ISME J.* 9:922–933.

Paschoalino, M., Groposo, C., Dias, F. et al. 2017. *Laboratory study and field trial validation of a glutaraldehyde/THNM blend for injection water at petrobras onshore field. Paper presented at the Reservoir Microbiology Forum*, Energy Institute, London.

Paschoalino, M., Groposo, C., Waldow, V. et al. 2019a. *Advanced diagnostics of the microbial contamination and field trial of glutaraldehyde/THNM in a water treatment plant in Brazil. Paper #128 presented at ISMOS 7 - International Symposium on Applied Microbiology and Molecular Biology in Oil Systems*, Halifax.

Paschoalino, M., Raymond, J., Sianawati, E. et al. 2019b. *Evaluation of a glutaraldehyde/THNM combination for microbial control in four conventional oilfields. SPE International Conference on Oilfield Chemistry*, Galveston. SPE-193594-MS.

Paschoalino, M., Silva, V., Dias, F. et al. 2018. *Field trial of a glutaraldehyde/THNM combination for production/injection water at an argentinian onshore field.* In *Rio Oil & Gas Expo and Conference 2018*, Rio de Janeiro, IBP1996_18.

Salgar-Chaparro, S. J., Lepkova, K., Pojtanabuntoeng, T., Darwin, A., Machucaa, L. L. 2020a. Nutrient level determines biofilm characteristics and subsequent impact on microbial corrosion and biocide effectiveness. *Appl. Environ. Microbiol.* 86:e02885–e02919.

Salgar-Chaparro, S. J., Lepkova, K., Pojtanabuntoeng, T., Darwin, A., Machucaa, L. L. 2020b. Microbiologically influenced corrosion as a function of environmental conditions: A laboratory study using oilfield multispecies biofilms. *Corros. Sci.* 169:108595–108614.

Sianawati, E., Curci, E., Lorca, M. L. et al. 2016. *Holistic approach of microbial control solution for YPF barrancas field – from laboratory evaluation to field trial.* In *Rio Oil & Gas Expo and Conference 2016*, Rio de Janeiro, IBP1536_16.

Slobodkin, A. I. 2005. Thermophilic microbial metal reduction. *Microbiology* 74:501–514.

Thornton, M. M. 2017. The role of Gammaproteobacteria in aerobic alkane degradation in oilfield production water from the Barnett Shale. MSc diss., University of Oklahoma.

Tidwell, T. J., Keasler, V., De Paula, R. 2017. How production chemicals can influence microbial susceptibility to biocides and impact mitigation strategies. In *Microbiologically Influenced Corrosion in the Upstream Oil and Gas Industry*, ed. T. L. Skovhus, D. Enning and J. S. Lee, 379–392. Boca Raton: Taylor & Francis, CRC Press.

Trivedi, R. 2018. *Biocide stability and efficacy in soured oilfield environments. Paper presented at the Reservoir Microbiology Forum*, Energy Institute, London.

Van der Kraan, G., Bruining, J., Lomans, B. P. et al. 2010. Microbial diversity of an oil-water processing site and its associated oil field: The possible role of microorganisms as information carriers from oil-associated environments. *FEMS Microbiol. Ecol.* 71:428–443.

Van der Kraan, G., Buijzen, F., Kuijvenhoven, A. T. et al. 2011. Which microbial communities are present? Application of PCR-DGGE: Case study on an oilfield core sample. In: *Applied Microbiology and Molecular Biology in Oilfield Systems*, ed. C. Whitby and T. L. Skovhus, 33–44. New York: Springer.

Van der Kraan, G., Silva, V., Paschoalino, M. et al. 2018. *Using integrated microbial auditing for an argentinian oilfield, enabled an improved complete chemical treatment regime development.* In *Rio Oil & Gas Expo and Conference 2018*, Rio de Janeiro, FM_3809_ 2013.

Van der Kraan, G. M., Morris, B. E. L., Widera, I. et al. 2014. *A microbiological audit of an FPSO*. In *Rio Oil & Gas Expo and Conference 2014*, Rio de Janeiro, IBP1652_14.

Vigneron, A., Alsop, E. B., Lomans, B. P., Kyrpides, N. C., Head, I. M., Tsesmetzis, N. 2017. Succession in the petroleum reservoir microbiome through an oil field production lifecycle. *ISME J.* 11:2141–2154.

Wang, Q., Shen, S., Badairy, H. et al. 2015. Laboratory assessment of tetrakis(hydroxymethyl) phosphonium sulfate as dissolver for scales formed in sour gas wells. *Int. J. Corros. Scale Inhib.* 4:235–254.

Wei, L., Ma, F., Zhao, G. 2010. Composition and dynamics of sulfate-reducing bacteria during the waterflooding process in the oil field application. *Bioresour. Technol.* 101: 2643–2650.

Widdel, F. and Pfenning, N. 1982. Studies on dissimilatory sulfate-reducing bacteria that decompose fatty acids II. Incomplete Oxidation of Propionate by *Desulfobulbus propionicusgen. nov., sp. nov*. *Arch. Microbiol.* 131:360–365.

Wrangham, J. B. and Summer E. J. 2013. *Planktonic microbial population profiles do not accurately represent same location sessile population profiles*. In *NACE International Corrosion 2013 Conference & Expo*. Orlando. C2013-0002780.

Wu, J., Zhang, D., Wang, P., Cheng, Y. et al. (2016). The influence of *Desulfovibrio sp.* and *Pseudoalteromonas sp.* on the corrosion of Q235 carbon steel in natural seawater. *Corros. Sci.* 112:552–562.

Zadjelovic, V., Chhun, A., Quareshy, M. et al. (2020). Beyond oil degradation: Enzymatic potential of Alcanivorax to degrade natural and synthetic polyesters. *Environ. Microbiol.* 22:1356–1369.

6 Microbial Control during Hydraulic Fracking Operations

Challenges, Options, and Outcomes

Renato De Paula, Irwan Yunus, and Conor Pierce

CONTENTS

6.1 Introduction 99
6.2 Microbial Control in Frac Operations: The Challenges 100
6.3 Microbial Control in Frac Operations: The Options 103
6.3.1 Oxidizing Biocides 105
6.3.2 Non-Oxidizing Biocides 108
6.4 Microbial Control during Fracking Operations: The Outcomes 111
6.4.1 Laboratory Evaluation 112
6.4.2 Field Evaluation 112
References 116

6.1 INTRODUCTION

The advent of horizontal drilling and hydraulic fracturing represented a milestone in oil and gas production, allowing the extraction of hydrocarbons from tight shale reservoirs. The economic impact of unconventional production has brought significant changes to many aspects in society, from increasing the availability of energy sources to raising further concerns around the environmental impact caused by oil & gas exploration. From 2009 to 2019, US production of crude oil and gas increased from 5.6 to 12.3 million of barrels/day, mostly driven by unconventional production in Texas (Permian basin) and North Dakota (Bakken basin). In the West Texas region only, oil & gas production from unconventional wells has increased 333% since 2010, from 1 mb/d to 3.9 mb/d (US Energy Information Administration).

One of the hallmarks of unconventional production is the hydraulic fracturing process, in which large volumes of water and additives are injected at high pressure via the horizontal well to create cracks in the tight shale rock allowing hydrocarbons to flow into the well (King, 2012). The use of injected water to create cracks in an oil reservoir is not unique to unconventional well production but has been optimized to allow production from a geological formation with low porosity. Significantly larger volumes of water are used to create fractures in unconventional wells compared to conventional wells. Horizontal wells can be drilled in different directions and span miles away from the well head, requiring that hydraulic fracturing to be performed in stages. Each stage can require between 3,000 to 12,000 of barrels of water. This creates a new set of challenges to unconventional producers such as water availability, societal pressure to conserve water, and the subsequent impact of water reuse on the lifespan of the well.

Due to the amount of water used for the hydraulic fracturing, one of the main problems that arises from is microbial contamination of the well at very early stages of production. This chapter will focus on the main challenges arising from microbial activity in unconventional production of oil and gas, the options available for mitigating these problems, and how using a combination of solutions can lead to better outcomes.

6.2 MICROBIAL CONTROL IN FRAC OPERATIONS: THE CHALLENGES

Hydraulic fracturing uses multiple sources of water for injection. In many instances, fresh water is utilized in the process, which can be obtained from different sources such as wells, ponds, rivers, and municipalities as treated water. Fresh water contains less than 3,000 ppm of total dissolved salts (TDS). Water containing between 3,000 and 10,000 ppm is defined as brackish water. Waters containing more than 10,000 and 35,000 ppm TDS are considered saline and brine, respectively (Stanton et al., 2017). Water is brought onsite mostly by trucking and stored either in open ponds or tank batteries until the time of use. Although the water can be treated proactively with a biocide before or during storage, it has been proven to be more economical to treat the water at the time of use. At time of use, the water is treated with biocides and scale inhibitors and mixed with other additives shown in Table 6.1 in a blender and immediately injected. Each stage of the fracturing process is completed by adding plugs along the length of the well. Once all the stages are completed, the plugs are removed, and the injected water flows back followed by the produced fluids. The recovered flow back water (about 20%–40% of the injected volume) can be subsequently treated and mixed with fresh water for fracking new wells.

Microorganisms are ubiquitous and can colonize any environment where water is present. Since microbial consortia will rapidly change in response to characteristics of the environment they are in, those microorganisms that are better equipped to sustain growth during fracking conditions will multiply and dominate the population. Characteristics such as temperature, water salinity, aerobiosis, presence of electron acceptors will determine whether certain microorganisms can grow. This is a key aspect during hydraulic fracturing where different sources of water are used, mixed, and injected into a formation that greatly differ from the surface characteristics.

Water salinity has a profound impact on the selection of halophilic organisms. Frequently, there is a significant change in chemical composition between the injected and produced fluids (Blauch et al., 2009) and that can define the microbial

TABLE 6.1
Frac Chemical Components, Use Estimates, and Associated Purpose (US Dept. of Energy)

Frac Fluid Components	Estimated Ratio	Active Compound	Purpose
Water	90%	Water	Carrier solvent
Proppant	9%	Silica, sand	Prop fractures and allow fluid flow
Acid	0.07%	Hydrochloric acid	Dissolves minerals and clean flow line
Corrosion Inhibitor	0.05%	Isothiazolin, N,n-dimethyl formamide	Prevent corrosion in metal pipe
Friction Reducer	0.05%	Polyacrylamide, mineral oil	Minimizes water friction in pipe to reduce power consumption during frac
Clay Control	0.03%	Potassium chloride salt	Stabilize surrounding soil
Crosslinker	0.03%	Borate salts	Maintain fluid viscosity as temperature increase
Scale Inhibitor	0.02%	Phosphonate	Prevent scale build up in the pipe
Breaker	0.02%	Ammonium persulfate	Delayed breakdown of the gel polymer
Iron Control	<0.01%	Citric acid	Prevent precipitate of metal oxide
Biocide	<0.01%	Glutaraldehyde, bleach	Eliminate bacteria in water
Gellant	0.50%	Guar gum, Hydroxyethyl cellulose	Thickens water to carry/suspend sand

problems that a specific well will face during its production lifetime. Early flowback water may show microbial communities that closely resemble the communities in fresh water used for fracking (Mohan et al., 2013). Nonetheless, the later flowback and produced fluids will frequently show an increase of halophilic anaerobic organisms. For example, *Halanaerobium sp* were found in early/late flowback and produced fluids in 9 out of 11 studies of the microbial ecology in fracked unconventional wells, although their abundance in fresh water is rather limited (Mouser et al., 2016, Booker et al. 2017, Lipus et al. 2017). Members of the *Halanerobiaceae* family seem to be one of the dominant sulfate-reducing (thiosulfate-reducing) organisms responsible for souring of unconventional wells (Booker et al., 2017). Similarly, members of the *Methanosarcineae* family, which are involved in methane production, seem to thrive in conditions of high salinity and anaerobiosis and are frequently found in flowback/produced water from unconventional wells (Cluff et al., 2014) but rarely found in fresh water. Nonetheless, the practice of comingling fresh and flowback water frequently leads to an increase in these halophilic, anaerobic organisms that will remain stagnant in water tanks for days or weeks until the water is used for another hydraulic fracture. Therefore, these organisms can be detected at the earlier flowback water of the subsequent fracking, creating cycles of enrichment for these organisms if proper microbial control is not accomplished.

Temperature is another factor that can impact the makeup of the microbial community during the fracking process. Some shale reservoirs can reach temperatures up to 310°F (155°C). While these high temperatures are mostly likely not conducive for microbial growth, the injection of cool water during the fracking can result in temporary cooling of the regions around the fractures, which may allow microorganisms to grow for a determined period until temperature equalization. For example, microorganisms such as *Thermoacetogenium, Thermovirga*, and *Thermoanaerobacter* have been identified in produced water from hot reservoirs like the Haynesville shale (Fichter et al., 2012), and a number of thermophilic anaerobes have been linked to corrosion failures in topside equipment during production due to poor practices during hydraulic fracturing (Tidwel et al., 2017).

Whether these organisms are native to reservoirs and are brought up during production or whether they are injected into the reservoir during the fracking process is the subject of much controversy. Although microorganisms are known to inhabit oil reservoirs and hydrocarbon seep sediments in the ocean (LaMontagne et al. 2004, Liu et al., 2018), the presence of native organisms in high salinity, high temperature (>120°C) shale reservoirs has not been proven. Collection of samples for microbial analysis during the drilling process is difficult to obtain and contamination of the samples with drilling fluids/mud cannot be excluded. But it is plausible to speculate that these organisms are likely introduced during the fracking process and enriched downhole by conditions of anaerobiosis, increased salinity, presence of frack additives (e.g. guar gum), and temporary cooling due to large volumes of water injected.

Poor hydraulic fracturing practices may lead to chronic microbial problems that span the entire lifecycle of an unconventional well. These will include early souring of the well, corrosion failures in topside equipment and pipelines due to Microbiological Influenced Corrosion (MIC), and limited flow due to biomass buildup in the fractures.

Souring is the production of toxic H_2S gas as a metabolic byproduct of sulfur-reducing prokaryotes (SRP). These organisms use sulfur species as electron acceptors for their conversion of carbon sources into energy. Souring is common in aging conventional wells when the water cut increases and the reservoir is waterflooded to maintain pressure (Basafa and Hawboldt, 2019). Nonetheless, the use of large volumes of untreated water during the hydraulic fracking can increase the risks of souring. Newly drilled and completed unconventional wells can show signs of souring as early as the initial onset of flowback water (Mohan et al., 2013). Once contaminated, a sour well may require constant care to minimize the production of toxic H_2S, either by performing clean-outs with biocides or treatment of the produced fluids with H_2S scavengers.

Early contamination of the reservoir during hydraulic fracturing can have a devastating impact in the operations and integrity of topside vessels and transmission pipelines. The contamination in the reservoir can seed the topside separation vessels with microorganisms that can colonize the surface of the metallurgy (pipes, tanks, valves, etc.) and lead to progressive corrosion. Thermophilic SRBs and Methanogens are known culprits in the process of microbial corrosion and can be enriched during the hydraulic fracking process as described earlier (Mouser et al., 2016).

In summary, the large amount of water used during the hydraulic fracturing process and the commingling of different water sources can have a critical impact in the life of an unconventional well. It can lead to enrichment of microbial populations that will thrive in reservoir conditions, even if in a transient period and it can be the source of contamination of topside production vessels, leaving producers to deal with chronic problems during the production phase.

6.3 MICROBIAL CONTROL IN FRAC OPERATIONS: THE OPTIONS

Water is the major component of hydraulic fracturing, comprising up to 90% of the content in the frack fluids. Sand represents around 9%[5] and the remaining 1% is composed of frac chemicals. Water acts as a carrier and the sand acts as a proppant to keep the fissures open to allow for the flow of hydrocarbons.

The type and amount of chemicals added to the fracking fluids varies according to the formation, water quality, and economics. A description of the chemicals usually present in a frack fluid and an explanation for its use is shown in Table 6.1.

The impact of fracking to the environment is a long and ongoing debate. The scarcity of fresh water and public pressure has pushed operators to reuse flowback and produced water for their fracking operations. This has created a new set of challenges to operators regarding microbial control, as the use of co-mingled fresh and flowback/produced water stimulate the growth of harmful anaerobic microorganisms and pose challenges to the biocide treatment. In these circumstances, the proper use of biocides has proved to be crucial to maintain the health of the well and avoid early contamination of the operations.

Biocides are affected by the water chemistry (fresh versus flowback/produced); therefore, finding the proper biocide product that can perform equally well in different waters is challenging. In the oil & gas industry, biocides are used to reduce and maintain low population of bacteria rather than the full eradication of microorganisms required in medical or food industries.

The pressure toward more sustainable use of water has advocated for the reuse of contaminated flowback and produced waters. Biocides are often considered a onetime "low cost insurance" to avoid the risk of perpetual high cost and safety risk expense of dealing with a sour well or reservoir. Most E&P (Exploration & Production) companies use biocides in their operations to reduce bacteria count in the water prior to use in hydraulic fracking, but identifying the proper biocide product is usually a question that must factor in several parameters such as compatibility of the product with other chemicals in the frac fluid, safety of the operation, and economics.

Many biocides are available in the market. The most used products have the following actives: glutaraldehyde, sodium hypochlorite (bleach), chlorine dioxide, quaternary ammonium, and peracetic acid. There are many factors influencing biocide selection such as cost, water quality, downhole well temperature, compatibility with other chemicals and equipment, and expected duration of efficacy. For example, quaternary ammonium-based biocides are very effective and cost competitive but are not suitable in waters with high hardness and may impact efficacy of anionic friction reducers. In addition, the impact of these surfactant products on the surface of the rock is not fully understood. Non-ionic glutaraldehyde usually shows good compatibility with frack fluids, but its inherent instability at high temperatures limits its use in high temperature reservoirs (Kahrilas et al., 2015). Quaternary phosphonium biocides such as THPS and TTPC have a higher thermal stability but also present their own set of challenges. Certain THPS products have been shown to interfere with the performance of friction reducers, although new formulations that are suitable for use with certain friction reducers have been developed (Jones et al., 2016).

Another aspect that is usually considered is when and where in the process should the water be treated to render optimal effectiveness while balancing the costs of the operation. Biocides can be applied at several locations of the frack operation including during transport of the water (truck treatment), in open holding ponds, battery tanks, or at the blenders, where the components of the frack chemical package are mixed. Operators balance the risks of microbial contamination of a well with costs to treat a barrel of water prior to fracking. A risk model is depicted in Figure 6.1. Fresh water can remain in open pits and storage tanks for several days or weeks before it is used for the fracking process. During this time microbial growth will occur, leading to waters with high microbial loads at the time of the hydraulic fracking. Microbial control treatments can be applied at different stages of the process to maintain acceptable levels of microorganisms and minimize the risks of microbial issues originating from the well during production. Nonetheless, this approach is expensive and requires multiple treatments. Usually, operators prefer to treat the water at the time of use ("on the fly"). When microbial counts are unusually high, a single treatment requires a highly efficient biocide that can achieve key performance indicators (KPI), such as the appropriate microbial reduction compared to the untreated water and/or the maximum number of microorganisms allowed be introduced into the reservoir. Failure to achieve control in highly contaminated water just prior to injecting the water downhole results in high risk of well contamination.

Frequently, the choice of when to treat the water for fracking operations will define the biocide product selected. Treating the water days or weeks ahead of the fracking process will require a product with a higher persistence to avoid recontamination of the water prior to the fracking and/or multiple biocide treatments. In situations where a single treatment on the fly is preferred, the use of a fast killing biocide,

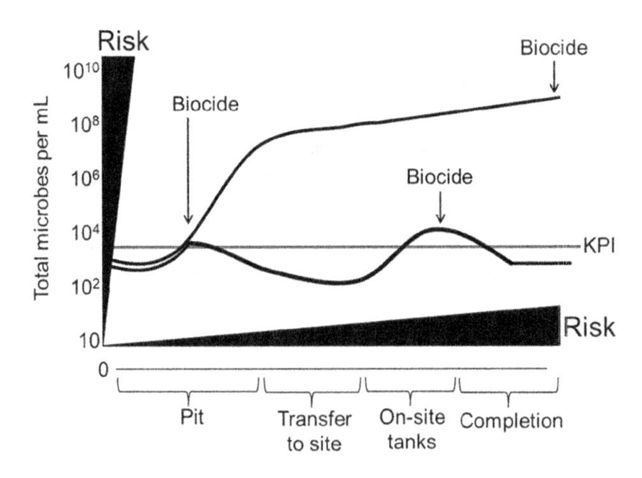

FIGURE 6.1 Risk model for microbial control treatment during hydraulic fracking operations.

alone or in combination with a more persistent biocide can provide the full spectrum of control for the entire duration of the fracking stages, drilling of the plugs, and flowback. Among the fast-acting biocides are oxidizing chemistries such as hypochlorous acid, sodium hypochlorite (bleach), chlorine dioxide, ozone, and peracetic acid. Hypochlorous acid and bleach are highly effective and cost competitive; however, waters with higher organic load will require greater volume of these products and in waters with low buffering capacity scaling tendency may occur, which can cause restricted production flow as the scale builds up.

The antimicrobial efficacy of each biocide is dependent on both the reservoir conditions as well as the other chemistries used in the fracturing fluid package. Just as reservoir temperature and pressure will affect microbial growth, extreme conditions can cause molecular degradation of less resilient biocides. The more reactive biocides may also suffer degradation due to incompatibly with other components of the fracturing fluid. Conversely, combinations of certain biocides in the fracturing fluid can generate synergistic effects, increasing the overall antimicrobial efficacy of the package and requiring a smaller dosage for sufficient treatment (Williams 2008, Xu et al., 2012). Considering the consumption of the biocide with components of the water and other chemicals in the frack fluid is critical to designing an optimized treatment that will render enough biocide to target the microbial organisms. For example, certain biocides such as THPS and quaternary ammonium compounds may be consumed by interactions with friction reducers and proppants, decreasing its availability to react with components of the microbial cells.

In the case of hydraulic fracturing fluids, treatment targets a reduction in bacterial counts sufficient to mitigate risks associated with uncontrolled microbial growth as mentioned above. To do this effectively, it is necessary to understand how the composition of the treated fluid will interact with any given biocide. The number of microorganisms present will directly affect the amount of biocide required. Equally important, however, is identifying the presence of nitrogen and sulfur sources or transition metals like iron or copper as each of these can have deleterious interactions with several biocidal molecules. Preliminary laboratory testing to determine the optimal biocide for use in fluids is typically performed according to NACE Standard TM0194 (Field Monitoring, 2004). This testing is used in tandem with parameters that include fluid compatibility with other fracturing chemicals used, the flow rates of those chemicals, and reservoir conditions like temperature and pressure (Kallmeyer and Boetius, 2004) to identify an optimal antimicrobial program.

Identification of biocide compatibility in both the treated fluid and fracturing package is important given the wide range of antimicrobial compounds available on the market. While structurally unique, biocides can generally be grouped by their mechanisms of action as either oxidizing or non-oxidizing molecules.

6.3.1 Oxidizing Biocides

Oxidizing biocides all share a common mechanism of action: removal of electrons from the molecular structures of cellular components resulting in the breaking of chemical bonds essential for cell survival (Finnegan et al., 2010). Due to the specifics of this mechanism, oxidizers are generally broad-spectrum antimicrobials that do not allow for the development of antimicrobial resistance given the prevalence of oxidizable functional groups like thiols and thioethers that play a crucial role in all living systems (Noszticzius et al., 2013). The highly reactive nature of oxidizing biocides also increases the risk of equipment corrosion due to reactions with some metals including copper, steel, and iron. Oxidizing biocides will also react with other compounds commonly found in treated fluids like iron, manganese, copper, and some organics.

Two widely used oxidizing biocides used in hydraulic fracturing operations are the chlorine-based species sodium hypochlorite, or bleach, and chlorine dioxide. Both compounds react readily with sulfur-containing amino and nucleic acids like cysteine and methionine leading to cell death (Hawkins et al., 2003). However, chlorine dioxide is more site-selective than hypochlorite and less likely to react with non-critical cellular components such as carbohydrates and lipids found in cellular membranes. This selectivity often translates to equivalent antimicrobial efficacy at a lower treatment rate.

When sodium hypochlorite is added to water, it creates a pH-dependent equilibrium mixture of hypochlorite and hypochlorous acid, with the active species existing as hypochlorous acid. While the mechanism of action for both is the same, hypochlorous acid is a much stronger oxidizer than the hypochlorite ion, thus a more efficient disinfectant.

Some of the more specific sodium hypochlorite mechanisms of action that have been proposed include inducement of cell death through metabolic dysfunction resulting from adenine nucleotide depletion and the inhibition of DNA replication, both due to the oxidation of key components involved (Tischler et al., 2009).

Hypochlorite is available as either a bulk solution of 11%–14% sodium hypochlorite or can be generated on-site through the electrolysis of a sodium chloride solution in water. Both methods have disadvantages. Bulk storage of sodium hypochlorite requires a climate-controlled environment to prevent hypochlorite degradation over time and on-site generation used to be limited to concentrations of <1.0%. More recent advances in the on-site generation of hypochlorite solutions allow for concentrations of up to 15% utilizing a mixture of hydroxide and chlorine instead of sodium chloride. It is also important to note that hypochlorite will generate by-products resulting from the reaction of chlorine with hydrocarbons in treated fluids. The most common of these by-products are trihalomethanes and haloacetic acids, both of which are listed as carcinogenic by the Environmental Protection Agency (Duong et al., 2003).

Unlike hypochlorite, chlorine dioxide cannot be stored in bulk solutions due to its inherent instability and explosive limit of 10% v/v in air so must be generated on-site. Generation methods for chlorine dioxide vary, but the most common are through an electrochemical system or the reaction of sodium chlorite with either hydrochloric acid or both hydrochloric and hypochlorous acids. Electrochemical generation yields a dilute solution of chlorine dioxide that is free of residual chlorine, thus carrying a lower risk of halogenated by-products like those observed during the application of hypochlorite. Generating chlorine dioxide via the two- and three-component reaction methods yields more concentrated solutions, but residual unreacted chlorite retains the risk of carcinogenic halogenated by-products.

Chlorine dioxide is an effective biocide in aqueous hydraulic fracturing applications in part due to its ability to accept up to five electrons versus chlorine's two. In other words, a single molecule of chlorine dioxide will react more times as an oxidizing biocide before being consumed than one of chlorine. The reaction pathway initially generates a chlorite ion after accepting one electron then proceeds through to a chloride ion after accepting four additional electrons. It is generally effective at pHs below 12 (Tischler et al., 2009).

An alternative to halogenated oxidizing biocides, and their associated risks, is hydrogen peroxide. Used throughout the oil and gas industry, hydrogen peroxide is an inorganic molecule available in a range of concentrations with long-term storage capability due to its relative stability when compared to the chlorine species discussed above. Prevailing theories regarding its mechanism of action state that hydrogen peroxide reacts with either a biological source of iron within the bacteria or iron/copper/manganese present in the treated fluid to generate a hydroxyl radical. This radical then causes cell death through the oxidation of DNA, proteins, and membrane lipids (Finnegan et al., 2010). More recent studies have proposed a DNA oxidation mechanism involving a ferryl radical formed from DNA-associated iron reacting with the peroxide, rather than from a hydroxyl radical. In either case, hydrogen peroxide can serve as an effective oxidizing biocide if the conditions to generate a reactive hydroxyl radical species are present (Linley et al., 2012).

Peroxy acids are related to hydrogen peroxide. Peroxy acids are synthesized by mixing hydrogen peroxide with carboxylic acids bearing various carbon chain lengths. Peracetic acid, from acetic acid and hydrogen peroxide, is the most prevalent of the peroxy acid oxidizing biocides in hydraulic fracturing operations. While it also generates hydroxyl radicals, peracetic acid is a more reactive molecule than hydrogen peroxide with a higher oxidation potential so is capable of oxidizing thiols and disulfide bonds in key cellular components without the need of a radical intermediate. An additional benefit of peracetic acid is the release of only environmentally benign byproducts: Acetic acid and water from the peracetic acid and oxygen from the hydrogen peroxide used to synthesize it (De Paula et al., 2013). Recent advances have introduced performic acid, from formic acid and hydrogen peroxide, as a more reactive peroxy acid variant. Mechanistically identical to peracetic acid but substantially more reactive, performic acid must be generated on-site. The byproducts of performic acid are also environmentally benign, forming only carbon dioxide and water due to the decreased amount of hydrogen peroxide required to generate it (Pierce and Peter, 2019).

Contrary to oxidizers chemistries that can be brought to the frack site in their final form of use, certain oxidizing chemistries must be generated *in situ*, at the time of use. Ozone must be generated on-site as its rapid reaction kinetics and relative instability cause it to quickly decay into diatomic oxygen. The half-life of ozone is only 15 minutes when dissolved in water at 25°C. Ozone is a gas molecule generated through the application of a high voltage electric discharge, often referred to as a corona discharge, to a source of either dry air or oxygen gas. When air is used, an ozone concentration of approximately 1% is produced in gas-phase and injected into the water stream. The use of more concentrated oxygen gas, often purified onsite via oxygen concentrators that remove nitrogen from ambient air, allows for higher concentrations of ozone but is still limited by the flow rate of the feed gas and efficiency of the corona cell. Specifically, slower flows and lower temperatures provide higher ozone concentrations. Maintaining suitable gas temperatures can be especially difficult given the inefficiencies of current corona discharge technologies that dissipate nearly 90% of the energy required to produce ozone as heat. Ozone will also introduce oxygen into the system which can lead to equipment corrosion if not dosed properly. This risk in diminished when treating ponds and storage tanks in hydraulic fracturing operations that are already open systems exposed to ambient air.

Electrochemically activated solutions, or electrochemically activated water, is another oxidizing biocide that requires on-site generation. Like on-site hypochlorite generation, water is passed through an electrochemical cell to form oxidizing species with antimicrobial activity. A key difference between electrochemically activated solutions and on-site hypochlorite generation is the ability to run the treated fluid itself through the electrochemical cell to generate biocidal species *in situ*. Depending on applied dosage and the composition of the fluid, a mixture of free chlorine and radical oxygen species such as ozone and hydroxyl radicals will form at low concentrations. The efficacy of these systems is dependent on the flow rate, to prevent dilution of the active species, and composition of the fluid, as the initial concentrations of chloride and reactive oxygen precursors will influence the amount of biocidal species formed (Ghebremichael et al. 2011, Robinson et al., 2012).

6.3.2 Non-Oxidizing Biocides

Contrary to oxidizing biocides, non-oxidizing biocides typically have lower reactivity and slower reaction kinetics as their mechanisms of action do not involve free radical species but rather bond-forming interactions between the biocidal molecular structure and cellular components on or within bacteria (McDonnell and Russell, 1999). These intermolecular interactions between non-oxidizing biocides and certain cellular components are categorized into two groups based on their mechanisms of action: Lytic and electrophilic.

Biocides categorized as lytic are typically amphiphilic surfactants, which contain both hydrophobic and hydrophilic components that disrupt the bacterial membrane through dissolution of the cell wall. This dissolution process proceeds via a binding of the cationic hydrophilic component of the lytic biocide with anionic functional groups on the cell wall followed by perturbation of the lipid bilayer, subsequent loss of osmotic regulation capacity, and finally cell death (Gilbert and Moore, 2005). The bacteria cell essentially loses structural integrity of the cytoplasmic membrane, inducing lysis.

Quaternary amine ammonium compounds, or QACs/quats, are the foremost example of lytic biocides used in hydraulic fracturing applications. Mechanistically, the cationic quaternary amine component of the QAC binds to the bacterial cell wall

TABLE 6.2
Biocides Used in Hydraulic Fracturing Operations by Frequency of Use (FracFocus, 2013)

Name	CAS No.	Chemical Structure	Frequency of Use
Glutaraldehyde	111-30-8	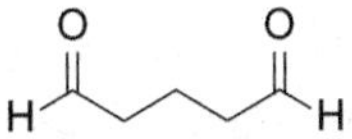 O, O, H, H	25.60%
Alkyl dimethyl benzyl ammonium chloride	68424-85-1	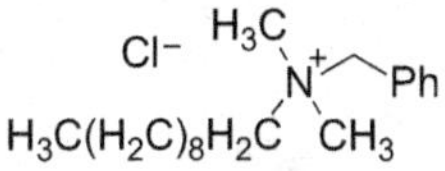 Cl^- H_3C N^+ Ph $H_3C(H_2C)_8H_2C$ CH_3	19.77%
Didecyl dimethyl ammonium chloride	7173-51-5	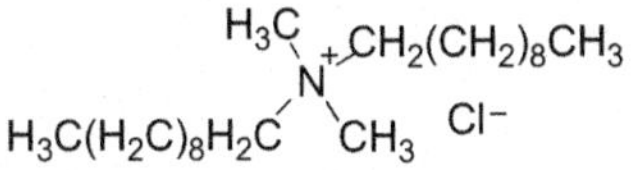 H_3C N^+ $CH_2(CH_2)_8CH_3$ $H_3C(H_2C)_8H_2C$ CH_3 Cl^-	17.93%
Tributyl tetradecyl phosphonium chloride	81741-28-8	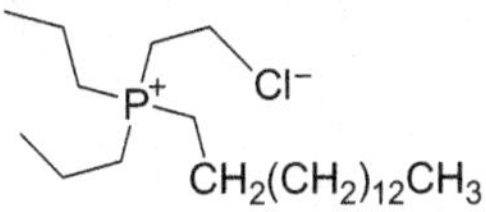 P^+ Cl^- $CH_2(CH_2)_{12}CH_3$	7.13%
Sodium Hypochlorite	7681-52-9	$Na^+ \; ^-O-Cl$	6.50%
Chlorine Dioxide	10049-04-4	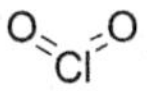 O, Cl, O	4.82%
Dibromonitrilopropionamide	10222-01-2	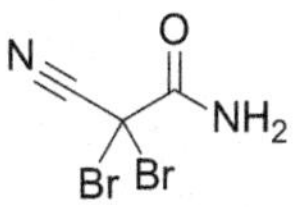 N, O, NH_2, Br Br	4.11%
Peracetic Acid	79-21-0	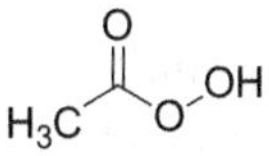 O, H_3C, O, OH	3.47%
Tetrakis hydroxymethyl phosphonium sulfate	55566-30-8	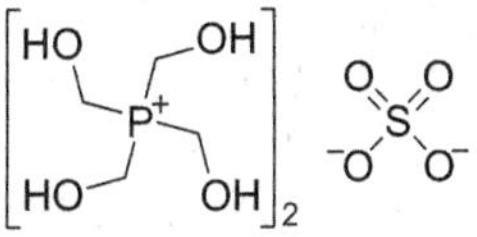 [HO, OH, P^+, HO, OH]$_2$ O, O, S, ^-O, O^-	2.48%
2-bromo-2-nitro-1,3-propanediol (Bronopol)	52-51-7	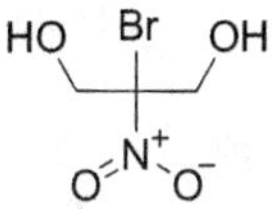 HO, Br, OH, N^+, O, O^-	1.54%
Methylisothiazolinone	2682-20-4	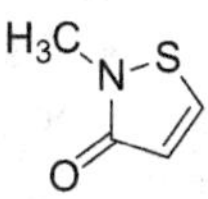 H_3C, N, S, O	1.37%
Chloromethylisothiazolinone	26172-55-4	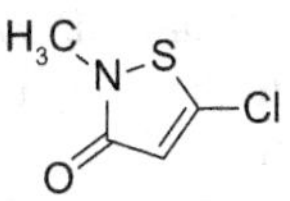 H_3C, N, S, Cl, O	1.37%
Dimethyloxazolidinone	51200-87-4	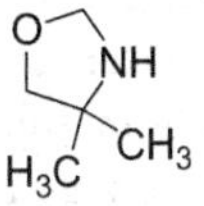 O, NH, CH_3, H_3C	1.34%

(Continued)

TABLE 6.2 (Continued)
Biocides Used in Hydraulic Fracturing Operations by Frequency of Use (FracFocus, 2013)

Name	CAS No.	Chemical Structure	Frequency of Use
Trimethyloxizolidinone	75673-43-7		1.32%
Dazomet	533-74-4		1.02%
Tris(hydroxymethyl) nitromethane	126-11-4		0.16%
1-(3-Chloroallyl)-3,5,7-triaza-1-azoniaadamantane chloride	4080-31-3		0.07%

while alkyl/carbon-containing substituents upset the lipid bilayer, leading eventually to lysis of the cell. The most common QACs found in frac applications are didecyl dimethylammonium chloride (DDAC, See Table 6.2) and alkyldimethylbenzylammonium chloride (ADBAC, Table 6.2). Synergistic antimicrobial effects have been observed when either of these two species is combined with an electrophilic biocide such as glutaraldehyde, increasing the efficiency of the chemical package and thus reducing the total amount of biocide necessary for effective bacterial control (Enzien et al., 2011).

Biocides categorized as electrophilic also bind cellular components as part of their mechanism of action but do so through the formation of covalent rather than ionic bonds, as is the case with lytic biocides. Electrophilic biocides contain an electrophilic functional group, typically carbonyl-based such as an aldehyde, capable of reacting with electron-rich nitrogen- and sulfur-containing nucleophiles found in key cellular components of bacteria (Maillard, 2002). This reaction and subsequent binding of the electrophilic biocide to the cellular components disrupts normal biological functions and leads to cell death.

The most common electrophilic biocide used in hydraulic fracturing operations is glutaraldehyde. It serves as an effective amino and nucleic acid cross-linker wherein sulfur- and nitrogen-containing moieties of the cellular components react with the aldehydes, causing damage and eventual cell death (Gorman et al.,

1980). Glutaraldehyde also inhibits DNA synthesis and can remove parts or all of the cell membrane. In application, Glutaraldehyde is most active at alkaline pHs but should be stored at acidic pH to prevent deactivation as a result of self-polymerization.

Not all electrophilic biocides are aldehydes, however. 2,2-dibromo-2-nitrilopropionamide (DBNPA) and 2-bromo-2-nitropropane-1,3-diol (bronopol) also react readily with electron-rich sulfur-containing functional groups such as those found in cysteine and glutathione, with similar results observed. Tetrakis(hydroxymethyl) phosphonium sulfate (THPS) has a unique mechanism of action and is another example of an electrophilic biocide lacking any aldehyde functional groups. THPS releases tris(hydroxymethyl)phosphine via deformylation which cleaves disulfide bonds in amino acids contained in the bacterial cell wall (Jones et al., 2010).

6.4 MICROBIAL CONTROL DURING FRACKING OPERATIONS: THE OUTCOMES

While the short-term microbial kill efficacy of the above biocides has been sufficiently demonstrated via traditional lab studies, their long-term performance under reservoir conditions is less known. The fast-acting and short-lived nature of oxidizing biocides makes them unsuitable for downhole use. Even lytic and electrophilic microbiocides with comparably longer half-lives such as glutaraldehyde, THPS, and some QACs yield reduced performance and stability under the temperatures and pressures typical of reservoir conditions. Fluid treatment with these biocides yields an initial reduction in microbial populations, but samples exposed to reservoir-simulated temperatures monitored over two months show near complete regrowth of acid-producing (APB) and sulfate-reducing (SRB) bacteria populations, even at increased treatment rates (Raymond et al., 2014).

These performance and stability concerns can be addressed by pairing the short-term antimicrobial efficacy of one of these biocides with the long-term bacterial growth control of a preservative. In hydraulic fracturing operations, a preservative is either a highly stable compound not consumed by its mechanism of action, so capable of longer-term efficacy, or a molecule that lacks inherent antimicrobial activity but releases formaldehyde over time through degradation. Some examples of preservatives used in the industry are benzylalkonium chloride, 1-(3-Chloroallyl)-3,5,7-triaza-1-azoniaadamantane chloride (CTAC), dimethyloxazolidine (DMO, and tris(hydroxymethyl)nitromethane (THNM).

When comparing the efficacy of glutaraldehyde to three of these preservatives (CTAC, DMO, and THNM), glutaraldehyde is proven most effective in the shortest time against SRBs. However, glutaraldehyde efficacy decreases as temperatures increase, requiring higher treatment rates to achieve equivalent bacterial reduction. Conversely, each of the three preservatives act more slowly but persist longer in treated fluids and their antimicrobial efficacy is enhanced at elevated temperatures (Yin et al., 2018). This data supports the idea that a combination of a faster acting biocide like an oxidizing biocide or a non-oxidizing biocide such as glutaraldehyde with a preservative would provide both an initial reduction in topside fluids and longer-term control downhole.

6.4.1 Laboratory Evaluation

One example of the benefits achieved with a treatment that combines short- and long-term efficacy is peracetic acid (PAA), an oxidizing biocide, dosed alone and in combination with a preservative in fluids collected in the Permian Basin from a hydraulic fracturing operation. Frac fluids were treated with a combination of peracetic acid, a quaternary ammonium compound, and a preservative chemistry (THNM). Samples were incubated and monitored over time for microbial reduction and potential regrowth. As shown in Figure 6.2, fluid treated with peracetic acid alone was compared to fluid treated with a combination of PAA and either ADBAC Quat or THNM, with samples monitored by ATP over 42 days. In all cases, PAA was shown to quickly reduce bacterial counts, represented as ATP MEq/mL (microbial equivalents/mL). However, regrowth of bacteria was seen over time, although the trend was less significant at higher dosages. When ADBAC was added to the biocide package, the initial reduction in microbial count by PAA was unchanged but the downward trend continued through the first three weeks of the study. After 42 days, a combination of PAA and ADBAC was shown to effectively control bacterial growth in both the short- and long-term. Interestingly, THNM did not show the same efficacy as ADBAC, with near equivalent microbial counts over the course of the full study (Tidwell, 2019). This discrepancy in preservative performance highlights that proper biocide selection is extremely important given the wide range of conditions present in any given fluid or reservoir.

Comparison of the metabolic profiles of microbial communities in hydraulic fracturing source water and produced water from the Marcellus Shale further underscore the importance of utilizing an effective preservative. While the source water bacteria are composed predominantly of aerobic *Alphaproteobacteria*, the produced water contained a majority of anaerobic *Gammaproteobacteria* capable of metabolizing sulfur compounds to sulfide. Initial biocide treatment would reduce the microbial count in the source water, but without use of a preservative the reservoir communities would increase risks associated with H_2S and FeS such as infrastructure corrosion, souring, and operator safety. This same study also found that the microbial communities in produced water showed an increased genetic ability to handle stress, making management of that water for reuse or disposal through disinfection more difficult when reservoir growth is uncontrolled (Mohan et al., 2014).

6.4.2 Field Evaluation

Lab studies like this can identify an optimal biocide program for use in the field. Nonetheless field data will ultimately be the decisive factor for the adoption of a proper biocide regime for hydraulic fracturing. In studies conducted in the field, following the outcome of hundreds of fracked wells using different biocide packages, numerous samples were collected from various locations and the microbial population in the collected samples were analyzed using techniques such as serial dilution, ATP, and molecular methods. When results indicated a combination of glutaraldehyde and THPS to be the optimal biocide package, field trials were conducted in four

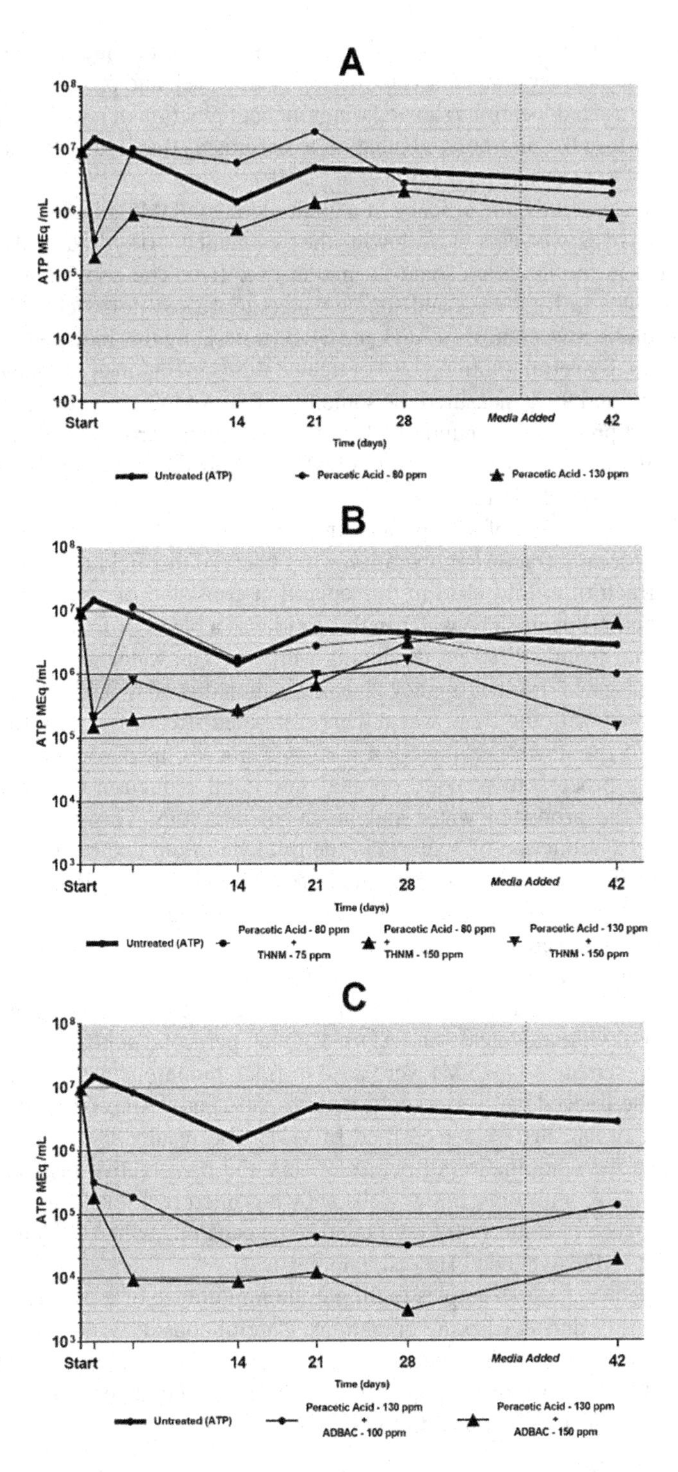

FIGURE 6.2 Peracetic acid and preservative biocide efficacy. Frac fluids were treated with peracetic acid alone or in combination with a preservative (THNM) or a quaternary ammonium compound. The number of microorganisms was monitored over 42 days.

separate locations using the lowest viable concentration of biocide as determined in the lab studies. Enumeration by serial dilution of flowback and produced water from all four of the treated locations showed a significant reduction in total bacterial count and even a shift in the microbial population to organisms that do not increase risk of MIC in two of the cases (Paschoalino et al., 2019).

In another case study, the efficacy of a glutaraldehyde/DMO biocide combination was evaluated over two years by, following the treatment of 70 hydraulically fractured wells located in the Niobrara shale formation over two. The comparison of initial microbial counts in frack tank and source waters to treated flowback and produced fluid showed effective control of APB and SRB bacteria by this biocide combination out to 90 days. By serial dilution, 93% of treated fluids maintained APB levels below 1000 cells/mL and 99% measured the same for SRB. Additionally, this combination of biocide and preservative required a lower total volume across all treated fluids to achieve a satisfactory reduction in microbial counts as compared to the incumbent programs (Corrin et al., 2015).

An effective outcome of the application of an oxidizing chemistry and a preservative to a hydraulic fracturing operation was observed in a field in the Fayetteville shale. The fracturing fluid used in this operation consisted of a mixture of water, chemicals, and proppant. The water in this fluid was a blend of fresh and produced/flowback water from a different producing well site. The water was brought onsite by truck, aggregated in a nearby recycle battery, then flowed to a second set of holding tanks to mix with the fresh water. This combined stream was sent to a blender tub which fed the downhole injection pumps. Peracetic acid was injected at two points in this process to provide optimal microbial reduction before downhole injection; in the produced water and in the blender tub. This biocide program yielded a >95% reduction of bacteria in the produced fluid and resulted in 85% of completed wells measuring under the goal of 100,000 MEq/mL. Bacteria counts in flowback/produced water measured over one year showed that initial treatments by peracetic acid allowed for continuous control of microbial growth in reservoir (Balasubramanian, 2016).

In a different field study in the Permian basin, different biocide regimes using a combination of Glutaraldehyde and ADBAC Quat, peracetic acid alone or peracetic acid and a preservative (THNM) were used to frack numerous wells. The flowback water from the fracked wells were collected 30-, 60-, and 90-days post-fracking and the levels of organisms were measured by ATP. The results shown in Figure 6.3 indicated that the combination of peracetic acid and preservative resulted in better microbial control, with only a few wells showing microbial levels above the KPIs established by the operator ($<10^5$ total microorganisms/mL) when compared to PAA alone or Glut+ADBAC Quat (Tidwell, unpublished).

Taken together, these examples highlight the importance of a proper strategy for microbial control during fracking operations. Careful consideration about the water sources used each step in the process, the efficiency of the microbial treatments, and the potential for survival of the population downhole will dictate the outcome of the process and the health of an unconventional well.

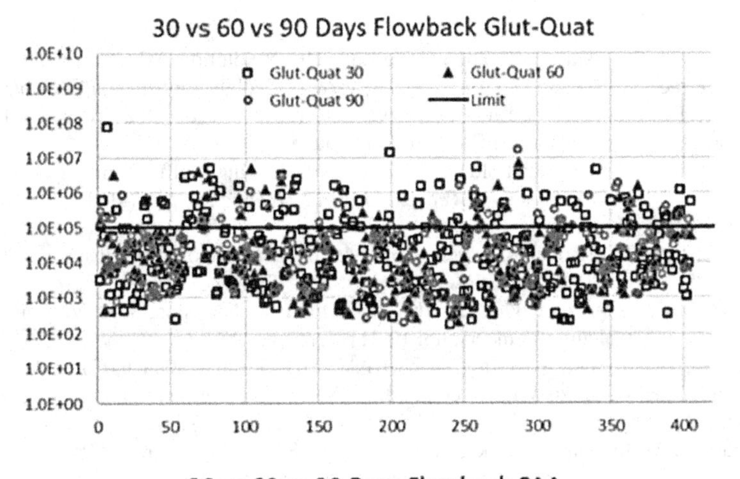

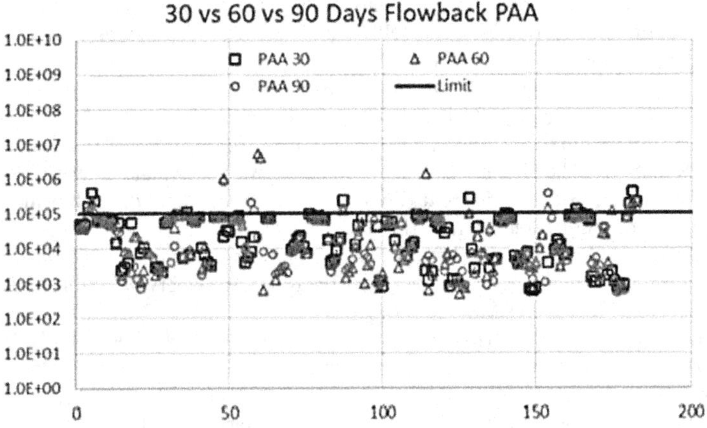

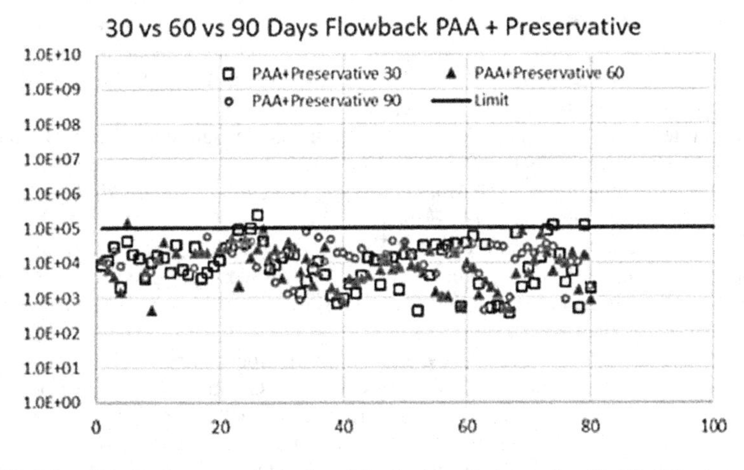

FIGURE 6.3 Flowback water analysis of fracked wells treated with Glut/Quat, peracetic acid, and preservative.

REFERENCES

Balasubramanian, Ramakrishnan. 2016. "The RenewIQ™ Solution: Microbial and iron control in frac water." *Case Study. Ecolab.* https://www.ecolab.com/-/media/Ecolab/Ecolab-Home/Documents/DocumentLibrary/CaseStudies/Nalco-Champion/TheRenewIQSolutionMicrobialandIronControlinFracWater.pdf.

Basafa, Mahsan, and Kelly Hawboldt. 2019. "Reservoir souring: Sulfur chemistry in offshore oil and gas reservoir fluids." *J. Pet. Explor. Prod. Technol.* 9: 1105–1118. doi:10.1007/s13202-018-0528-2.

Blauch, Matthew Eric, Roger R. Myers, Tom Moore, Brian Andrew Lipinski, and Nathan Allan Houston. 2009. *Marcellus Shale Post-Frac Flowback Waters - Where is All the Salt Coming from and What are the Implications?* Charleston, West Virginia, USA: Society of Petroleum Engineers. doi:10.2118/125740-MS.

Booker, Anne E., Mikayla A. Borton, Rebecca A. Daly, Susan A. Welch, Carrie D. Nicora, David W. Hoyt, Travis Wilson, et al. 2017. "Sulfide generation by dominant halanaerobium microorganisms in hydraulically fractured shales." *mSphere* 2 (4).

Cluff, Maryam A., Angela Hartsock, Jean D. MacRae, Kimberly Carter, and Paula J. Mouser. 2014. "Temporal changes in microbial ecology and geochemistry in produced water from hydraulically fractured marcellus shale gas wells." *Environ. Sci. Technol.* 48: 6508–6517. doi:10.1021/es501173p.

Corrin, E., C. Rodriguez, and T. M. Williams. 2015. "A case study evaluating a co-injection biocide treatment of hydraulic fracturing fluids utilized in oil and gas production." *NACE - International Corrosion Conference Series 2015*, Dallas, TX, USA.

De Paula, Renato M., Vic Keasler, Junzhong Li, David McSherry, and Richard Staub. 2013. "Development of Peracetic Acid (PAA) as an environmentally safe biocide for water treatment during hydraulic fracturing applications." In *Development of Peracetic Acid (PAA) as an Environmentally Safe Biocide for Water Treatment During Hydraulic Fracturing Applications*. The Woodlands, Texas, USA: Society of Petroleum Engineers. 10. doi:10.2118/164088-MS.

Duong, H. A., M. Berg, M. H. Hoang, H. V. Pham, H. Gallard, W. Giger, and U. V. Gunten. 2003. "Trihalomethane formation by chlorination of ammonium- and bromide-containing groundwater in water supplies of Hanoi, Vietnam." *Water Res.* 37: 3242.

Enzien, Michael V., Bei Yin, Donald Love, Michael Harless, and Edward Corrin. 2011. "Improved microbial control programs for hydraulic fracturing fluids used during unconventional shale-gas exploration and production." In *Improved Microbial Control Programs for Hydraulic Fracturing Fluids used during Unconventional Shale-Gas Exploration and Production*. The Woodlands, Texas, USA: Society of Petroleum Engineers. 10. doi:10.2118/141409-MS.

Fichter, Jennifer, Kenneth Wunch, Robert Moore, Elizabeth Summer, Shelby Braman, and Phineas Holmes. 2012. *How Hot is Too Hot For Bacteria? A Technical Study Assessing Bacterial Establishment in Downhole Drilling, Fracturing and Stimulation Operations*. Salt Lake City, Utah, USA: NACE International.

Finnegan, Michelle, Ezra Linley, Stephen P. Denyer, Gerald McDonnell, Claire Simons, and Jean-Yves Maillard. 2010. "Mode of action of hydrogen peroxide and other oxidizing agents: Differences between liquid and gas forms." *J. Antimicrob. Chemother.* 65: 2108–2115. doi:10.1093/jac/dkq308.

FracFocus Chemical Disclosure Registry. 2013. http://fracfocus.org/.

Ghebremichael, K., E. Muchelemba, B. Petrusevski, and G. Amy. 2011. "Electrochemically activated water as an alternative to chlorine for decentralized disinfection." *J. Water Supply: Res. Technol - Aqua* 60: 210–218. doi:10.2166/aqua.2011.034.

Gilbert, P., and L. E. Moore. 2005. "Cationic antiseptics: Diversity of action under a common epithet." *J. Appl. Microbiol.* 99: 703.

Gorman, S. P., E. M. Scott, and A. D. Russell 1980. "Antimicrobial activity, uses and mechanism of action of glutaraldehyde." *J. Appl. Bacteriol.* 48(2):161–190. doi:10.1111/j.1365-2672.1980.tb01217.x. PMID: 6780502.

Hawkins, C. L., D. I. Pattison, and M. J. Davies 2003. Hypochlorite-induced oxidation of amino acids, peptides and proteins. *Amino Acids* 25(3–4):259–274. doi:10.1007/s00726-003-0016-x.

Jones, C. R., B. L. Downward, Kansas Hernandez, Tim Curtis, and Francis Smith. 2010. *Extending Performance Boundaries with Third Generation THPS Formulations.* San Antonio, Texas, USA: NACE International.

Jones, Chris, Stephanie Edmunds, Francis Smith, Gareth Collins, and Jean Molina. 2016. *Enhanced THPS for Challenging Environments.* Vancouver, British Columbia, Canada: NACE International.

Kahrilas, Genevieve A., Jens Blotevogel, Philip S. Stewart, and Thomas Borch. 2015. "Biocides in hydraulic fracturing fluids: A critical review of their usage, mobility, degradation, and toxicity." *Environ. Sci. Technol.* 49: 16–32. doi:10.1021/es503724k.

Kallmeyer, J., and A. Boetius. 2004. "Effects of temperature and pressure on sulfate reduction and anaerobic oxidation of methane in hydrothermal sediments of Guaymas Basin." *Appl. Environ. Microbiol.* 70: 1231.

King, George Everette. 2012. *Hydraulic fracturing 101: What Every Representative, Environmentalist, Regulator, Reporter, Investor, University Researcher, Neighbor and Engineer Should Know About Estimating Frac Risk and Improving Frac Performance in Unconventional Gas and Oil Wells.* The Woodlands, Texas, USA: Society of Petroleum Engineers. doi:10.2118/152596-MS.

LaMontagne, Michael G., Ira Leifer, Sandra Bergmann, Laurie C. Van De Werfhorst, and Patricia A. Holden. 2004. "Bacterial diversity in marine hydrocarbon seep sediments." *Environ. Microbiol.* 6 (8): 799–808.

Linley, Ezra, Stephen P. Denyer, Gerald McDonnell, Claire Simons, and Jean-Yves Maillard. 2012. "Use of hydrogen peroxide as a biocide: New consideration of its mechanisms of biocidal action." *J. Antimicrob. Chemother.* 67 (7): 1589–1596.

Lipus, Daniel, Amit Vikram, Daniel Ross, Daniel Bain, Djuna Gulliver, Richard Hammack, and Kyle Bibby. 2017. "Predominance and metabolic potential of halanaerobium spp. in produced water from hydraulically fractured marcellus shale wells." *Appl. Environ. Microbiol.* 83 (8).

Liu, Yi-Fan, Daniela Domingos Galzerani, Serge Maurice Mbadinga, Livia S. Zaramela, Ji-Dong Gu, Bo-Zhong Mu, and Karsten Zengler. 2018. "Metabolic capability and in situ activity of microorganisms in an oil reservoir." *Microbiome* 6 (1): 5.

Maillard, J. 2002. "Bacterial target sites for biocide action." *J. Appl. Microbiol.* 92: 16S.

McDonnell, G., and A. Denver Russell 1999. "Antiseptics and disinfectants: Activity, action, and resistance." *Clin. Microbiol. Rev.* 12 (1): 147–179. doi:10.1128/CMR.12.1.147.

Mohan, Arvind Murali, Angela Hartsock, Richard W. Hammack, Radisav D. Vidic, and Kelvin B. Gregory. 2013 "Microbial communities in flowback water impoundments from hydraulic fracturing for recovery of shale gas." *FEMS Microbiol. Ecol.* 86(3): 567–580. doi:10.1111/1574-6941.12183.

Mohan, Arvind Murali, Kyle J. Bibby, Daniel Lipus, Richard W. Hammack, and Kelvin B. Gregory. 2014. "The functional potential of microbial communities in hydraulic fracturing source water and produced water from natural gas extraction characterized by metagenomic sequencing." *PLoS One* 9: e107682.

Mouser, Paula J., Mikayla Borton, Thomas H. Darrah, Angela Hartsock, and Kelly C. Wrighton. 2016. "Hydraulic fracturing offers view of microbial life in the deep terrestrial subsurface." *FEMS Microbiol.* 92 (11).

Noszticzius, Zoltán, Maria Wittmann, Kristóf Kály-Kullai, Zoltán Beregvári, István Kiss, László Rosivall, and János Szegedi. 2013. "Chlorine dioxide is a size-selective antimicrobial agent." *PLoS One* 8: e79157.

Paschoalino, Matheus, Jon Raymond, Emerentiana Sianawati, and Veronica Silva. 2019. *Evaluation of a glutaraldehyde/THNM Combination for Microbial Control in Four Conventional Oilfields*. Galveston, Texas, USA: Society of Petroleum Engineers. doi:10.2118/193594-MS.

Pierce, Conor, and Cruz St. Peter. 2019. *Performic Acid for MIC Prevention in Treated Water Systems: Balancing Biocidal Efficacy Against Oxidizer Induced Corrosion*. Nashville, Tennessee, USA: NACE International.

Raymond, Jon, Earl Parnell, and Jennifer Fichter. 2014. *Determining Effective Antimicrobial Treatments for Long-Term Protection of Hydrocarbon Reservoirs*. San Antonio, Texas, USA: NACE International.

Robinson, G., R. Thorn, and D. Reynolds. 2012. "The effect of long-term storage on the physiochemical and bactericidal properties of electrochemically activated solutions." *Int. J. Mol. Sci.* 14: 457.

Stanton, J.S., D.W. Anning, C.J. Brown, R.B. Moore, V.L. McGuire, S.L. Qi, A.C. Harris, K.F. Dennehy, P.B. McMahon, J.R. Degnan, and J.K. Böhlke. 2017. "Brackish groundwater in the United States: U.S. Geological Survey." Professional Paper 1833, 185 p., doi:10.3133/pp1833.

Tidwell, Timothy. 2019. *Implementation of Microbial Control Strategies that Work: Lab to Frac, Lab to Production*. Halifax, Nova Scotia, Canada: ISMOS-7.

Tidwell, Timothy J., Renato De Paula, Zach Broussard, and Victor V. Keasler. 2017. *Mitigation of Severe Pitting Corrosion Caused by MIC in a CDC Biofilm Reactor*. New Orleans, Louisiana, USA: NACE International.

Tischler, Alfred, Tyrel Roy Woodworth, Sheril Dale Burton, and Robert Dean Richards. 2009. "Controlling bacteria in recycled production water for completion and workover operations." In *Controlling Bacteria in Recycled Production Water for Completion and Workover Operations*. Denver, Colorado: Society of Petroleum Engineers. 13. doi:10.2118/123450-MS.

Williams, Terry M. 2008. "Optimizing and improving biocide performance in water treatment systems." In *Optimizing and Improving Biocide Performance in Water Treatment Systems*. New Orleans, Louisiana, USA: NACE International.

Xu, D., Y. Li, and T. Gu. 2012. "A synergistic D-tyrosine and tetrakis hydroxymethyl phosphonium sulfate biocide combination for the mitigation of an SRB biofilm." *World J. Microbiol. Biotechnol.* 28: 3067.

Yin, Bei, Terry Williams, Thomas Koehler, Brandon Morris, and Kathleen Manna. 2018. "Targeted microbial control for hydrocarbon reservoir: Identify new biocide offerings for souring control using thermophile testing capabilities." *Int. Biodeterior. Biodegrad.* 126: 204–207.

7 The Application of Bioinformatics as a Diagnostic Tool for Microbiologically Influenced Corrosion

Nora Eibergen, Geert M. van der Kraan, and Kenneth Wunch

CONTENTS

7.1 Introduction 120
7.1.1 Microbiologically Influenced Corrosion 120
7.1.2 MIC Monitoring Techniques 120
7.2 Case Study 122
7.2.1 Materials and Methods 123
7.2.1.1 Site Sampling & DNA Extraction 123
7.2.1.2 Total Cell Determination and Subgroups via qPCR 123
7.2.1.3 Microbial Diversity Studies via NGS and Basic Metabolic Mapping 124
7.2.1.4 Bug Bottles 124
7.2.2 Applied Sampling Plan 125
7.2.3 Case Study Results 125
7.2.3.1 Visual Observation of Water Samples 126
7.2.3.2 Bug Bottle Analysis 126
7.2.3.3 qPCR Analysis 126
7.2.3.4 Next Generation Sequencing Diversity Study 129
7.2.4 Case Study Discussion and Interpretation 129
7.2.4.1 Interpretation of qPCR Results 129
7.2.4.2 Interpretation of NGS Results 131
7.2.4.3 Comparison of Coupon Corrosion Data with Microbial Data 132
7.3 Conclusions 133
References 135

7.1 INTRODUCTION

7.1.1 Microbiologically Influenced Corrosion

The vast infrastructure that is required for the production and transport of oil and gas is largely constructed of carbon steel. While this material is less expensive and easier to fabricate than stainless steel, it is highly susceptible to corrosion. Indeed, it has been estimated that the annual global cost of corrosion was $2.5 trillion in 2013 (Koch et al. 2016) and that 25% of the failures that occur in the oil and gas industry can be attributed to some type of corrosion (Kermani and Harrop 1998). In particular, microbiologically influenced corrosion (MIC) has been estimated to account for 10–20% of this costly damage to steel oilfield assets (Skovhus, Eckert and Rodrigues 2017).

MIC is characterized by the acceleration of corrosion rates in the presence of microorganisms and has been traditionally associated with microbial metabolic processes that result in the production of acid and the reduction of sulfate to hydrogen sulfide. For this reason, topside corrosion monitoring programs typically include efforts to detect both sulfate-reducing bacteria (SRB) and acid-producing bacteria (APB) (Sharma and Voordouw 2017). These groups of microorganisms can grow in sessile communities on metallic surfaces and can directly or indirectly contribute to the corrosion of the metal surface on which they grow (Enning et al. 2012, Enning and Garrelfs 2014, Gu 2012, Iverson 1987, Pak et al. 2003). Through the production of biogenic hydrogen sulfide and acids near the metallic surface, SRB and APB have been implicated as the primary contributors to biocorrosion. However, the advent of bioinformatics now allows for the application of more comprehensive microbiological field monitoring programs to elucidate the complex MIC mechanisms that threaten asset integrity.

7.1.2 MIC Monitoring Techniques

MIC monitoring has been historically achieved using a combination of two tools: corrosion coupons and bug bottles. The installation of corrosion coupons allows one to detect corrosion events and the rate at which these events cause damage. Corrosion coupons detect corrosion damage that results from both abiotic and microbiological induced mechanisms. Bug bottles are used to analyze water samples for the presence of problematic classes of microorganisms such as SRB and APB that are known to cause corrosion (Little, Lee, and Ray 2006). Bug bottles are an effective, albeit slow, tool for detecting culturable microorganisms of the microbial classes for which bug bottles are commercially available. However, it has been estimated that over 85% of viable microorganisms cannot be cultured (Amann, Ludwig, and Schleifer 1995), and bug bottles will not detect potentially problematic microorganisms within a field water sample that cannot be cultured under bug bottle growth conditions. Also, because of their design, bug bottles are typically only used to analyze liquid samples. It is difficult to use bug bottles to anoxically analyze solid field samples such as scrapings from corrosion coupons or schmoo collected from pigging runs. These samples are potentially more informative to what conditions and microorganisms are present at the surface where corrosive events are occurring (Wrangham and Summer 2013).

Alternative tools are available that not only detect unculturable microorganisms but can do so in minutes to hours instead of the days that are required to receive a result from bug bottle testing. One of these tools, ATP measurement, is facilitated using any one of a variety of commercial kits. These kits allow one to estimate the number of metabolically active cells in a solid or liquid sample based on the amount of intact ATP, a molecule produced by metabolically active cells, that is present in that sample (Little, Lee, and Ray 2006). One of the biggest advantages to using this method is the speed at which results can be obtained. Because the method uses a rapid enzymatic reaction to convert ATP molecules to a product that can be detected using a luminometer, results can be obtained within a half hour after sample collection if portable luminometers are used. However, while this method can quickly detect metabolically active microorganisms, including those that are unculturable, it does not provide the user with any information about the microorganisms it detects. From a diagnostics perspective, it is important to understand not only how many cells are present within a sample, but also what type of microorganisms are present or what metabolic processes may be active in the microbial communities within the sample.

Molecular microbiological methods allow one to realize the benefits of both testing speed *and* information about the microbial profile of a sample. The utilization of these methods requires the extraction and isolation of DNA from a microbial sample, which can be done at the site of collection if a microcentrifuge is available. Otherwise, liquid and solid samples can be preserved at low temperature or using any one of a variety of commercially available chemical preservatives. These methods have been shown to keep genetic material intact for several days to accommodate the shipment or transport of samples to a laboratory (Rachel and Gieg 2020). Thus, the analysis of a sample that has been adequately preserved can provide accurate information about the microbial profile of that sample at its time of collection.

Extracted DNA is typically analyzed via one of two methods: Quantitative Polymerase Chain Reaction (qPCR) or Next Generation Sequencing (NGS). These two methods provide different types of information about the microbial profile of a given sample in hours to days. The use of the qPCR technique provides the user with a quantitative measurement of functional genes that are known to equip microorganisms with metabolic processes that can contribute to corrosion. This is achieved through amplification of specific functional genes from sample DNA and an estimation of the number of copies of a specific functional gene in the sample from the rate at which amplification occurs. Functional gene copy numbers can then be used to enumerate the microbial cells within a sample that contain that functional gene. Primer DNA sequences have been published that target and amplify genes specific to sulfate-reducing microorganisms (*dsrA*), nitrate reducers (*narG*), methanogens (*mcrA*), bacteria, and archaea (Sørenson, Skovhus and Larsen 2011; Zhu et al. 2005). By analyzing a sample with the appropriate primers and qPCR program, one can estimate the risk of asset failure from the number of cells that have the potential to cause deleterious effects.

NGS provides a list of microbial genera and their relative abundances within an analyzed sample. This is achieved by sequencing the ribosomal 16S region that has

TABLE 7.1
Summary of MIC Monitoring Methods

	Speed	Cost	Ease of Analysis at Sampling Site	Provides Quantitative Data	Detects Unculturable Organisms	Provides Info about Microbial Profile	Provides Live/Dead Cell Data
Bug bottles	–	++	++	++	–	+	++
ATP	++	+	+	++	++	–	+
qPCR	+	+	–	++	++	+	–
NGS	+	–	–	+	++	++	–

Monitoring methods and their strongly advantageous (++), slightly advantageous (+), and disadvantageous (–) qualities.

been amplified from sample DNA. Obtained sequences are then matched to those of known microorganisms in established databases. Sample sequences that match database entries can be classified as belonging to a particular microbial family or genus. Unlike qPCR analysis where quantitative results are obtained for cells containing certain functional genes of interest, NGS analysis will only provide relative abundances of genera within a sample. In addition, this tool is only as powerful as the database that is used to analyze the data, as sample sequences that cannot be matched with sufficient confidence to sequences in the database will remain unclassified. Despite these limitations, however, the NGS technology provides one advantage that many others lack. While tools such as bug bottles and qPCR provide quantitative information about particular types of cells, they provide the user with little information about the rest of the microbial community. In contrast, NGS allows an analyst to obtain detailed information about a sample's entire microbial profile without looking for any particular microbial or metabolic class.

In summary, the various methods that can be used to monitor for MIC each have their own set of advantages and disadvantages, which are summarized in Table 7.1. Therefore, the case can be made that the most holistic representation of a system's risk-causing microbial characteristics can be achieved when several complementary methods of monitoring are used in tandem. In this chapter, a case study will be presented that illustrates the use of the bug bottle, qPCR, and NGS techniques to monitor the corrosion risk of a topside oilfield system. This case study highlights how three monitoring techniques can be used not only to identify potential problem areas within a system but also to diagnose the underlying cause of the corrosion so that it can be addressed.

7.2 CASE STUDY

One of the largest in North America, the oilfield described in this case study produced 1.5 million barrels of oil per day at its peak production. Even though inspection and maintenance programs for infrastructure were in place at this site, severe

MIC failure events were reported for its facilities. The financial and environmental impact of these failures, which mostly occured in low-flow areas of the water injection system, varied in severity.

This study focused predominantly on the water treatment facilities, which consisted of two main systems: the Seawater Treatment Plant (STP) and the Seawater Injection Plant (SIP). The seawater was pumped in at the STP where it was passed through one of several deaerators to strip it of oxygen. In addition, sulfite was added to strip the last bit of oxygen from the seawater. From there, the water was pumped through an 18 km (11.2 mile) pipeline to the SIP, where the pressure was increased to above 170 bars. The water was subsequently distributed to several locations via manifolds, where it was then pumped downhole into the oil reservoir via several injection wells. Water and coupon samples were collected from the injection water and production systems, and data for these samples are included in this chapter; however, discussion in this chapter will be focused on samples collected from the injection water system consisting of STP and SIP.

7.2.1 Materials and Methods

7.2.1.1 Site Sampling & DNA Extraction

Equipment was transported to the sampling site and temporarily setup in a field laboratory. Over the course of two days, different topside facility locations were visited where water and coupon swabs were collected. Water samples were collected in 1L bottles that were filled to the brim in order to minimize oxygen intrusion. Coupons were sampled by collecting coupon swabs which were obtained by swabbing a 1 cm^2 area of coupon. The samples were subsequently transported to the field laboratory where the collected waters and coupon swabs were subjected to DNA extraction (MoBio Power Water), and bug bottles were inoculated. The isolated DNA and bug bottles were then shipped offsite for incubation and further analysis.

7.2.1.2 Total Cell Determination and Subgroups via qPCR

To determine the total bacterial number via quantitative PCR (qPCR), the 16S rRNA gene was targeted. The 16S rRNA gene is present in all bacteria, and each bacterial cell contains an average of two to four copies (3.6) on its genome (Klappenbach et al. 2001). Two primer sets for the determination of the total bacterial number in the samples retrieved have been applied (typically listed as cells/mL). Primer set 1 included primers Bac8F and Bac338 (Gittel et al. 2009), and primer set 2 includes primers that anneal to gene locations 325–343 and 791–766 (Nadkarni et al. 2002). The thermocycler programs used for this study have been optimized from those reported in the original publications.

Archaea are a different kingdom of unicellular microorganisms frequently found in oil fields and oil-associated ecosystems. To determine the total number of archaea, the 16S rRNA gene was targeted. The 16S rRNA gene is present in all archaea, and each archaeal cell contains an average of one to two copies (1.77) on its genome (Gittel et al. 2009). One primer set, Arch806F and Arch958R (Gittel et al. 2009), was applied to quantify archaea in the collected samples. The thermocycler programs used for this study have been optimized from those reported in the original publications.

Sulfate-reducing prokaryotes (SRP) are organisms that produce H_2S as part of their metabolism. SRP include both sulfate-reducing bacteria (SRB) and sulfate-reducing archaea (SRA). To determine the total number of SRP in the samples, the dissimilatory sulfate reduction (*dsr*) gene was targeted by primer set DSRp2060F and DSR4R (Dar et al. 2007, Foti et al. 2007). This gene encodes a protein involved in the sulfate reduction pathway of all SRP. The *dsr* gene appears only once in the genome of each SRP and can therefore be directly linked to cell amounts. The thermocycler programs used for this study have been optimized from those reported in the original publications.

7.2.1.3 Microbial Diversity Studies via NGS and Basic Metabolic Mapping

DNA isolated from collected samples were subjected to next generation sequencing (NGS) of 16S rDNA gene amplicons utilizing the Illumina platform. The diversity of both bacteria and archaea was explored by amplifying the V3-V5 region of the 16S rDNA using the primers in Table 7.2.

The returned sequences were aligned using the MOTHUR program (Version 1.38.0), chimeras were removed, and the optimized sequences were compared to the SILVA V-128 database. Returned sequence alignments were retrieved and their basic metabolism and isolation sources were gleaned based on existing information in the literature.

7.2.1.4 Bug Bottles

Water samples and selected coupon swab samples were used to inoculate bug bottles filled with either Modified Postgate's B Media (to detect the presence of SRB) or Phenol Red Dextrose Media (to detect the presence of APB). These bug bottles each contained 9 mL media and were purchased from Biotechnology Solutions USA. Bug bottles were injected with 1 mL of each water sample. Coupon swab inoculums were prepared by immersing each swab in a small amount (~1–2 mL) of its corresponding water sample and shaking gently. Bug bottles were then inoculated with 1 mL of the resulting water. After six days of shipping and handling, bug bottles were received in the laboratory and incubated at 30°C for 28 days. At the 28-day time point, growth was assessed and recorded.

TABLE 7.2
Primers Used for NGS Analysis

Bacteria/Archaea	Forward/Reverse	Primer Sequence (5′-3′)
Bacteria	Forward	GCCTACGGGAGGCAGCAG
Bacteria	Reverse	CTACCAGGGTATCTAATCC
Archaea	Forward	CCCTAYGGGGYGCASCAG
Archaea	Reverse	GTGCTCCCCCGCCAATTCCT

MiSeq v3 reagent kits were used and the read length was 2×300 base pairs. This delivered 5000 to 25000 sequences per sample.

7.2.2 Applied Sampling Plan

Coupon and water sampling occurred at designated coupon sampling locations across the system (see Table 7.3 for sampling locations and Figure 7.1 for a map of these locations). Coupons of a representative metal were placed on a holder which allows them to be exposed to the liquids and flow conditions of the pipe. Upon collection, coupons were cleaned and weighed such that corrosion rates could be calculated. Water samples were either used to inoculate bug bottles designed to detect SRB and APB or they were filtered such that DNA could be isolated onsite and shipped away for analysis by qPCR and NGS.

7.2.3 Case Study Results

The aim of this study was to estimate the MIC risk for various assets within the topside facility. These risk estimates were then used to optimize the microbial control program for the oilfield. The results of the analyses are described in the sections below.

TABLE 7.3
Locations Where Water and Coupon Samples Were Collected

Sample Number	Description of Location
1	Seawater reservoir upstream of chemical treatment
2	STP; after deaerator
3	STP, after deaerator
4	STP, after deaerator
5	STP, after malfunctioning deaerator
6	Sales line - last sampling location before SIP
7	SIP inlet
8	SIP, injection pump outlet
9	SIP outlet
10	SIP, outlet of tank with oxygen ingress
11	SIP outlet to injection water distribution station
12	Production water line
13	Production water line

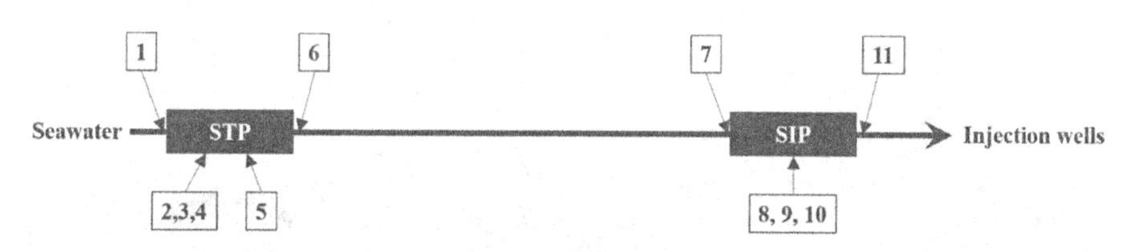

FIGURE 7.1 Map of sampling locations for water & coupons. Sample numbers are in white boxes. Production water samples (labeled 12 and 13) were also collected.

7.2.3.1 Visual Observation of Water Samples

The water collected from the different sampling points varied significantly in appearance. A photo of most of the collected water samples is shown in Figure 7.2. A short description of the physical appearances of the water samples along with other observations that were made is provided in Table 7.4.

7.2.3.2 Bug Bottle Analysis

Bug bottles were inoculated with both water samples and microorganisms eluted from coupon swabs samples. For most samples, the resulting cultures were then serially diluted such that samples could be enumerated by MPN analysis. This analysis revealed that samples taken from sampling points at SIP were more contaminated with SRB and APB than those taken from STP (Figure 7.3). While water samples taken from STP had APB counts ranging from 10^1 – 10^4 cells/mL and SRB counts ranging from 10 – 10^2 cells/mL, those taken from SIP had APB counts ranging from 10^4–10^6 cells/mL and 10^4 SRB/mL. Swab samples also fit this trend, as the sample taken from STP contained 10^3 APB/cm^2 and 10 SRB/cm^2 while samples taken at STP had APB counts of 10^6 cells/cm^2 and SRB counts ranging from 10^2–10^3 cells/cm^2.

7.2.3.3 qPCR Analysis

While bug bottle analysis is a useful tool for enumerating the number of living organisms with sulfate-reducing and acid-producing metabolic processes, it does suffer from the limitation that only organisms that can be cultured in the growth medium chosen for the analysis are counted. To ensure that all organisms present in the system were accounted for, qPCR analysis was performed on DNA isolated from both water samples and coupon swabs. In this analysis, total bacterial counts, total archaeal counts, and counts of corrosion influencing SRP were determined in the system. qPCR analysis of the collected water samples (Figure 7.4) revealed that while the microbial load of seawater flowing into STP was low (10^1–10^3 cells/mL), the load increased substantially within STP. All of the water samples collected within STP were contaminated with similar amounts of bacteria (~10^6 cells/mL), archaea (~5×10^3 cells/mL), and SRP (~5×10^5 cells/mL). This microbial load decreased during the transport of water from STP to SIP, as planktonic cell counts at the SIP inlet dropped to less than 10^3 cells/mL. However, populations of bacteria, archaea, and SRP rebounded within SIP to densities that were at least as high as those observed at STP. Samples 10 and 11, both of which were collected downstream of a tank with known oxygen ingress, were exceptions to this trend.

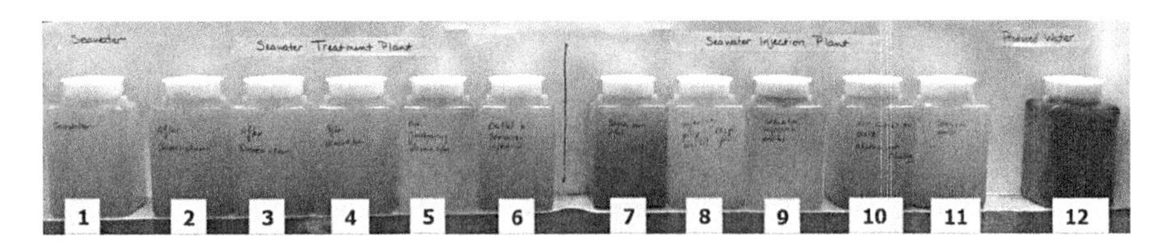

FIGURE 7.2 Overview of the collected water samples.

TABLE 7.4
Water Sample Details

Sample #	Description	Temperature °F (°C)	Appearance Description	Pressure (psi)	Volume of Water Filtered (mL)
1	Seawater reservoir upstream of chemical treatment	42.6 (5.5)	Brown with little particulates	–	230
2	STP; after deaerator	66.6 (19.5)	Gray with little particulates	–	350
3	STP, after deaerator	68.2 (20)	Gray with little particulates	–	200
4	STP, after deaerator	66.9 (19.8)	Gray with little particulates	–	250
5	STP, after malfunctioning deaerator	70 (21)	Light gray with little particulates	–	370
6	Sales line - last sampling location before SIP	55 (12.5)	Brown/gray with little particulates	–	250
7	SIP inlet	55 (12.5)	Dark gray with moderate particulates	15	220
8	SIP, injection pump outlet	55 (12.5)	Light gray with little particulates	2245	190
9	SIP outlet	–	Small blue particulates	–	100
10	SIP, outlet of tank with oxygen ingress	55 (12.5)	Brown with moderate particulates	20	300
11	SIP outlet to injection water distribution station	55 (12.5)	Dark gray with moderate particulates	–	260
12	Production water line	102.2 (39)	Black with moderate particulates	–	200
13	Production water line	102.2 (39)	Black with moderate particulates	–	180

A similar analysis of coupon swab samples by qPCR revealed consistently high loads of sessile microbes (Figure 7.5). Coupon samples collected from STP carried >10^5 cells/cm^2 of both bacteria and archaea. SRP were present on these coupon samples at densities >10^4 cells/cm^2. While a slight decrease in the population of sessile archaea and SRP was observed at the SIP inlet relative to STP sampling points, bacterial counts remained high (6×10^6 cells/cm^2). The counts for all three organism

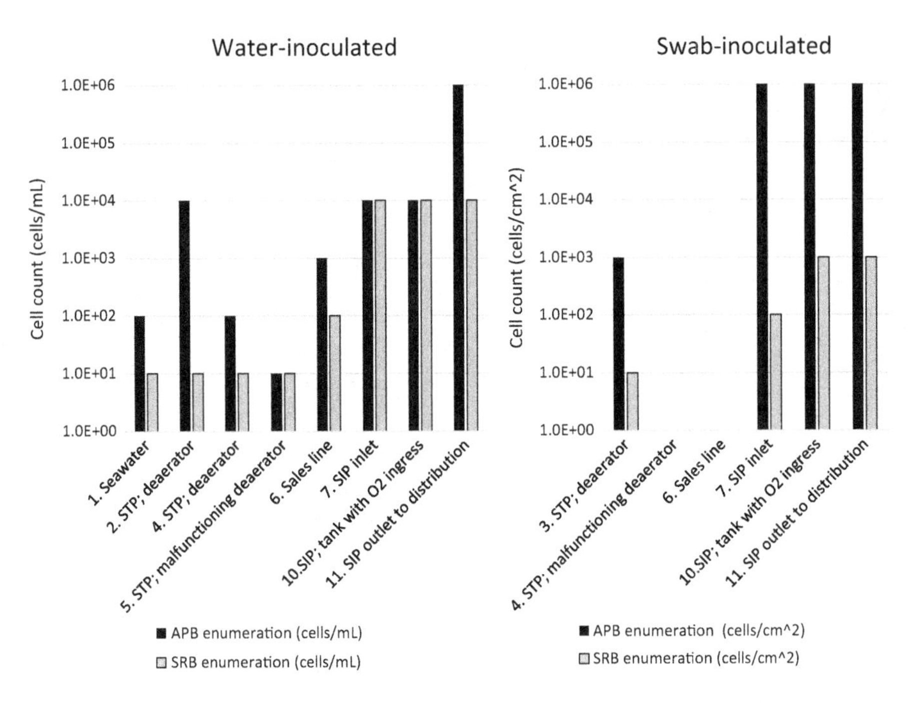

FIGURE 7.3 Results from bug bottles inoculated with water and coupon swabs.

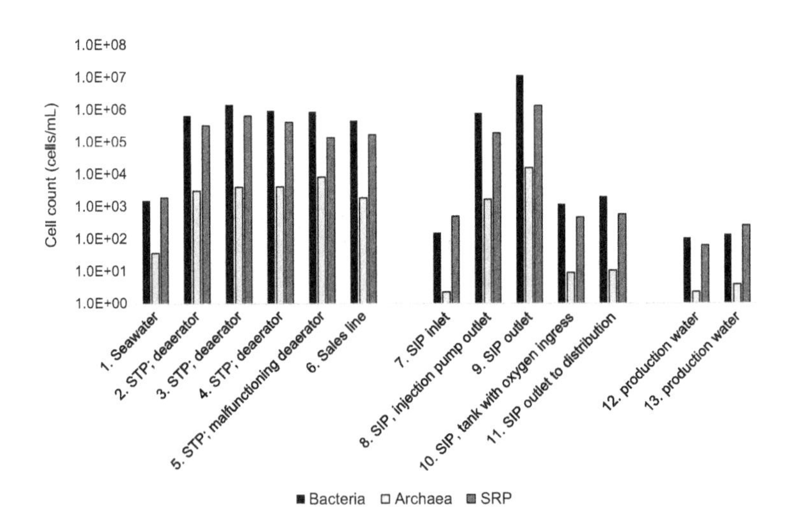

FIGURE 7.4 qPCR results for filtered water samples.

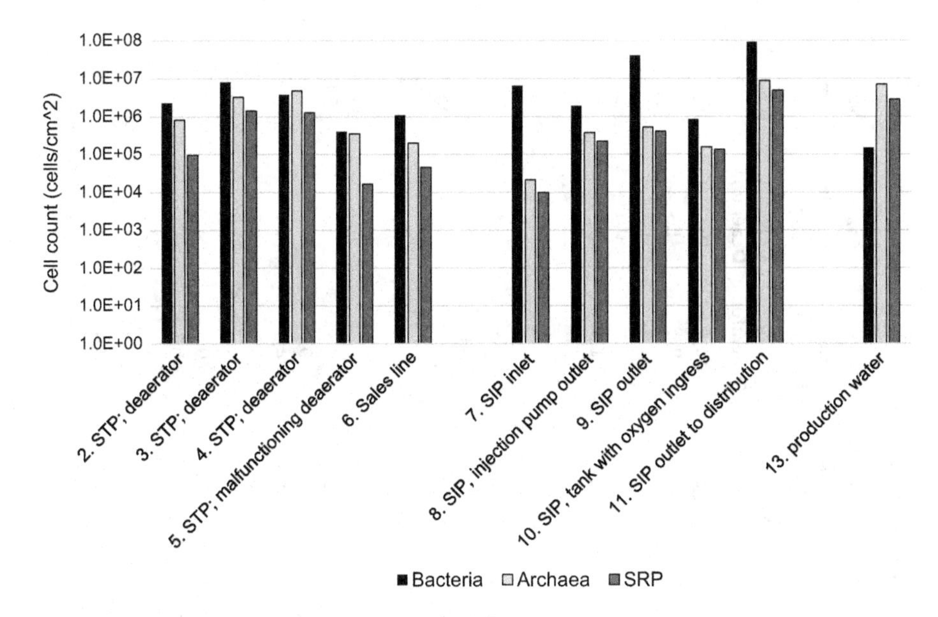

FIGURE 7.5 qPCR results for coupon swabs.

classes remained high (>10^5 cells/cm^2) for all remaining samples including those collected from SIP locations known to have oxygen ingress and those taken from production water lines.

7.2.3.4 Next Generation Sequencing Diversity Study

The objective of the qPCR program in combination with the bug bottle program was to estimate the absolute quantification of contamination in the water and coupon samples obtained from the different locations. Also, a selective set of samples was chosen for a more detailed next generation sequencing (NGS) analysis. The selection of these samples was based on qPCR data and their system relevance. NGS uses the 16S rRNA gene as a common classifier for the determination of microbial diversity. The extracted DNA for the selected sample locations was subjected to MiSeq Illumina sequencing and the retrieved sequences were aligned using the open source software MOTHUR against the available SILVA database. To discuss all the results in detail is beyond the scope of this chapter, and a summary of the most relevant results is provided in Table 7.5. Taxon levels 5 and 6 are shown; these levels corresponded to two sequential depth levels of the phylogenetic tree, class, and genera, respectively. Where further taxonomic information was available, the classification of taxon level 5 was further refined by division into taxon level 6.

7.2.4 Case Study Discussion and Interpretation

7.2.4.1 Interpretation of qPCR Results

In general, cell counts obtained by qPCR followed the same trends as those observed from bug bottle analysis. The microbial load of water increased to concerning levels

TABLE 7.5
Relative Abundance of Most Abundant Classes and Genera of Bacteria Identified through NGS*

Class	Coupon Sample #3 STP; Deaerator	Coupon Sample #7 SIP Inlet	Coupon Sample #9 SIP Outlet	Coupon Sample #10 Tank with O_2 Ingress	Coupon Sample #13 Production water	Water Sample #1 Seawater Inlet	Water Sample #3 STP; Deaerator	Water Sample #5 Defective Deaerator	Water Sample #10 Tank with O_2 Ingress
Acidimicrobiales	10.0	6.4	11.5	6.7	0.0	24.7	14.1	0.6	11.1
Actinomycetales	5.7	4.2	7.4	11.0	1.7	10.2	9.2	1.3	9.4
Rhodobacteraceae	46.0	45.5	31.2	36.4	0.1	32.1	41.4	31.8	34.8
SAR116	1.8	1.3	1.1	4.3	0.0	2.7	1.1	0.0	3.6
Sphingomonadaceae	1.4	1.4	2.0	1.5	0.5	2.1	1.3	0.3	1.7
Unclassified Class	3.5	3.4	3.7	4.2	0.0	2.4	3.0	22.0	3.3
Campylobacteraceae	9.3	13.3	8.1	11.6	1.3	0.1	0.5	0.5	9.1
Helicobacteraceae	3.0	4.8	6.7	7.6	0.0	2.2	1.4	1.9	4.0
Synergistaceae	0.0	0.0	0.0	0.1	44.8	0.0	0.0	0.0	0.0
Unclassified Class	2.1	1.5	2.6	0.8	43.1	2.7	4.0	3.7	2.2
Others	17.3	18.2	25.9	15.9	8.5	20.8	23.9	37.9	20.8
Genus									
Acidimicrobineae	10.0	6.4	11.5	6.7	0.0	24.7	14.1	0.6	11.1
Micrococcineae	1.8	2.6	4.7	7.5	0.8	5.6	4.3	0.3	3.8
Kordiimonas	0.3	0.0	0.0	0.6	0.0	0.0	0.0	7.3	0.0
Roseobacter	31.3	33.5	22.4	26.3	0.0	19.9	24.5	19.4	22.4
Unclassified Class	12.5	10.5	6.8	8.6	0.0	10.4	14.2	11.5	10.2
Unclassified Class	3.5	3.4	3.7	4.2	0.0	2.4	3.0	22.0	3.3
Arcobacter	9.3	13.3	7.9	8.5	0.0	0.0	0.5	0.5	9.1
Sulfurimonas	1.1	0.6	3.9	5.9	0.0	0.2	0.2	1.7	1.2
Sulfurovum	1.8	2.6	2.0	1.6	0.0	1.9	1.1	0.1	2.9
Unclassified Class	0.0	0.0	0.0	0.1	44.7	0.0	0.0	0.0	0.0
Unclassified Class	2.1	1.5	2.6	0.8	43.1	2.7	4.0	3.7	2.2
Others	26.3	25.7	34.4	29.3	11.2	32.1	34.0	32.9	34.0

* Sequences for which a matching database entry could not be found with at least 95% confidence are listed as “Unclassified Class”

(~10^6 cells/mL) as it flowed deeper into the treatment plant. In particular, counts of SRP, organisms known to contribute to biocorrosion, reached levels of 10^5–10^6 cells/mL. Water entering the injection plant contained significantly lower counts of microorganisms than that collected from STP. This decrease in cell count could be attributed to a drop in water temperature in the pipeline between STP and SIP and/or by an enhancement in the O_2 scavenging potential of the pipeline. The cell levels, however, quickly increased within the injection plant, and by the time the water was pumped to the injection wells, were again elevated.

Analysis of coupons by qPCR revealed a consistently high number of sessile bacteria, archaea, and SRP throughout the system. This is particularly concerning, as biomass can prevent the diffusion of corrosive metabolites and waste products, significantly increasing the effective concentration of these chemicals at the surface of the pipeline. Maximum sessile cell counts (for bacteria, archaea, and SRP) were measured on the coupon collected from the SIP outlet. The coupon at this last sampling point prior to downhole injection was covered in 10^8 bacterial cells/cm^2 and nearly 10^7 cells/cm^2 of both archaea and SRP.

7.2.4.2 Interpretation of NGS Results

The microbial profiles for water samples collected from the seawater inlet, one of the functioning deaerators, and the seawater injection outlet are described in Table 7.5. These locations empirically have been the most informative indicators of the microbial profile of a system as a whole. Overall, the population distribution of microorganisms was fairly consistent throughout the treatment and injection plants. Only in sampling locations where oxygen ingress was clearly measured, were significant changes in the microbial distribution observed. However, these changes manifest as the emergence of a cluster of DNA sequences that could not be classified, so information could not be obtained for what species might be responsible for the emergence of this cluster. In addition, the communities identified in the analyzed water and coupon samples were similar.

The identified communities display signatures of an environment which is marine and O_2 limited but not O_2 depleted. A similar environment can, for example, be found in a low tide marine sediment. The identified community is also indicative of sulfur cycling, utilization of organic components, and metallic iron utilization. In particular, organisms belonging to the following microbial classes were identified:

- **Actinobacteria:** The genera contained in the phylum of the Actinobacteria (more specifically, the Acidimicrobineae) contain members which are involved in iron cycling. A well-described member of this genus, *Acidimicrobium ferrooxidans*, is a ferrous-iron-oxidizing, acidophilic bacterium (Barka et al. 2016, Clark and Norris 1996)
- **Micrococcineae:** The suborder Micrococcineae contains several genera which can be found in a variety of terrestrial and aquatic (marine) ecosystems and are often associated with hydrocarbon-contaminated environments. They are known for their ability to degrade difficult substrates like aromatic components and polycyclic aromatic hydrocarbons (PAH) (Hennessee and Li 2010). The presence of microorganisms belonging to this suborder in a sample can be

indicative of organic matter utilization; however, no other evidence of organic matter was observed for these particular injection water samples.

- ***Roseobacter*:** The presence of the *Roseobacter* genus was not surprising. Members of this genus are involved in many biochemical elemental cycles and represent a significant part of the marine coastal bacteria pool. Roseobacters are typically mesophilic and survive under anoxic conditions. In their natural habitat, they are responsible for the turnover of organic material and are known as decomposers (Buchan, Gonzalez and LeCleir 2014). *Roseobacter* members can be mobile.
- **Alphaproteobacteria:** Alphaproteobacteria are known for their cycling and oxidation of organic matter, their ability to oxidize metal, and their ability to perform sulfur cycling (Bergauer et al. 2018, Zhang et al. 2017).
- **Epsilonproteobacteria:** Epsilonproteobacteria are a highly specialized class of proteobacteria, which is primarily involved in sulfur cycling. Organisms of this class predominantly appear around thermal vents in the ocean floor where they oxidize the released sulfur species like H_2S to sulfate (or elemental sulfur) (Akerman, Butterfield and Huber 2013, Djurhuus et al. 2017). Their presence in a water treatment system indicates reduced, *but not depleted*, O_2 levels. Epsilonproteobacteria can therefore be a sign that oxygen management improvements are needed in the system.
    - ***Arcobacter*:** Members of the *Arcobacter* genera are Epsilonproteobacteria, and they inhabit a variety of ecosystems (Wang et al. 2014). In particular, many of them are microaerophilic and live in locations with reduced O_2 levels (Evans et al. 2018). They are involved in sulfur cycling and can reduce nitrate. Hence, they can be related to nitrate-related corrosion processes, where nitrate is amended but still may lead to MIC.
    - ***Sulfurimonas* and *Sulfurovum*:** Like the *Arcobacter* genus, these genera can be classified as Epsilonproteobacteria. The presence of bacteria belonging to the *Sulfurimonas* and *Sulfurovum* genera is an indicator of both sulfur cycling processes and an oxygen-reduced environment. Members of these genera can reduce nitrate and can oxidize both sulfur and hydrogen (An et al. 2016). They can also use a variety of organic substrates.

Even though numerous SRB bug bottles turned positive (see Figure 7.3), the Deltaproteobacteria class, which contains most of the traditional sulfate-reducing bacteria, was not among the most abundant genera in any sample taken from the water injection system. This observation highlights a drawback of bug bottle testing in that it provides a rather narrow view of the microbial population.

7.2.4.3 Comparison of Coupon Corrosion Data with Microbial Data

Corrosion data for the pulled coupons was provided by the operator (Table 7.6). Some overlap was observed between locations with high measured corrosion rates and the samples that were highly contaminated with problematic organisms. The occurrence of severe MIC is difficult to predict and comparing the generated data from one coupon pull may be speculative. It is however clear that the highest corrosion/pitting rates were found at locations where oxygen ingress had been

TABLE 7.6
Corrosion Data for Collected Coupons (Data Supplied by Operator)

Sample Number	Description	Corrosion Rate (mpy)	Pit Depth (mil)	Pit Rate (mpy)
2	STP, after deaerator	4.99*	0.005*	11.79
3	STP, after deaerator	4.45*	0.004*	10.48
4	STP, after deaerator	3.73*	0*	0
5	STP, after malfunctioning deaerator	5.82*	0.004*	9.17
6	Sales line - last sample collection location before SIP	0.82*	0*	0
7	SIP inlet	3.23	0.005	14.104
8	SIP; injection pump outlet	0.52	0	0
9	SIP outlet	0.47	0	0
10	SIP, outlet of tank with oxygen ingress	3.64*	0.004*	9.87
11	SIP outlet to injection water distribution station	0.83*	0*	0
12	production water line	0.03681	0	0
13	production water line	0.04*	0*	0

* Two coupons were pulled from sampling location, and the average of measurements from two coupons is reported.

reported. Namely the corrosion rates reported for the non-functioning deaerator (sample #5) and the seawater injection plant tank inlet where oxygen ingress had been reported (sample #10) were both between 3 and 10 mpy. One simple explanation for these elevated corrosion rates could be that oxygen ingress has a big influence on original pit formation. The O_2 reaction with metal creates the first onset of anodic and cathodic zones on a metal surface, and this effect may be evident from the pitting data reported by the operator. However, NGS analysis also revealed the presence of sulfur and sulfide oxidizers in these oxygen-contaminated assets, so it is also possible that low levels of oxygen are promoting the growth of a microbial population with greater corrosive potential.

7.3 CONCLUSIONS

In the above described case study, three methods of MIC monitoring were used to assess the microbial status of an oilfield water injection system. Bug bottle analysis, qPCR analysis, and NGS were complementary in their ability to detect live, dead, and unculturable microorganisms with various metabolic functions. The analysis of water and coupon samples using these three methods revealed that the microbial load of the water increases as it moves through the water injection system of this particular oilfield. Concerning levels of planktonic and sessile microbial counts were measured at the last sampling point prior to downhole injection. When the total cell levels and cell types detected throughout the system are considered, a qualitative estimate

TABLE 7.7
Concern Level Given to the Tested Environments

	Coupon Swabs				Water Samples					
Sample Number and Description	**Total Cells (qPCR)**	**Total SRP (qPCR)**	**SRB (Bug Bottle)**	**APB (Bug Bottle)**	**Total Cells (qPCR)**	**Total SRP (qPCR)**	**SRB (Bug Bottle)**	**APB (Bug Bottle)**	**NGS**	**Total**
1. Seawater reservoir					++	++	++	++	+	++
2. STP; after deaerator	–	+	+	+	+	+	++	+		+
3. STP, after deaerator	–	–	+	+	–	+	+	+	–	+
4. STP, after deaerator	–	–	++	+	+	+	++	++		++
5. STP, after malfunctioning deaerator	+	+	++	++	+	+	++	++	–	+
6. Sales line - last sampling location before SIP	+	+	++	++	+	+	+	+		+
7. SIP inlet	–	+	+	–	++	++	–	+	–	+
8. SIP, injection pump outlet	–	–	+	+	+	+	+	+	+	+
9. SIP outlet	–	–	+	+	–	–	+	+	–	–
10. SIP, outlet of tank with oxygen ingress	+	–	+	–	+	+	–	–	–	–
11. SIP outlet to injection water distribution station	–	–	+	–	+	+	–	–	–	–

Gray: Not tested.
++ : Low concern.
+ : Moderate concern.
– : High concern.

of the risk of each asset for MIC can be made (Table 7.7). In estimating the overall risk for a particular asset, greater weight is given to measurements of corrosion-influencing microbes relative to measurements of total cells. Also, greater weight is given to sessile cell data, as these populations have greater contact with asset interior surfaces.

Using NGS, it was found that the microbial communities identified in the water and on the coupons throughout the water injection system were similar and were primarily composed of organisms of the Alphaproteobacteria and Epsilonproteobacteria classes. Microorganisms of these classes promote the cycling of organic matter and suggest the presence of an O_2-reduced environment, along with sulfur/H_2S cycling and oxidation processes that can potentially result in polysulfur formation. Polysulfur is detrimental to metals as it reacts with metallic iron. In addition, our findings from

NGS analysis indicate that while the water injection system is operating under reduced oxygen levels, some oxygen does remain in the system. The presence of oxygen can contribute significantly to corrosion, and optimization of the O_2 management system was recommended to deplete oxygen from the system.

From this holistic data set, we determined that the microbial control program employed at the time of analysis was not optimized and needed improvement. It was recommended that this include the injection of biocide at multiple points in the water injection system. This improvement could provide protection to assets that are far downstream from the current site of biocide application within the seawater treatment plant. In addition, the dosing strategy for functional chemicals, including biocides, needs to be evaluated to ensure the compatibility of chemicals during co-application.

REFERENCES

Akerman, N.H., D.A. Butterfield, and J.A. Huber. 2013. Phylogenetic Diversity and Functional Gene Patterns of Sulfur-Oxidizing Subseafloor *Epsilonproeobacteria* in Diffuse Hydrothermal Vent Fluids. *Frontiers in Microbiology* 4:185.

Amann, R., W. Ludwig, and K.-H. Schleifer. 1995. Phylogenetic Identification and *In Situ* Detection of Individual Microbial Cells without Cultivation. *Microbiological Reviews* 59 (1995):143.

An, D., et al. 2016. Metagenomic Analysis Indicates Epsilonproteobacteria as a Potential Cause of Microbial Corrosion in Pipelines Injected with Bisulfite. *Frontiers in Microbiology* 7:28.

Barka, E.A., et al. 2016. Taxonomy, Physiology, and Natural Products of *Actinobacteria*. *Microbiology and Molecular Biology Reviews* 80:1.

Bergauer, K., et al. 2018. Organic Matter Processing by Microbial Communities Throughout the Atlantic Water Column as Revealed by Metaproteomics. *PNAS* 115:E400.

Buchan, A., J.M. Gonzalez, and G.R. LeCleir. 2014. Master Recyclers: Features and Functions of Bacteria Associated with Phytoplankton Blooms. *Nature Reviews Microbiology* 12:686.

Clark, D.A., and P.R. Norris. 1996. *Acidmicrobium ferrooxidans* gen. nov., sp. nov.: Mixed-Culture Ferrous Iron Oxidation with *Sulfobacillus* Species. *Microbiology* 142:785.

Dar, S.A., et al. 2007. Analysis of Diversity and Activity of Sulfate-Reducing Bacterial Communities in Sulfidogenic Bioreactors Using 16s rRNA and *dsrB* Genes as Molecular Markers. *Applied and Environmental Microbiology* 73:594.

Djurhuus, A., et al. 2017. Cutting Through the Smoke: The Diversity of Microorganisms in Deep-Sea Hydrothermal Plumes. *Royal Society Open Science* 4:160829.

Enning, D., and J. Garrelfs. 2014. Corrosion of Iron by Sulfate-Reducing Bacteria: New Views of an Old Problem. *Applied and Environmental Microbiology* 80:1226.

Enning, D., et al. 2012. Marine Sulfate-Reducing Bacteria Cause Serious Corrosion of Iron Under Electroconductive Biogenic Mineral Crust. *Environmental Microbiology* 14:1772.

Evans, M.V., et al. 2018. Members of *Marinobacter* and *Arcobacter* Influence System Biogeochemistry During Early Production of Hydraulically Fractured Natural Gas Wells in the Appalachian Basin. *Frontiers in Microbiology* 9:2646.

Foti, M., et al. 2007. Diversity, Activity, and Abundance of Sulfate-Reducing Bacteria in Saline and Hypersaline Soda Lakes. *Applied and Environmental Microbiology* 73:2093.

Gittel, A., et al. 2009. Prokaryotic Community Structure and Sulfate Reducer Activity in Water from High-temperature Oil Reservoirs with and without Nitrate Treatment. *Applied and Environmental Microbiology* 75:7086.

Gu, T. 2012. Can Acid-Producing Bacteria Be Responsible for Very Fast MIC Pitting? *NACE CORROSION, paper no. 1214*. Salt Lake City, UT.

Hennessee, C.T., and Q.X. Li. 2010. Micrococcineae: Arthrobacter and Relatives. In *Handbook of Hydrocarbon and Lipid Microbiology*, ed. K.N. Timmis, 1854–1861, New York: Springer.

Iverson, W.P. 1987. Microbial Corrosion of Metals. *Advances in Applied Microbiology* 32:1.

Kermani, M.B., and D. Harrop. 1998. The Impact of Corrosion on Oil and Gas Industry. *SPE Production & Facilities* 11:186.

Klappenbach, J.A., et al. 2001. Rrndb: The Ribosomal RNA Operon Copy Number Database. *Nucleic Acids Research* 29:181.

Koch, G., et al. 2016. International Measures of Prevention, Application, and Economics of Corrosion Technologies (IMPACT). *Report to NACE by DNVGL and APQC* http://impact.nace.org/economic-impact.aspx (accessed December 18, 2020).

Little, B.J., J.S. Lee, and R.I. Ray. 2006. Diagnosing Microbiologically Influenced Corrosion: A State-of-the-Art Review. *CORROSION* 62:1006.

Nadkarni, M.A., et al. 2002. Determination of Bacterial Load by Real-Time PCR Using a Broad-Range (Universal) Probe and Primers Set. *Microbiology* 148:257.

Pak, K.R., et al. 2003. Involvement of Organic Acid During Corrosion of Iron Coupon by *Desulfovibrio desulfuricans*. *Journal of Microbiology and Biotechnology* 13:937.

Rachel, N.M., and L.M. Gieg. 2020. Preserving Microbial Community Integrity in Oilfield Produced Water. *Frontiers in Microbiology* 11:2536.

Sharma, M., and G. Voordouw. 2017. MIC Detection and Assessment: A Holistic Approach. In *Microbiologically Influenced Corrosion in the Upstream Oil and Gas Industry*, eds. T. Skovhus, D. Enning, and J. Lee. Boca Rotan: CRC Press.

Skovhus, T.L., R.B. Eckert, and E. Rodrigues. 2017. Management and Control of Microbiologically Influenced Corrosion (MIC) in the Oil and Gas Industry – Overview and a North Sea Case Study. *Journal of Biotechnology* 256:31.

Sørenson, K., T.L. Skovhus, and J. Larsen. 2011. Techniques for Enumerating Microorganisms in Oilfields. In *Applied Microbiology and Molecular Biology in Oilfield Systems*, eds. C. Whitby and T.L. Skovhus. New York: Springer.

Wang, L-Y., et al. 2014. Comparison of Bacterial Community in Aqueous and Oil Phases of Water-Flooded Petroleum Reservoirs Using Pyrosequencing and Clone Library Approaches. *Applied Microbiology and Biotechnology* 98:4209.

Wrangham, J.B., and E.J. Summer. 2013. Planktonic Microbial Population Profiles Do Not Accurately Represent Same Location Sessile Population Profiles. *NACE CORROSION, paper no. 2780*. Orlando, FL.

Zhang, Y., et al. 2017. Microbial Diversity and Community Structure of Sulfate-Reducing and Sulfur-Oxidizing Bacteria in Sediment Cores from the East China Sea. *Frontiers in Microbiology* 8:2133.

Zhu, X.Y., et al. 2005. Applications of Quantitative, Real-Time PCR in Monitoring Microbiologically Influenced Corrosion (MIC) in Gas Pipelines. *NACE CORROSION, paper no 05493*. Houston, TX.

8 Methanogens and MIC

Leveraging Bioinformatics to Expose an Underappreciated Corrosive Threat to the Oil and Gas Industry

Timothy J. Tidwell and Zachary R. Broussard

CONTENTS

8.1 Introduction 138
8.1.1 Working with Methanogens 138
8.2 We've Come a Long Way from Bug Bottles 140
8.3 Archaeal Methanogens 141
8.3.1 Role in the Biosphere 141
8.3.1.1 Environments 141
8.3.1.2 The Human Archaeome 142
8.3.1.3 Temperature 142
8.3.1.4 Salinity and pH 142
8.3.1.5 Osmoprotectants 142
8.3.1.6 High-Salt-In Strategy 143
8.3.1.7 pH 143
8.3.2 Morphology 143
8.3.3 Archaeal Genetics 143
8.3.3.1 Lateral Gene Transfer (LGT) 144
8.3.3.2 Taxonomy 146
8.3.3.3 Metabolism/Methanogenesis 147
8.3.3.4 Syntrophic Relationships 151
8.4 Considerations for Oil and Gas 152
8.4.1 Crude-Oil Biodegradation via Methanogenesis 152
8.4.2 Methanogen Mediated Iron Corrosion 152
8.4.2.1 Syntrophic Relationships 153
8.4.2.2 Direct Iron Oxidation 153
8.4.2.3 Discovery of MIC Island and a Binary MIC Factor 153
8.4.2.4 Archaeoglobus 155
8.4.3 Mitigation of Methanogens 155

8.5 Methanogen Oilfield Corrosion Examples....156
8.5.1 Dry Gas Pipeline....157
8.5.2 West Texas Shale Production....158
8.5.3 Biofilm Reactor Coupons – Deepwater Brazil....159
8.5.4 West Texas Salt Water Disposal Facility....160
8.5.5 South America Pipeline....160
8.6 Conclusion....163
References....163

8.1 INTRODUCTION

The Archaea are a distinct phylogenetic lineage of microorganisms representing the third domain of life next to Bacteria and Eukarya. This group of organisms exhibits a wide range of physiological and morphological diversity, including extreme halophiles, sulfur-metabolizing hyperthermophiles, and methanogens. Archaea are often found thriving in conditions that have periodically required redefining the limits for life as we know it. These limits include surviving salinities at saturation, highly reduced anoxic environments, and extremely high temperatures of up to 130°C (Lovey 2003). Archaea are commonly found in petroleum reservoirs, pipeline deposits, and tubing failure investigations. Research completed in the last few years has shown that the involvement of Archaea in MIC, particularly the methanogens, has been historically underestimated. One reason for that is the historical dependence on culture-based methods for microbial enumeration. However, even since the introduction and adoption of molecular biology techniques, which are currently required in practice for detecting methanogens, interest in identifying the corrosion risk from methanogens has lagged. Published research on methanogenic corrosion is dwarfed in comparison to SRB- or APB-associated MIC research, as seen in Figure 8.1. Recent studies on methanogens have discovered novel corrosion mechanisms that elegantly facilitate severe iron corrosion by the oxidation of metallic iron (Deutzmann, et al. 2015, Tsurumaru, et al. 2018, Lahme, et al. 2020).

8.1.1 Working with Methanogens

Considering all the exciting aspects of Archaea, why is there such a disproportionate amount of attention given to this domain of life? Archaea's tend to thrive in conditions that are difficult to replicate in laboratory conditions such as high temperature, salinity, pressure, and pH. Perhaps one of the biggest challenges to routinely working with methanogens is their extreme oxygen sensitivity. Labs that study methanogens must be equipped with efficient anaerobic apparatuses capable of reducing oxygen to ppb levels and create a stable, reducing environment with a redox potential below 50 mV (Bhattad 2012). Essential to the study of methanogens is a mechanism for scrubbing traces amounts of oxygen from commercial carbon dioxide sources. Keep in mind that the quality control specification for oxygen in research-grade carbon dioxide is 2ppm (99.998% purity). Industrial grade, cryogenic carbon dioxide may contain as much as 0.5% oxygen.

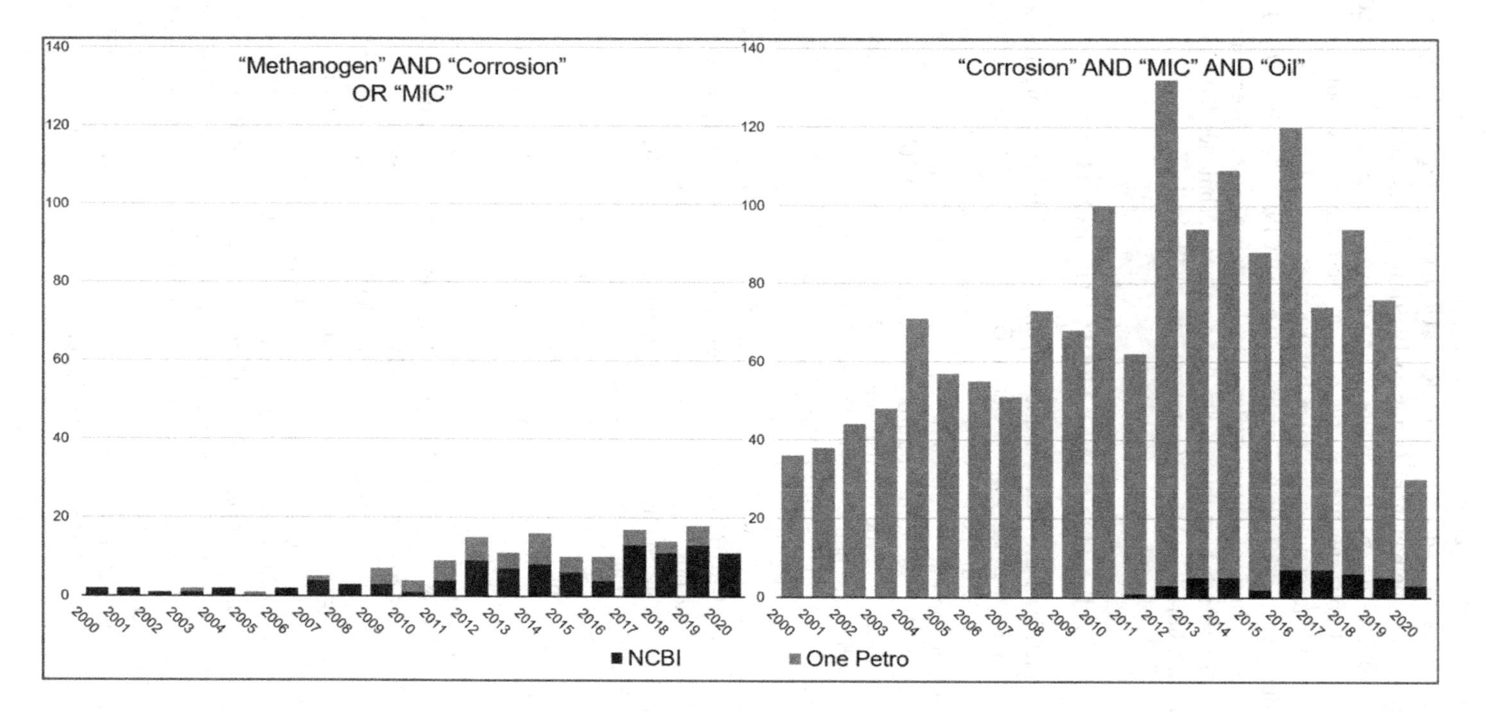

FIGURE 8.1 Publications from NCBI and One Petro as of July 2020 (original).

These technical limitations explain, in some part, the lag in archaea research. It was not until the 1950s and 1960s when Robert Hungate published blueprints for how to reliably grow methanogens (Hungate 1969). In 1974, a new era was beginning as the use of pressurized atmosphere chambers was adopted. These allowed for a greater degree of consistency when research on these organisms could progress. Previous work with strict anaerobic microbes was met with set back after set back when procedures were not followed perfectly (Hungate 1950, Wolfe 1976). These anaerobic culture techniques have contributed to the study of many diverse microbes that could otherwise not be cultured, but these culture techniques have not been transferred to monitor oil and gas infrastructure on a wide scale.

8.2 WE'VE COME A LONG WAY FROM BUG BOTTLES

The first introduction of the idea that microorganisms may be the etiological agent for metal corrosion was in 1891 by JH Garret (Garrett, et al. 1891). Garret hypothesized that the increase in the corrosive action of lead could be due to ammonia, nitrites, and nitrates produced by bacterial action. From there, numerous reports were published linking MIC to sulfur and iron bacteria (Gaines 1910). However, most people did not understand the mechanisms of MIC until von Wolzogen Kuehr and van der Vlugt published their Cathodic Depolarization Theory (CDT). The original CDT theory attempted to describe a MIC mechanism with the electrochemical reasoning of cathodic hydrogen release kinetics. But the idea did not address all the factors involved in the corrosion process, such as the presence of the cathodically active H_2S (Costello 1974). CDT was used to explain the relationship to known SRB in oil production to the observed MIC in practice. Thus, the focus was given to SRB detection with available methods of the time, which led to the development of Postgate medium (Postgate 1963). Today, the CDT is accepted as a classic/historical explanation rather than a completely accurate mechanism (von Wolzogen Kuhr and van der Vlugt, 1934, Thierry and Sand, 2002).

In 1985, Saiki *et al.* published a method for diagnosing sickle cell anemia by detecting specific alleles in the beta-globulin gene using a precursor method of qPCR (Saiki, et al. 1985). There are very few inventions that can compete against the importance of PCR over the past 100 years as it revolutionized biological and genetic research. qPCR is now commonly used for oil and gas risk assessments to enumerate organisms and genes of interest. In the early 2000's, next generation 16S rRNA sequencing bombarded producers with new information about the microbial communities involved and by using this approach to characterize MIC risk has brought the importance of methanogens to the forefront. As qPCR and next-generation sequencing have become more prevalent as diagnostic tools in the oilfield, the importance of procedural selection has become abundantly clear, especially when considering the identification of Archaeal populations. Utilizing universal primers in oilfield applications (Geissler 2014; Lomans 2016), we recognize the frequency that archaea or methanogens is present in areas where corrosion is problematic. Our analysis of almost 14,000 DNA samples from oil and gas production and completions has found methanogens present in 31% of all samples.

This chapter hopes to serve as a reference to those in oil and gas that are learning about methanogens and how they may impact pipelines, production tubing, water treatment, or reservoir conditions. Here we summarize some of the latest developments in methanogen research concerning MIC and present an opportunity to view microbial communities associated with severe corrosion and or failures.

8.3 ARCHAEAL METHANOGENS

8.3.1 Role in the Biosphere

The observation that methane emissions were biogenic in origin was originally attributed to Italian scientist Alessandro Volta in 1776, who first described flammable freshwater swamp gas and postulated it was derived from decaying organic material. However, methanogens would not be isolated in the lab for another 157 years (Stephenson, and Strickland 1933, Wolfe 1993).

Methanogens are abundant in a wide variety of anaerobic environments, where they catalyze the terminal step in the anaerobic food chain by converting methanogenic substrates, such as CO2, acetate, and other C1-compunds, into methane. The complexity of methanogenesis pathways suggests an ancient monophyletic origin of methanogens, a hypothesis supported by phylogenetic analyses based upon DNA sequences. The only enzyme present in all types of methanogenesis is methyl-coenzyme M reductase (*Mcr*), a Ni-corrinoid protein catalyzing the last step of methyl group reduction to methane (Berghuis, et al. 2019).

Methanogenic microbes may be one of the most primitive organisms. Evidence of microbial methanogenesis has been dated to 3.46 billion years. However, it is uncertain when methanogens first appeared on Earth (Ueno 2006). During the Archaean era (before 2.5 Gyr ago), methanogens may have been influential in regulating climate because they could have provided enough greenhouse gas methane to warm a frozen earth (Canfield 2006).

Methanogens play critical roles in the carbon cycle by producing 900 million tons of methane annually, contributing to 16% of the total emission of global warming gases (Li, et al. 2014). Estimates show that <1 % of biological methane from the subsurface is released into the atmosphere (Thauer et al., 2008a). Overall, it is estimated that up to 2 % of carbon in the global carbon cycle (or 450 Tg annually) is mineralized by methanogens per year (Dlugokencky, et al. 2011). The biosynthesis of methane is the ubiquitous, defining characteristic of methanogens.

8.3.1.1 Environments

The archaea are notorious for inhabiting some of the harshest places on Earth and thrive under conditions that few bacteria and no eukaryotes would tolerate. It is hypothesized that these organisms are well suited to conditions that might have existed on the early *Archaean* Earth. Methanogens can be isolated from across a wide range of thermochemical gradients, from acidophilic to alkaliphilic (pH 3.0–10.2), from (hyper)thermophilic to psychrophilic temperature (from −2°C to 122°C), and from freshwater estuarine to halophilic environments up to 3 M NaCl (Buan 2018). However, it is wrong to presume that all archaea are extremophiles. Methanogens also colonize non-extreme environments. Methanogens can be found in most anaerobic habitats, and especially in low-sulfate environments such as in freshwater pond sediment. They can be isolated from anoxic soil sediments such as rice fields, peat bogs, marshland or wetlands, and the digestive tracts of animals (ruminants and humans), insects, and marine sediment, and terrestrial subsurface environments (Thauer et al., 2008b).

8.3.1.2 The Human Archaeome

Methanogens have been associated with human polymicrobial diseases, such as intestinal dysbiosis and periodontal inflammation; however, there are no reports of methanogens being directly involved in pathogenesis using classical virulence factors or toxins. There is growing evidence that methanogens may behave as opportunistic pathogens and necessary commensal residents (Lurie-Weinberger and Gophna, 2015). Molecular studies have indicated the presence of members of the orders *Methanosarcinales, Thermoplasmatales, Methanomicrobiales*, and *Nitrososphaerales* in the human gut microbiota; however, to date, these microbes have not been isolated. Archaeal abundance is associated with obesity, colon cancer, and diverticulosis. Confoundingly, the absence of methanogens is also associated with Crohn's disease and ulcerative colitis (Macario, et al. 2009, Vemuri, et al. 2020).

8.3.1.3 Temperature

Many methanogens have a mesophilic temperature spectrum, such as the *Methanosarcina*, *Methanobacterium*, or most *Methanococcus*. However, thermophilic and even hyperthermophilic methanogens, like *Methanothermobacter thermautotrophicus* or *Methanocaldococcus jannaschii*, grow at temperatures up to 75°C and 86°C, respectively. Even growth up to 110°C is possible, as shown for the hyperthermophilic strain *Methanopyrus kandleri* (Takai, et al. 2008). The most extreme example of a hyperthermophile was published in 2003, where Lovely incubated a yet uncharacterized Archaea, *strain 121* at 130°C for 2 hours, the strain still grew when transferred back to fresh medium at 103°C (Lovey 2003). Cold-loving methanogenic psychrophiles have been isolated, such as the methanol-converting archaeon *Methanolobus psychrophilus*, which grows optimally at 18°C and shows still metabolic activity at 0°C (Enzmann, et al. 2018).

8.3.1.4 Salinity and pH

Salt concentration is an important physiological parameter for methanogens. A few methanogens have colonized niches such as hypersaline lakes, due to the extremity of the environments for microbial growth. Because biological membranes are permeable to water, a higher solute concentration outside the cell, as in environments with high salinity, would drag water out of the cell leading to cell death. Microorganisms living under such salty conditions must adapt to prevent desiccation. This is achieved in two ways:

8.3.1.5 Osmoprotectants

The first mechanism for halotolerant/moderately halophilic methanogens are the synthesis and accumulation of osmoprotectants with a small molecular mass and high solubility. For example, *Methanosarcina mazei* synthesizes glutamate (MW: 147.13 g/mol) when NaCl reaches 400 mM and produces N-acetyl-β-lysine (MW: 188.22 g/mol) above 800 mM NaCl (Pflüger and Müller, 2003). The osmolyte *N*-acetyl-β-lysine is commonly found in halotolerant methanogen species. Interestingly, *N*-acetyl-β-lysine was originally discovered in Archaea and was once thought to be unique to methanogens, but some reports have shown certain green sulfur bacteria and *Bacillus cereus* CECT148 also produce this osmolyte under high salt stress. β-Amino acids, specifically, the accumulation of β-amino acids (and derivates) for osmoadaptation, are relatively rare in biological systems. β-Glutamate and *N*-acetyl-β-lysine have only been detected in a few organisms to date, and *N*-acetyl-β-lysine had previously been considered unique to methanogenic Archaea (Empadhinhas and da Costa 2008). It has been found in several species belonging to the Methanococcales, Methanomicrobiales, and Methanosarcinales (Enzmann, et al. 2018).

8.3.1.6 High-Salt-In Strategy

An influx of potassium and chloride into the cytoplasm to balance the cytoplasm osmotically with the high-salt environment. This *high-salt-in strategy* may also be used by the recently discovered *Methanonatronarchaeia* (Sorokin, et al. 2017). They appear to be extremely halophilic, methyl-reducing methanogens related to the haloarchaea (Oren 2008).

8.3.1.7 pH

Most methanogens grow optimally around neutral pH. Many halophilic or halotolerant species of methanogens show adaptation to alkaline pH as well. *Methanocalculus alkaliphilus* grows alkaliphilically with an optimum at pH 9.5 and a moderate salinity up to 2 M of total Na^+, whereas *Methanosalsum natronophilum* can even tolerate higher salinities, up to 3.5 M of total Na^+, at the same alkaline pH (Sorokin, et al. 2017). Methanogens can also inhabit moderately acidic environments as, for example, *Methanoregula booneii*, which was isolated from an acidic peat bog and had an optimum growth pH of 5.1 (Bräuer, et al. 2006).

8.3.2 Morphology

Methanogens present as a diverse group. Observations of methanogens have shown considerable variations in cell dimension and organization. Methanogens may be in the form of rods (long and short), cocci (often arranged to sarcina), spheres, and spirals (Zeikus 1977).

8.3.3 Archaeal Genetics

In 1977, Carl Woese published, "these *Bacteria* appear to be no more related to typical bacteria than they are to eukaryotic cytoplasms." He found that a group of anaerobic *bacteria*, which had been studied for years due to their ability to produce methane, were not bacteria at all. There had already been observations that these microbes had some notable "nonbacterial" aspects, such as the presence of N-linked glycoproteins and a distinctive spectrum of antibiotic sensitivity. Subsequent studies of rRNA phylogeny revealed that archaea are no more related to conventional bacteria than to eukaryotes. This group of organisms was reorganized and renamed Archaebacteria (Woese and Fox, 1977). Archaea have demonstrated to be a medley of molecular features, which are encoded by three distinct groups of genes: a eukaryotic lineage that codes for information processing, a lineage that codes for operational (housekeeping) functions with a bacterial aspect (Rivera, et al. 1998), and a unique lineage that is poorly understood. For example, the subunit structure of RNA polymerases of Archaea is more like those of Eukarya than to the bacterial polymerase, but both in Bacteria and Archaea, there is only one polymerase. Additionally, the methanogenic RNA polymerase activity depends strongly on the presence of a promoter sequence of high homology to the TATA box recognized by one of the eukaryal RNA polymerases (Allers and Mevarech 2005). However, like bacteria, archaea use polycistronic Shine-Dalgarno sequence to recognize mRNA translation. Archaea also possesses a unique translation initiation mechanism that operates on leaderless mRNAs that hints at a eukaryotic origin. It has been suggested that informational genes are less prone to lateral gene transfer than operational genes as they are usually part of complex and highly regulated systems (Schmitt, et al. 2020).

Bioinformaticians should find fertile territory of discovery as the Archaea are not merely a eukaryotic/bacterial mash-up. As much as 50 % of the archaeal genome is characterized with no known function (Allers and Mevarech 2005). As opposed to bacterial phylogenetics, which relies on 16S rRNA, the evolutionary tree of methanogens is designated by the evolution of McrA. Figure 8.2 shows the structure of methanogen groups based on the sequence homology of McrA from Boyd, 2019.

8.3.3.1 Lateral Gene Transfer (LGT)

It is commonly understood that only the more complex eukaryotes indulge in sex, and prokaryotes rely on vertical inheritance to acquire novel gene functions. To the contrary, prokaryotes are highly promiscuous, and the role of lateral gene transfer (LGT) has played in muddying up the phylogenetic tree in prokaryotic evolution has been underestimated (Boucher, et al. 2003). Methanogenic and haloarchaea are uninhibited in their quest to acquire new and share old gene functions with their community. The opportunity often arises because Achaea often cohabits with bacteria. Some estimates conclude that the portion of "foreign" (mostly bacterial) genes in archaea might be as high as 20%–30% (Koonin, et al. 2001). Much of this bacterial DNA influx can be attributed to orthologous replacement or acquisition of a paralogous gene. As a result, archaeal genomes resemble genetic mosaics (Doolittle 2003). Figure 8.3 shows an analysis by Nelson-Sathi in 2015, looking at the rRNA phylogenetic tree's linear progression with the phylogenomic analysis of essential informational genes. The balance of gene transfers between archaea and bacteria is highly asymmetric. Transfers from bacteria to archaea occur 5-times more frequently than archaea to bacteria (Nelson-Sathi, et al. 2015).

Occasionally, LGT leads to a novel gain-of-function event. For example, it has been suggested that the switch from an anaerobic to an aerobic lifestyle by the (methanogenic) ancestor of haloarchaea was facilitated by LGT of respiratory-chain genes from bacteria (Kennedy, et al. 2001).

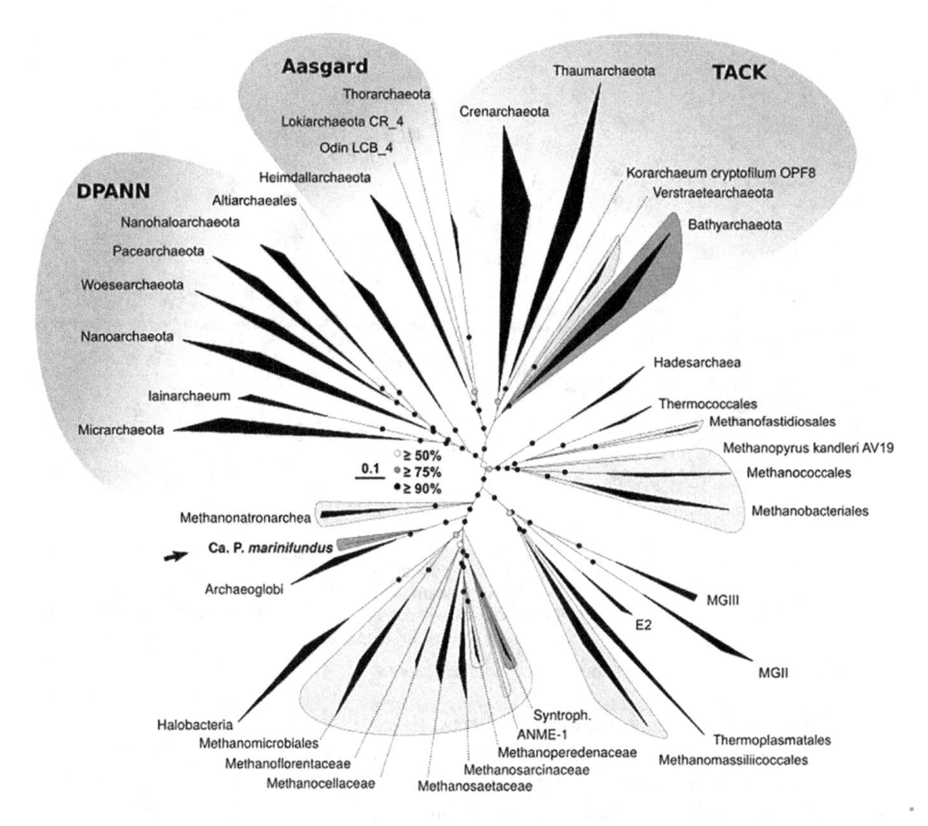

FIGURE 8.2 McrA genome tree phylogeny. A concatenated alignment of 122 single copy archaeal marker genes from high-quality archaeal RefSeq genomes (release 80). Hydrogenotrophic and acetoclastic methanogens are designated with †, H_2-dependent methylotrophic methanogens are marked with ‡, known/putative methane oxidizers are marked ¤, and lineages encoding divergent McrA are marked Ψ. Bootstrap support was generated from 100 replicates, and white, gray, and black nodes represent ≥50%, ≥75%, and ≥90% support, respectively (adapted from Boyd et al., 2019).

Archaea, like bacteria, often group co-regulated genes into operons, which can promote co-inheritance by LGT. There is a high degree of variability between species, but much of the archaeal genome is polluted with bacterial genes. LGT may account for some variation in archaeal genome sizes (Allers and Mevarech 2005). For example, *Methanosarcina mazeihas* a bloated genome of 4.10 Mb, 30% of which is bacterial in origin (Deppenmeier, et al. 2002). Because of rampant LGT, phylogenetic trees based on individual genes differ significantly from the rRNA tree (Brown and Doolittle 1997). Although it seems possible that there exists a "core" of non-transferable genes between bacteria and archaea, studies continue to find exceptions (Deppenmeier, et al. 2002, Doolittle 2003).

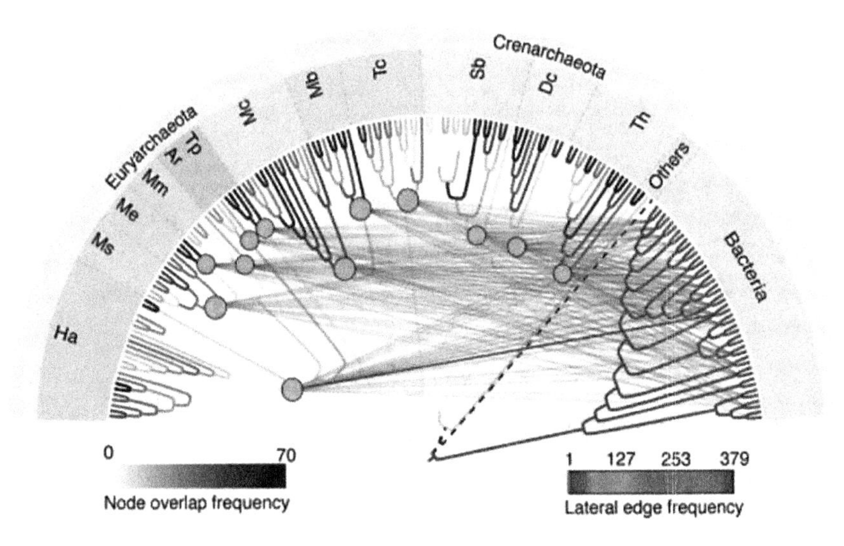

Ha - Haloarchaea (1047)
Ms - Methanosarcinales (338)
Me - Methanocellales (83)
Mm - Methanomicrobiales (85)

Ar - Archaeoglobus (51)
Tp - Thermoplasma (49)
Mc - Methanococcales (100)
Mb - Methanobacteriales (128)

Tc - Thermococcales (101)
Sb - Sulfolobales (129)
Dc - Desulfurococcales (40)
Th - Thermoproteales (59)

Others - Korarchaeota, Nanoarchaeota and Thaumarchaeota

FIGURE 8.3 Archaeal gene acquisition network: Based on 70 concatenated genes, gray shading indicates how often the branch was recovered by the 70 genes analyzed individually. The vertical edge weight of each branch in the reference tree (scale bar at left) was calculated as the number of times associated node was present within the single gene trees. Lateral edges indicate 2,264 bacterial acquisitions in archaea. The number of acquisitions per group is indicated in parentheses, the number of times the bacterial taxon appeared within the inferred donor clade is color coded (scale bar at right). The strongest lateral edge links Haloarchaea with Actinobacteria. Archaea were arbitrarily rooted on the *Korarchaeota* branch (dotted line). Bacterial taxon labels are (from left to right) *Chlorobi, Bacteroidetes, Acidobacteria, Chlamydiae, Planctomycetes, Spirochaetes, ε-Proteobacteria, δ-Proteobacteria, β-Proteobacteria, γ-Proteobacteria, α-Proteobacteria, Actinobacteria, Bacilli, Tenericutes, Negativicutes, Clostridia, Cyanobacteria, Chloroflexi, Deinococcus-Thermococcus, Fusobacteria, Aquificae, Thermotogae* (adapted from Nelson-Sathi, *et al.* 2015).

8.3.3.2 Taxonomy

The archaeal taxonomical tree is currently in a constant state of evolution with new branches, lineages, phyla, classes, and orders continually being added and proposed. The number of validated archaeal species increased by almost 31% between 2012 and 2016, while the cumulative number of described archaeal genomes increased by over 500% (Adam, et al. 2017). Traditionally, methanogens have been classified into eight orders within the phylum Euryarchaeota: *Methanobacteriales, Methanomassiliicoccales, Methanococcales, Methanomicrobiales, Methanosarcinales, Methanocellales, Methanopyrales*, and *Candidatus Methanophagales* (Evans, et al. 2019).

Based on previous taxonomical classifications, it was suggested that, because modern methanogens are monophyletic (a claim only recently challenged), it is likely that methanogenesis evolved only once and that all modern methanogens share a single ancestor (Kim and Whitman 2014). However, within the past few years, the discovery of *mcr* and divergent *mcr*-like genes in not only new euryarchaeotal lineages but novel archaeal phyla such as the *Bathyarchaeota* and *Verstraetearchaeota* as well as *Korarchaeota* has challenged long-held views of the evolutionary origin of this metabolism within the Euryarchaeota (Evans, et al. 2019; Söllinger and Urich 2019). Several outstanding hypotheses exist as to the origin of *mcr*-containing archaea, and there seems to some evidence that 1) methanogenesis may have evolved independently more than once in Earth's history, 2) a common methanogenic ancestor for the Archaea exists, and 3) the origin of mcr-containing archaea is characterized by lateral gene transfer and neofunctionalization (Adam, et al. 2017, Buan 2018, Evans, et al. 2019).

In the coming years, molecular and synthetic biology will continue to modify and challenge the existing archaeal taxonomy, including the methanogens.

8.3.3.3 Metabolism/Methanogenesis

All organisms have essential requirements for life. Carbon, hydrogen, oxygen, nitrogen, phosphorus, and sulfur are considered the six elements required for life. The needs of prokaryotes can be further simplified to three requirements: 1) a carbon source that can be used to build its biomass, 2) an electron donor (such as hydrogen or acetate) for the energy it needs to survive, and 3) a terminal electron acceptor (such as oxygen, nitrate, sulfate, CO_2) to receive the electrons the organisms use for energy.

The carbon source will often serve as the electron donor and is required in macro proportions compared to the nutrients. Nitrogen, phosphorus, and sulfur can also fulfill the role of electron donors or acceptors. However, they are more often considered as nutrients and are required in smaller proportions than carbon, hydrogen, and oxygen. In each of these pathways, the electrons' energy state from the electron donor is higher than that of the electrons donated to the electron acceptor (Buan 2018). Figure 8.4 shows the dominant electron-accepting mechanism as redox conditions decrease (Bouwer and McCarty 1984).

When inorganic electron acceptors such as nitrate, ferric iron, or sulfate are depleted, methanogenesis serves as the terminal step in the degradation of organic matter. Substrates that methanogens utilize are limited. In the industrial process of anaerobic digestion, the usual substrates for methane formation are only H_2, CO_2, or formate for hydrogenotrophic methanogens, acetate for acetoclastic methanogens via acidogenesis and acetogenesis, and C1 compounds such as methanol for methylotrophic methanogens produced (Enzmann, et al. 2018).

$$e^{-}\text{donor} + e^{-}\text{acceptor} \rightarrow CH_4 + \text{byproduct}$$

For methanogens grown in axenic culture, the electron donor is either molecular hydrogen (for hydrogenotrophic or methyl respiration pathways) or the carbon source itself (for methylotrophic, carboxydotrophic, or acetoclastic fermentation or respiration pathways) (Buan 2018).

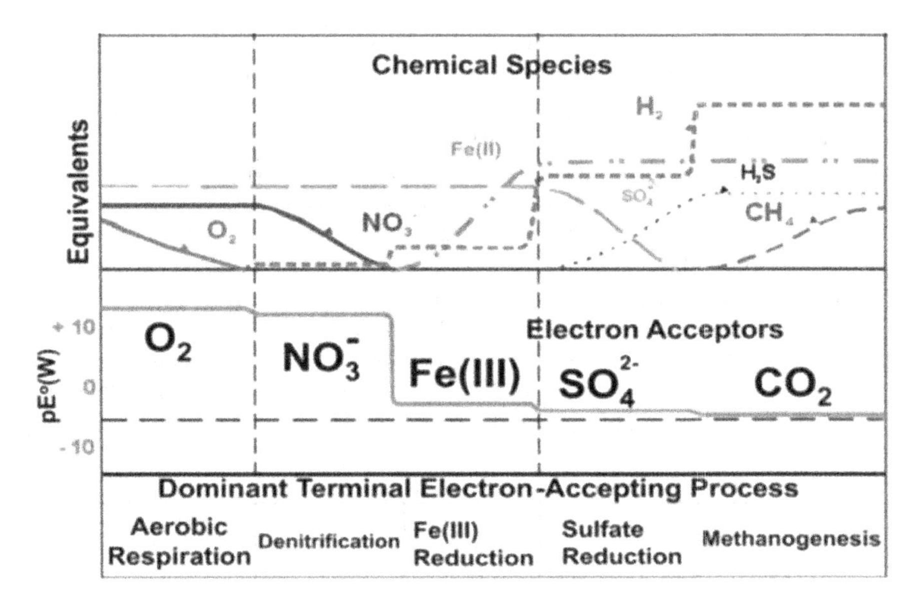

FIGURE 8.4 Dominant electron-accepting process (from Bouwer and McCarty 1984).

8.3.3.3.1 Methyl Coenzyme M Reductase (mcrA)

mcrA is a popular target for methanogen quantification (Steinberg and Regan 2009). The *mcrA* gene encodes for methyl-coenzyme M reductase (mcr), the final and rate-limiting step in methanogenesis (Thauer 1998). Mcr is the only enzyme that is present in all types of methanogens (Berghuis, et al. 2019). Mcr contains a nickel hydrocorphinate F_{430} at its active site (Meyerdierks, et al. 2003). The nickel-tetrapyrrole coenzyme F430 is only found in methanogens and evolved into a strong enough reducing agent to activate carbon to accept electrons to form methane. In the final step of methanogenesis, coenzyme B thiol (the terminal e^- donor) inserts into the mcr active site and donates an electron to the methyl radical producing methane. Subsequently, a coenzyme M-coenzyme B heterodisulfide is formed to regenerate the Ni(I)F430 cofactor which releases the methane (Wongnate, et al. 2016, Buan 2018).

8.3.3.3.2 Reverse Methanogenesis

Some methanogens can also reverse the methanogenesis pathway to oxidize methane, where mcr serves as the first step. Anaerobic methane-oxidizing archaea (ANME) are responsible for the *anaerobic oxidation of methane* (AOM) (Shima and Thauer 2005). AOM is catalyzed via a modified, reverse methanogenesis pathway, but only during net methane production or *trace methane oxidation* (TMO). TMO produces CO_2 in trace amounts relative to the methane produced. The Gibbs free energy change of the forward mcr reaction under standard conditions is $-30\,\text{kJ mol}^{-1}$ (Scheller, et al. 2010). Therefore, the reverse mechanism is endergonic under standard conditions, but in high methane concentrations, the reaction can become endergonic. Net AOM can also be exergonic when coupled to an external electron acceptor such as sulfate, nitrate, or metal oxides (Timmers, et al. 2017). No studies show that methanogens can conserve energy from TMO, even under thermodynamically favorable conditions. Although more questions remain about these divergent *mcrA* variants' functions, it seems the most likely role of AOM is a mechanism to deal with back flux of individual enzymes of the methanogenic pathway (Holler, et al. 2011).

8.3.3.3.3 Three Pathways of Methanogenesis

Figure 8.5 shows the three pathways of biological methane production: 1) the hydrogenotrophic pathway, the 2) acetoclastic pathway, and the 3) methylotrophic pathway (Wolfe 1993) figure adapted from (Buan 2018).

8.3.3.3.3.1 Hydrogenotrophic Methanogens Hydrogenotrophic methanogens grow on H_2 and CO_2 to generate methane as a waste product. The hydrogenotrophic pathway is the most broadly distributed mechanism for methanogenesis. Due to its vast phylogenetic diversity, it is considered the most ancient of the three

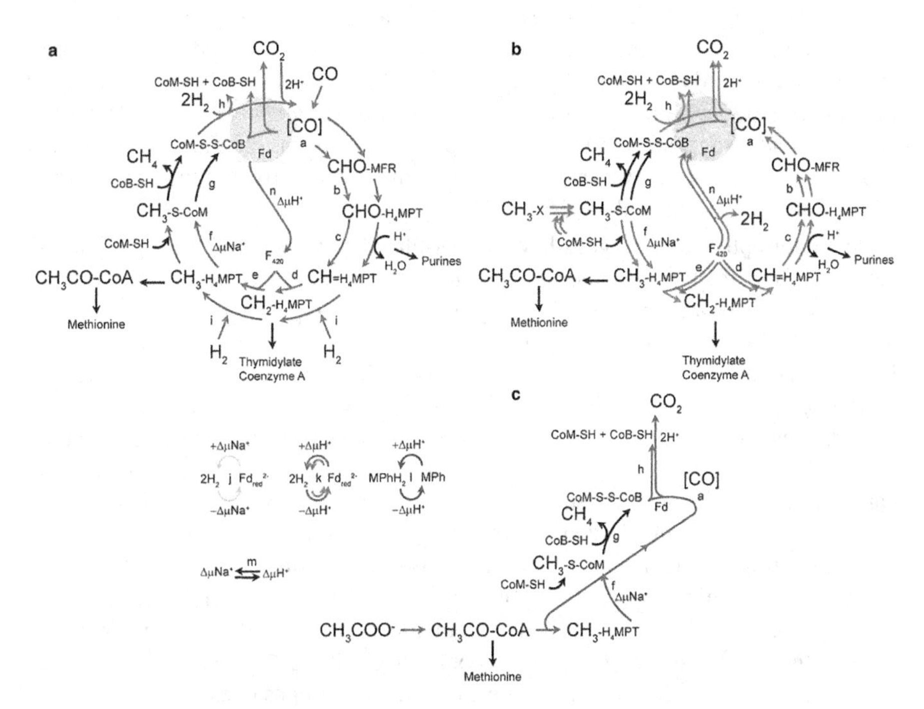

FIGURE 8.5 The Wolfe Cycle. Arrows represent direction of biochemical reactions. Black, reaction steps and directions common to all five methanogenesis pathways from C1 compounds or acetate. (**a**) Hydrogenotrophic and carboxydotrophic, methanogenesis pathways. Formic acid and primary or secondary alcohols are oxidized to CO_2 and hence methanogens that grow on these substrates use the hydrogenotrophic pathway. (**b**) Methyl respiration pathway and methylotrophic pathway. (**c**) Acetoclastic pathway. (From Buan 2018).

methanogenesis pathways (Bapteste, et al. 2005). Class I methanogens (Methanopyrales, Methanococcales, and Methanobacteriales) and most class II methanogens (*Methanomicrobiales, Methanocellales,* and *Methanosarcinales,* except for *Methanomassiliicoccales*) are hydrogenotrophs. They reduce CO_2 to CH_4 in six steps via the reductive acetyl-CoA or Wood–Ljungdahl pathway (WLP). To conserve energy, hydrogenotrophs couple the WLP to methanogenesis (Berghuis, et al. 2019). Hydrogenotrophic methanogens have evolved to grow by apparently "cheating' thermodynamics, in a sense. In the first step of the hydrogenotrophic pathway, CO_2 must be reduced by a low-potential Fe/S ferredoxin (−530 mV) to produce formyl-methanofuran in an "uphill" unfavorable reaction (Buan 2018).

Hydrogenotrophic methanogens are suspected to be directly involved in MIC because of their tendency to consume hydrogen; however, recent studies have shown it is not that simple (Lahme, et al. 2020).

8.3.3.3.3.2 Methylotrophic Methanogens Methylotrophic methanogenesis occurs when methylated substrates such as methanol, methylamines, or methylated sulfur compounds like methanethiol, dimethyl sulfide, or methylated thiols are reduced to methane. Previously, this pathway was thought to be limited to the orders *Methanomassiliicoccales*, *Methanobacteriales*, and *Methanosarcinales;* however, putative methylotrophic methanogenesis has recently been described in the proposed archaeal phylum *Verstraetearchaeota*, *Candidatus* Methanofastidiosa, *Candidatus Bathyarchaeota*, within members of the phylum *Korarchaeota*, and *Methanonatronarchaeia* suggesting that methylotrophic methanogens are much more diverse and widespread than previously thought (Nobu, et al. 2016, Söllinger and Urich 2019, Vanwonterghem et al., 2016).

Methylotrophic methanogens have been isolated from marine sediments, animal gastrointestinal tracts, wetlands, landfill/waste streams, and extreme environments. In marine sediments where sulfate-reducing bacteria outcompete hydrogenotrophic and acetoclastic methanogens and in ruminant animals, methylotrophic methanogens are likely among the significant methane producers (Söllinger and Urich 2019). Together, *Methanomassiliicoccales*, *Methanobacteriales*, and *Methanosarcinales* make up 44 % of Archaea found in our oilfield samples (Figure 8.7). It has been proposed that the high level of ammonium found in flow-back and produced waters from shale oil fields can be attributed to halophilic methylotrophic methanogens, such as *Methanohalophilus*, that grow on methylamines added to fracturing fluids (An, et al. 2019).

Methylotrophic methanogenesis starts with the activation of one-carbon compounds (methylamines, methanol, methylated sulfides, or methoxylated compounds) to methyl coenzyme M (CH_3-CoM). Methylotrophs are further broken into two groups, those that are obligately H_2 dependent and those that are H_2 independent (Vanwonterghem, et al. 2016). H_2 independent methylotrophs consist of members of *Methanosarcinales* that possess cytochromes. These cytochromes allow for the oxidation of methyl groups to CO_2 via a membrane-bound electron transport chain (Lang, et al. 2015). Within the H_2-independent pathway present in *Methanosarcinales*, about 75 % of the CH_3-CoM is reduced via mcr to produce CH_4, but about 25 % of CH3-CoM is oxidized to CO_2 via AOM (Timmers, et al. 2017). The oxidation pathway is required to provide reducing agents needed for the reduction of CH_3-CoM by mcr. The H_2-dependent methylotrophs include *Methanomassiliicoccales* and many of the recently proposed methylotrophs use H_2 as the electron donor for the reduction of methyl-groups.

8.3.3.3.3.3 Aceticlastic Methanogenesis Acetate provides a significant growth substrate for methanogens. Aceticlastic methanogenesis is the most recent addition to the methanogen repertoire of methane metabolism. Methane formation from acetate, called acetoclastic methanogenesis, can be found only in the order *Methanosarcinales* (Rotherman, et al. 2014). According to molecular phylogeny, global nickel deposition, data from volcanic activity, C isotope ratios, and the fossil record, it has been proposed that *Methanosarcina* likely acquired the ability to use acetate as a substrate from anaerobic *Clostridia* in an LGT event. Studies have shown that the high-efficiency pathway in *Methanosarcina* evolved via a single horizontal gene transfer event after the mid-Ordovician evolution of terrestrial plants. Cellulolytic bacteria from the Clostridia are the likely donors of the function. Considering its narrow phylogenetic distribution, it also appears to be a recent event within the evolution of methanogenesis (Fournier and Gogarten 2008).

The conversion of acetate to methane by acetoclastic methanogenesis begins by activating acetate to acetyl-CoA. Carbon monoxide dehydrogenase (CODH) then catalyzes acetyl-CoA's cleavage, after which, in common with the utilization of all methanogenic substrates, mcr catalyzes the reduction of a methyl group to methane. The CODH enzyme complex found in acetoclastic methanogens also contains a nickel cofactor (Ferry 1992).

The activation of acetyl-CoA within acetoclastic methanogens occurs via two different pathways in two distinct groups of organisms:

8.3.3.3.3.3.1 ACS Pathway *Methanosaetaceae* – Single Step Acetyl-CoA Synthase. Requires two ATP per acetate. Growth on acetate using the low-efficiency ACS pathway within Methanosaeta is thermodynamically possible because of poorly understood and enigmatic functions in their electron transport chain; this limitation may be responsible for their observed slow rate of growth (Welte and Deppenmeier, 2011).

8.3.3.3.3.3.2 AckA/Pta Pathway *Methanosarcina* – Two-step acetate kinase/phosphoacetyl transferase pathway requires 1 ATP per acetate.

8.3.3.4 Syntrophic Relationships

Syntrophy is the critical interdependency between producer and consumer (Dolfing 2014). Before it was understood that many methanogens are lithoautotrophs that use C1 compounds or acetate to grow, the addition of heterotrophic carbon sources enriched syntrophic bacterial/methanogen communities that were impossible to culture separately. Early methanogenic enrichments were considered "contaminated" and lost or discarded before an appreciation for coupled metabolism or an understanding of syntrophy. Syntrophy is a particular case of mutualistic metabolism between two or more organisms. In the strictest sense, a syntroph is an organism that cannot conserve energy on its own (Buan 2018). However, it requires a second organism, such as a methanogen, to maintain the concentrations of a secondary substrate (Schink and Stams 2006). The syntrophic relationship of hydrogen producers and hydrogen consumers is *called interspecies hydrogen transfer*. For example, methanogens can enhance the metabolism of fermenting bacteria by consuming hydrogen or acetate fermentation byproducts that eventually inhibit growth if concentrations are too high. When taken to the extreme, the removal of H_2 by hydrogenotrophic methanogens is vital for synthetic acetogenesis of higher organic acids. Without the coupling provided by the methanogen, the acetogenesis reaction is not exergonic (Nitschke and Russell 2012). Syntrophic communities of H_2-producing acetogenic bacteria and methanogens degrade fatty acids coupled to growth. However, neither the methanogens nor the acetogens alone can degrade these compounds. Intermicrobial distances between acetogens and methanogens can influence these microorganisms' specific growth rates (Batstone, et al. 2006).

Other mechanisms that facilitate syntrophy include interspecies formate transfer and direct interspecies electron transfer (DIET) (Stams and Plugge 2009).

8.3.3.4.1 Kind of a Common Ancestor

As further evidence of archaea's propensity to create synergistic relationships, in 2016, it was reported that *Lokiarchaeon* of the Asgard Group is hydrogen dependent. This is significant because such an organism was thought to exist as it satisfies the hydrogen hypotheses' phylogenetic gap. This ancient archaeon became embedded in the eukaryotic genome through symbiogenesis and may represent modern eukaryotic relative's closest prokaryotic relative. Such a symbiosis event would have been the origin of eukaryotic life (Sousa et al., 2016, López-García and Moreira, 2020).

8.4 CONSIDERATIONS FOR OIL AND GAS

8.4.1 Crude-Oil Biodegradation via Methanogenesis

Biodegradation of crude oil in subsurface petroleum reservoirs has adversely affected most of the world's oil, making recovery and refining of that oil more costly. The prevalent occurrence of biodegradation in shallow subsurface petroleum reservoirs has been attributed to aerobic bacterial hydrocarbon degradation stimulated by surface recharge of oxygen-bearing meteoric waters. However, a growing body of work suggests that methanogen-driven anaerobic biodegradation is altering hydrocarbon deposits (Head, et al. 2003). Hydrogenotrophic methanogens preferentially remove *n*-alkanes and generate near stoichiometric amounts of methane to show (Jones, et al. 2008). This biogenic conversion to heavy oil has certain advantages as a method of generating producible energy from unrecoverable oil (Strąpoć 2017).

8.4.2 Methanogen Mediated Iron Corrosion

Methanogens have been implicated in corrosion failures since their initial discovery. In 1987, Dainels observed that methanogenesis and cell division were capable with CO_2 when the sole electron donor was iron (Fe^0). Iron is an inexpensive metal and is widely used in many industrial processes and industrial/commercial products. When iron contacts an aqueous electrolyte, it readily corrodes. Metallurgical and environmental heterogeneities are not evenly distributed across the metal's surface, and, consequently, the electric potential is also unevenly distributed (Uchiyama, et al. 2010). Therefore, electrons flow within the metal from higher electrical potential (the anode) to an area of lower electrical potential (the cathode). At the anode, iron atoms lose electrons and dissolve into ferrous ions (Fe^{2+}), whereas cations or elements dissolved in solution, H^+ under anaerobic conditions or O_2 under aerobic conditions, are reduced by electrons at the cathode.

Sulfate-reducing bacteria (SRB) are considered the main causal agents of MIC under anaerobic conditions through the formation of hydrogen sulfide, which is a very corrosive compound. MIC induced by chemical reactions with biogenic compounds such as hydrogen sulfide is referred to as "chemical microbiologically influenced corrosion" or cMIC.

The much-criticized cathodic depolarization theory assumes that the anaerobic oxidation of Fe^0 ($Fe^0 \rightleftarrows Fe^{2+} + 2e^-$; anodic reaction) coupled with the reduction of H^+ ($2H^+ + 2e^- \rightleftarrows H_2$; cathodic reaction) is limited by the diffusion of H_2 from the cathode and that hydrogenotrophic microorganisms including SRB and methanogens accelerate the rate-limiting cathodic reaction by consuming H_2 built up around Fe^0 surfaces. However, this theory's validity has been thoroughly reconsidered, as not all hydrogenotrophic SRB and methanogens promoted Fe^0 corrosion. Currently, there are two mechanisms for methanogen MIC.

8.4.2.1 Syntrophic Relationships

Methanogens play a crucial role in the breakdown of organic matter in anaerobic environments. MIC induced by the metal surface's chemical reactions with biogenic compounds such as hydrogen sulfide, acetate, and other organic acids that cause corrosion is known as cMIC.

8.4.2.2 Direct Iron Oxidation

Metabolic activities of certain microorganisms catalyze the oxidation of metallic iron (Fe^0) through a process called electrochemical microbiologically influenced corrosion (eMIC). This mechanism has been experimentally verified, but direct genetic markers have not been identified in SRB (Enning and Garrelfs 2014). Recent reports have shown SRB can cause severe corrosion rates as high as 0.71mm/yr (Enning, et al. 2012). Using a multiport flow column, An, *et al.* showed that under flow conditions, the average corrosion rates of *Methanobacterium IM1* ranged between 0.2 and 0.3 mm/yr, with maximum corrosion rates above 0.5 mm/yr, which is roughly 25 times higher than previously reported methanogenic-corrosion rates and much higher than the *D. ferrophilus IS5* control in the study (Annie An et al., 2020).

8.4.2.3 Discovery of MIC Island and a Binary MIC Factor

The direct oxidation of metallic iron by microbes has only recently been proven (Hai-Yan et al., 2004) and accepted by oil and gas operators and researchers as a predominant issue (Enning and Garrelfs 2014). Studies demonstrated that the exception of corrosion rates produced by some bacteria are strain specific in SRB.

In particular, *Desulfovibrio ferrophilus* IS4 and *Desulfopila corrodens* IS5 caused corrosion rates as high as 0.31 mm/year and 0.71 mm/year, where Fe^0 was the only electron donor available. At the same time, the hydrogenotrophic control *Desulfopila inferna* managed only a 0.01 mm/year corrosion rate (Enning and Garrelfs 2014). Similarly, APB studies have demonstrated a similar phenomenon for acetogenesis-dependent MIC, particularly with *Clostridium* (Mand, et al. 2014, Kato, et al. 2015). These, along with other research, have debunked the idea that any hydrogen consuming organism could induce a cathodic depolarization and directly accelerate corrosion (Tori, et al. 2010, Enning and Garrelfs 2014). However, a direct mechanism has yet to be discovered. Like SRB and APB, it has been shown that methanogens can generate methane utilizing metallic iron as a sole source of electrons (Dainels 1987) but was initially attributed to the classic model of cathodic depolarization.

Acetogens are often associated with methanogens and methanogenesis. MIC in non-sulfidic environments is often linked to the presence of a community of acetogens such as *Clostridium* and methanogens such as *Methanosarcinales* (Mand, et al. 2014).

8.4.2.3.1 MIC Hydrogenase

Recent studies have demonstrated that cell-free filtrates, *Methanococcus maripaludis* MM901, accelerated H_2 generation from Fe^0, while filtrates of spent cultures of a live hydrogenase-knockout did not induce corrosion (Deutzmann, et al. 2015). It was found that the filtrates were thermolabile and proteinase sensitive, which indicated that the mechanism, which generated a corrosion rate of greater than 0.15mm Fe^0 yr^{-1}, was likely secreted extracellular hydrogenases that were capable of catalyzing the cathodic removal of hydrogen ions (Deutzmann, et al. 2015). Subsequent work by Tsurumaru et al. (2018) showed genetic evidence of an unstable genetic island in *Methanococcus maripaludis* that encodes an [NiFe] extracellular hydrogenase and a secretion system that allows *Methanococcus* to draw electrons from Fe^0, generating H_2: $2H^+ + Fe^0 \rightleftarrows H_2 + Fe^{2+}$ (Tsurumaru, et al. 2018), Figure 8.6. The gene responsible for is *micH*. With the sequence of *micH* in hand, bioinformatical studies on *Methanococcus maripaludis* MM901 and *Methanobacterium congolense—Buetzberg* revealed sequence homology with some exceptions in limited regions, which indicates that this *MIC island* and its orthologs have spread beyond the genus level (Tsurumaru, et al. 2018).

In 2020, Lahme & Enning published work applying this understanding of *micH* to a metagenomic analysis of observed pipeline corrosion. A simple qPCR assay was developed to test for the presence and abundance of the large subunit of the *micH* gene in nine separate petroleum pipeline systems. Their study shows that micH is found on at least four continents in its wild type. The *micH* assay characterized biofilms that caused technically significant corrosion (> 0.13 mm Fe^0 yr^{-1}) from those that did not (< 0.08 mm Fe^0 yr^{-1}) in experiments with both produced and synthetic water samples. For the first time in the history of oil and gas industries, a biomarker was identified and deployed to serve as a binary genetic marker for technically relevant corrosion. Additionally, the study also included the full genome characterization of oilfield samples in West Africa. They found a near-identical version of *micH* on a separate continent from the other isolates found to have *micH* orthologs.

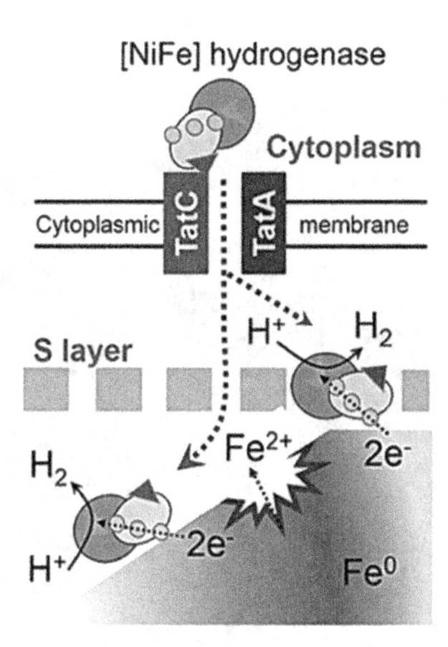

FIGURE 8.6 [NiFe] Hydrogenase and secretion system in *Methanococcus maripaludis* (from Tsurumaru et al., 2018).

This indicates that this mechanism is not only widespread but also highly conserved (Lahme, et al. 2020).

8.4.2.4 Archaeoglobus

Archaeoglobi is a class of thermophilic Euryarchaeota that is abundant in subsurface hydrothermal environments, where they likely play a role in carbon and nutrient cycling (Brileya 2014). The Archaeoglobi are split into three genera: 1) *Archaeoglobus*, which are all heterotrophic or chemolithotrophic sulfate reducers. 2) *Geoglobus* and 3) *Ferroglobus*, which both reduce nitrate and ferric iron. Studies have shown that some Archaeoglobus can oxidize alkanes. Archaeoglobus are distant cousins of methanogens and were once capable of biomethanogensis, but there are no known *Archaeoglobi* that encode the Mcr complex (Boyd, et al. 2019).

Archaeoglobus are of particular interest because they are commonly implicated in corrosion failures in high-temperature oil production systems. They can grow lithotrophically from elemental iron and form chimney-like structures of iron carbonate (siderite). Interestingly, *Archaeoglobus fulgidus* doesn't consume sulfate when iron is the electron donor (Amin Ali et al., 2020). Archaeoglobales represent 12 % of the total archaea identified in our database (Figure 8.7).

8.4.3 Mitigation of Methanogens

Studies of biocide efficacy against methanogens are scarce; however, (Okoro, et al. 2016) found that THPS concentrations less than 0.5% (5,000 ppm) did not effectively inhibit methane production and corrosion rates from a perceived methanogen infested production system. The impact on methanogen-mediated corrosion and methane

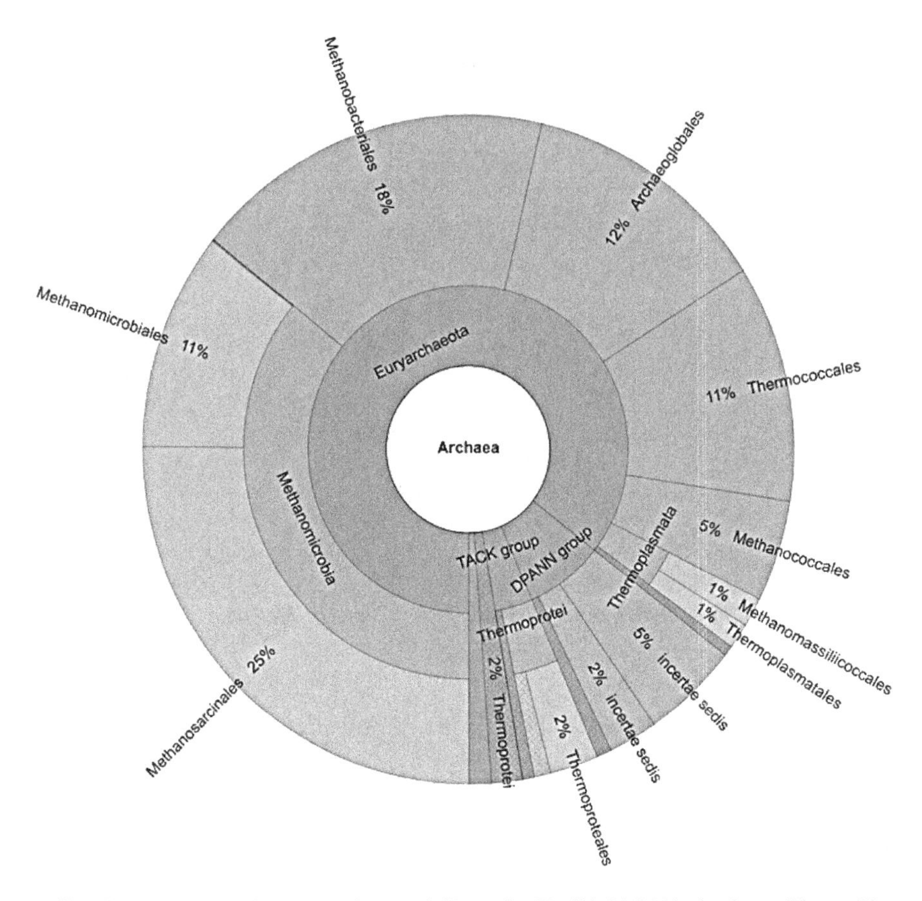

FIGURE 8.7 Krona plot metanalyses of ChampionX oilfield DNA database (Krona Plot, Ondov et al., 2011).

production by THPS became more apparent when the THPS concentration further increased, and methane production became negligible with 1 % THPS. Subsequent studies seemed to suggest that methanogens may be intrinsically more tolerant of THPS or may have a mechanism to develop resistance (Okoro and Lin 2017). This theory is interesting because there are still many underlying mechanisms we do not know about archaeal genetics and regulation.

8.5 METHANOGEN OILFIELD CORROSION EXAMPLES

As qPCR and next-generation sequencing has become more prevalent as a diagnostic tool in the oilfield, the importance of identifying Archaeal populations has grown. The development of standardized universal primers has now allowed research to be more uniform in its bias.

Our lab has processed almost 15,000 samples from oil and gas production facilities—the Krona plot in Figure 8.7 details the taxonomical breakdown of identified Archaea (Ondov, et al. 2011). We found genetic evidence from every established phylum of Archaea except the Asgard group. Euryarchaeota represented 86 % of archaea identified. The DPANN group represented 8 %, followed by the TACK group at 4 %, Thermoprotei at 2 %, and leaving 1 % as unclassified.

We found that methanogens were present in 31 % of oilfield samples. The three most abundant genera of methanogens were *Methanolobus sp., Methanohalophilus sp.*, and *Methanothermobacter sp.* These genera comprised ~46 % of samples containing methanogens. Given the implications from the before mentioned studies on *micH*, *Methanolobus* and *Methanohalophilus* are probably not indicators alone of highly corrosive populations. However, they are also present in samples associated with failures. Although methanogens lacking in a specialized extracellular hydrogenase may also contribute to MIC in less obvious ways.

Further assessment of 22 failure analysis samples, consisting of solids or swabs, found that all but one contained methanogens. Figure 8.8 shows the percent relative abundance of each of the samples highlighting the presented Archaeal classes. Methanogens represented more than 50 % of the population in six samples.

8.5.1 Dry Gas Pipeline

In this example of MIC in a dry gas pipeline, a 24-inch pipe failed at the 9 O'clock position after 10-year service life. Previous monitoring for MIC involved serial dilution at water drops. Historical testing did not show an elevated risk of MIC as the KPI of less than two SRB and/or APB turns was permissible. When the pipeline depressurized, the damaged section was immediately isolated from the service and shipped for failure analysis. When the pipeline arrived, solids were removed from the primary failure area, and the second large put approx. 10cm away. Not surprisingly, a large population of SRB, APB, and methanogens was identified. Figure 8.9 shows the failure (**A**) 11cm above another large corrosion pit. qPCR and DNA sequencing were run on the debris mechanically removed from each location. Microbes identified at the failure site consisted primarily of bacteria. The SRB, *Thermoanaerobacter* (53 %),

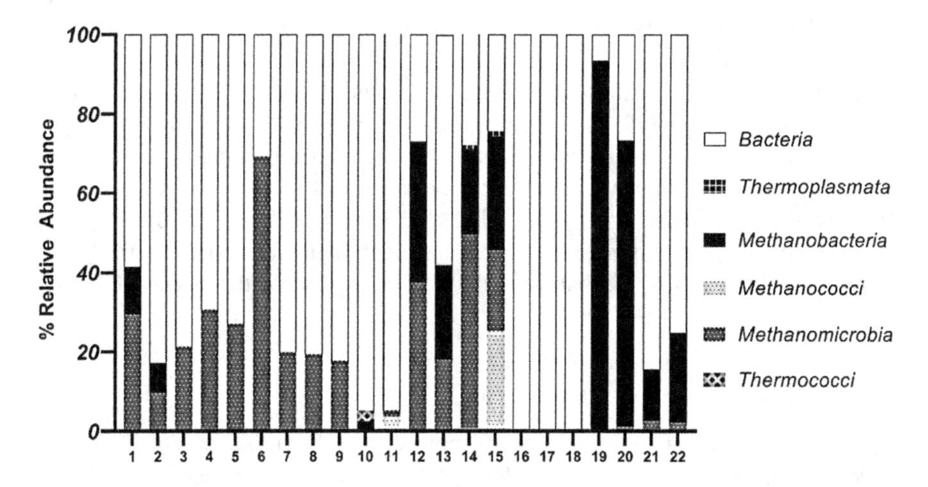

FIGURE 8.8 Relative abundance of methanogens in oilfield failure samples (original).

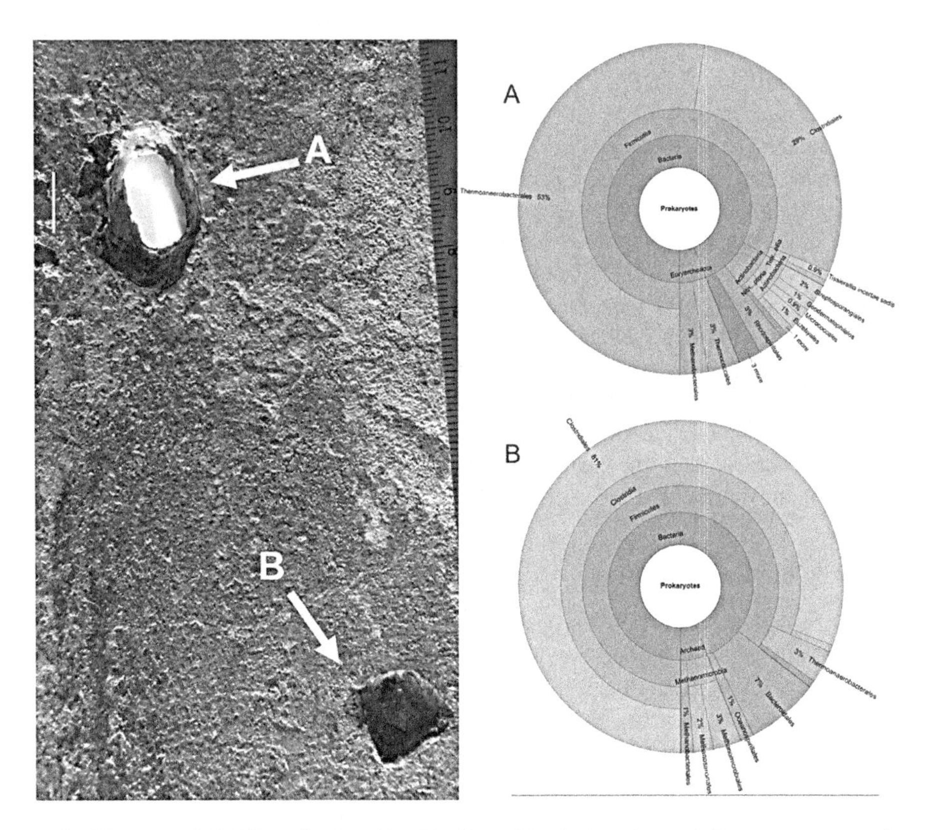

FIGURE 8.9 MIC Failure in Dry Gas Pipeline—North America and Krona metagenomic analysis (Krona Plot, Ondov et al., 2011) (photograph by TJ Tidwell).

was the most abundant organism, followed by the acid-producing *Clostridiales*, *Alkalibater* (15 %), and *Acetobacterium* (12 %). Methanogens included *Thermococcus* and *Methanothermobacter* species and were present in a small proportion (6 %) of the total population in the primary failure.

In the large pit (**B**), a similar population was found consisting of a higher proportion of *Acetobacterium* (79 %) and a much lower proportion of *Thermoanaerobacter* (3 %). The population of methanogens included the before-mentioned *Methanothermobacter* (1 %); however, *Thermococcus* was not found, and members of the *Methanomicrobia*, *Methanocalculus* (3 %), *Methanohalophilus* (1 %), and *Methanococcoides* (0.7 %) were found. Presumably, a similar population existed in the failure pit but were likely mostly discharged with the depressurization that occurred through the opening.

8.5.2 West Texas Shale Production

Figure 8.10 shows corrosion from failed tubing strings. The primary organisms identified in the corrosion failures were *Methanothermococcusx*, and various SRB, such as *Desulfohalbium*, *Thermodesulforhabdus*, and fermentative *Halanaerobium*. These

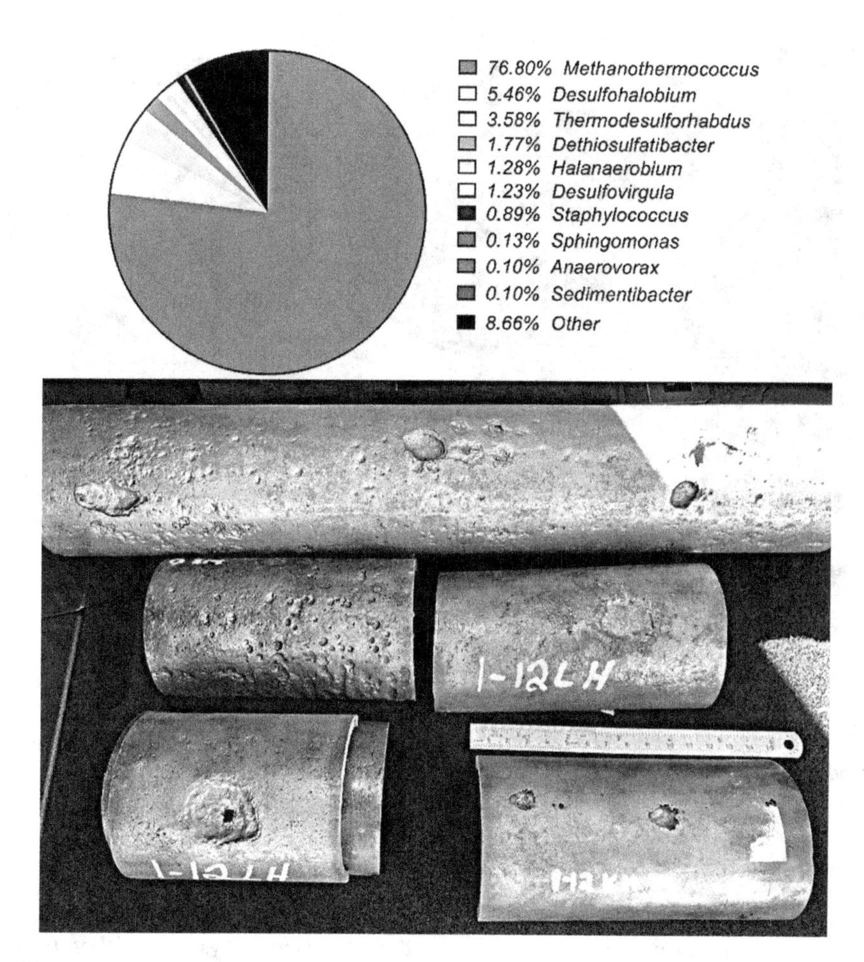

FIGURE 8.10 West Texas Tubing Failures and metagenomic relative abundance (photograph by TJ Tidwell).

failures continued until aggressive biocide treatments were implemented on a weekly basis until planktonic and sessile microbial counts fell below 50,000/mL (or cm^2).

8.5.3 Biofilm Reactor Coupons – Deepwater Brazil

In this example of methanogen-associated MIC, produced water samples from an offshore production FPSO off the coast of Brazil were cultivated in the lab to determine the efficacy of biocides against the corrosive biofilms. Figure 8.11 shows carbon steel coupons after 90 days of incubation in a biofilm reactor simulating field conditions. Macroscopic pitting was observed along with a diverse population of SRB, methanogens, and APB were identified. The SRB *Pelobacter* and *Desulfovibrio* made up almost 40 % of the community, followed by *Methanocalculus* at 10 %. The maximum pit depth was measured at 321 μm.

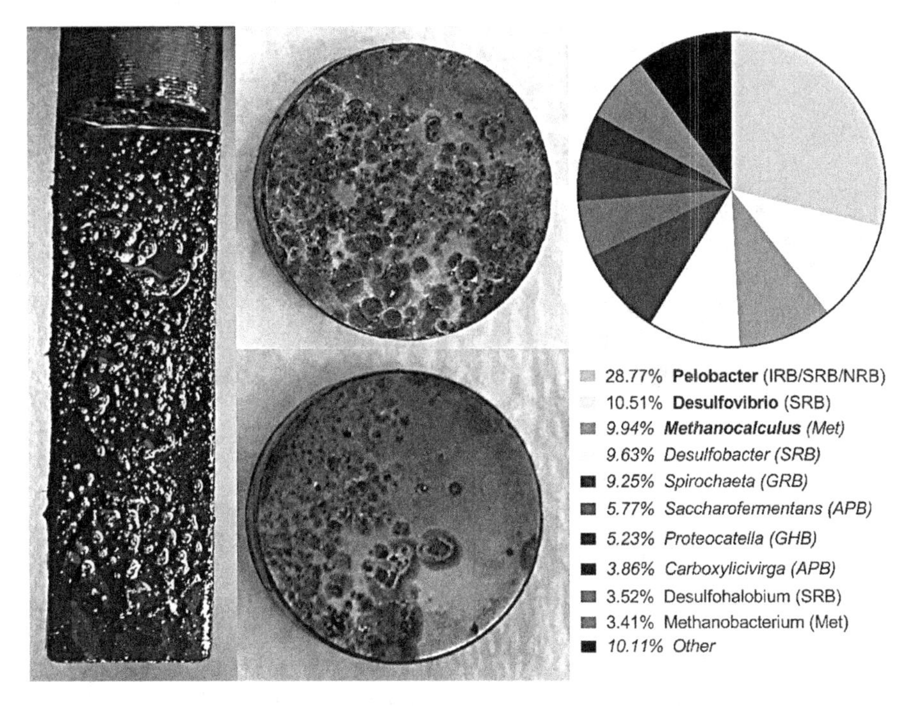

FIGURE 8.11 1018 CS coupons from biofilm reactor with produced water from deep-water production—South America (photograph by TJ Tidwell).

8.5.4 West Texas Salt Water Disposal Facility

In this study (Tidwell, et al. 2017), a highly corrosive (0.5mm/yr), methanogen-dominated biofilm was cultivated from a saltwater disposal system in West Texas. Figure 8.12 shows white-light-interferometry (WLI) scans and of two adjacent coupons. Despite the coupons coming from the same coupon holder and the near-identical microbial community, the pitting morphology was dramatically different. In practice, you often hear numerous references to "SRB or APB pitting" by engineers. This example of MIC is evidence that you cannot distinguish or diagnose a MIC mechanism from pit morphology alone.

8.5.5 South America Pipeline

In this example, an oil pipeline running through a sensitive tropical area was internally scanned (ILI) after a pinhole leak was discovered. These scans revealed 75 % wall loss throughout most of the pipeline in mostly the 6 o'clock position. The producer's asset management team had run numerous corrosion simulation models and had not come close to predicting the corrosion rates necessary for the corrosion observed. This operator had a minimal microbial monitoring program and had never observed more than a two-bottle turn of standard SRB media, indicating a low risk. However, a field visit and a day of testing ATP on pig debris showed at least

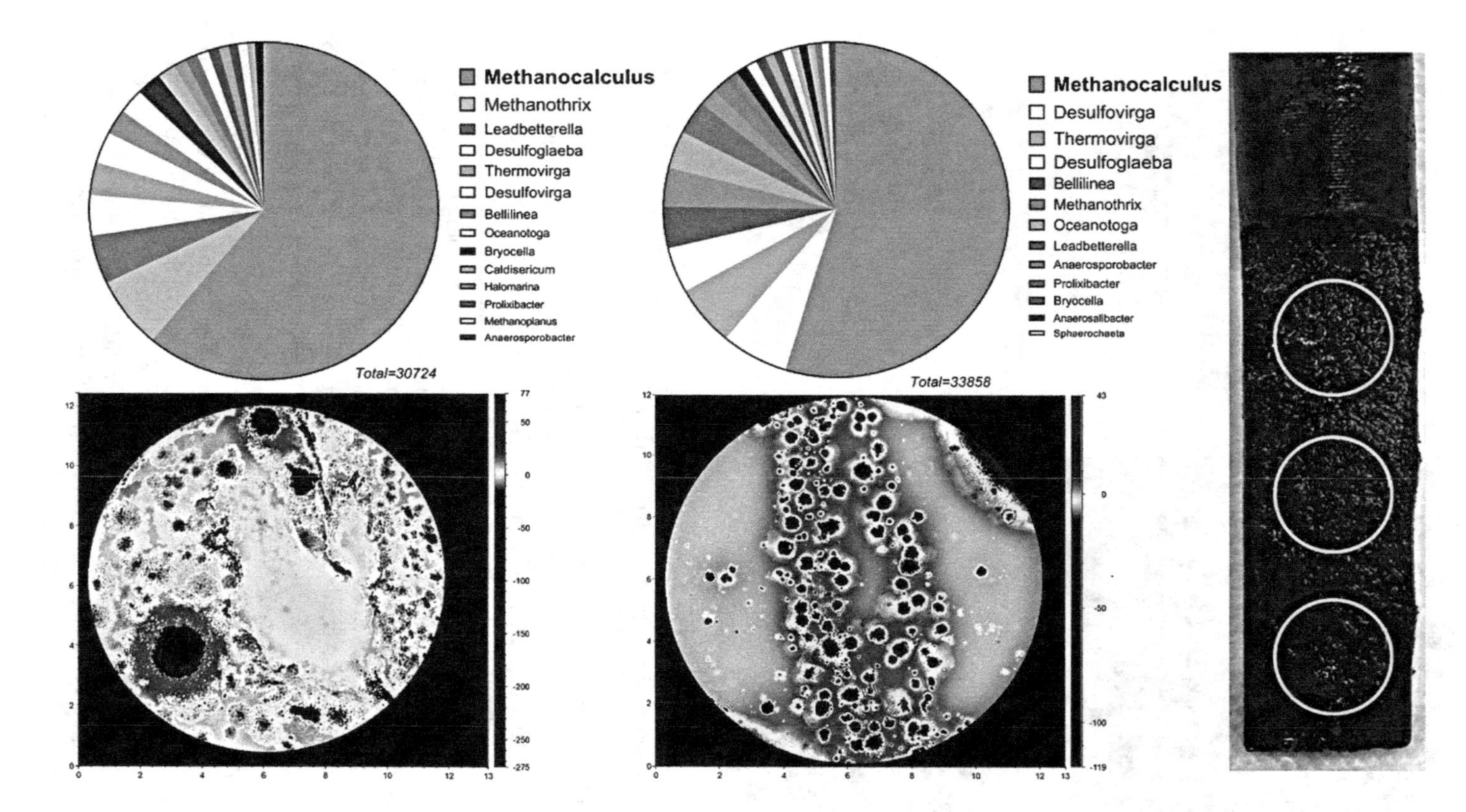

FIGURE 8.12 1018 CS coupons from biofilm reactor with produced water from a west Texas Saltwater disposal (photograph by TJ Tidwell).

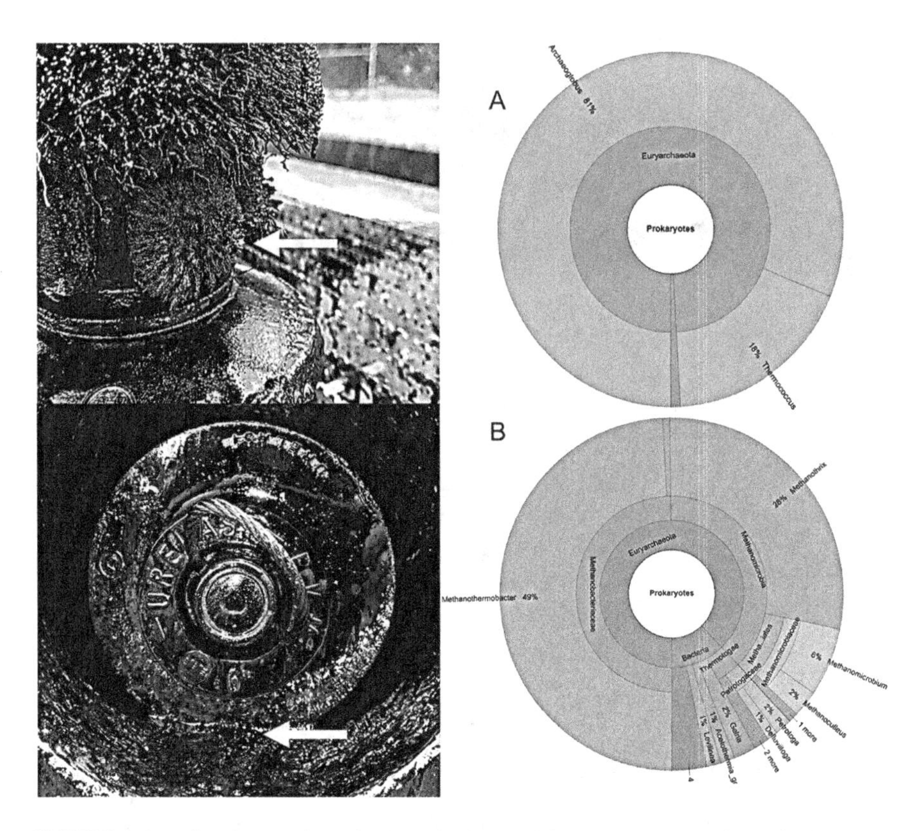

FIGURE 8.13 South America pipeline pigging and Krona metagenomic analysis (Krona Plot, Ondov et al., 2011) (photograph by TJ Tidwell).

2.5 million active microbes per gram of debris. The pig used was fitted with wire brushes to scour the pipe surface and strong rare-earth magnets to collect metallic debris for sampling. DNA sequencing was performed on two samples from the pig. Figure 8.13 shows the location of each sample and the Krona plot of the microbial community. This example shows many things, including some exciting aspects of collecting samples from pipelines. The debris' magnetic metallic components showed a completely different population than the oily debris ahead of the pig.

Sample A was taken from the pig's magnets and contained almost entirely the sulfate-reducing and methanogenic *Archaeoglobus* (81 %), as well as the SRB *Thermococcus* (18 %). The remaining 1% was composed of bacteria from the *Thermoanaerobacteraceae family*.

Sample B was taken from the slimy, oily debris in front of the front polyurethane disk. Interestingly, the population was almost 100% different. DNA sequencing showed that 12% of the total population was composed of thermophilic bacteria from the *Thermotogae* (4 %), as well as *Geiria* (2 %), *Acetothermia* (1 %), *Levilinea* (1 %), and less than 1 % of *Parvibacter*, *Thermovirga*, *Smithella*, and *Hydrogenothermus*. The remaining 88 % were members of the Euryarchaeota *Methanothermobacter*

(49 %), *Methanothrix* (28 %), *Methanomicrobium* (6 %), and less than 1% of *Methanolinea* and *Methanobacterium*.

In this case, the operator had little idea there were billions of Achaea chewing away at their pipeline.

Given this insight, it was easy for the operator to adopt an aggressive remediation approach and expand its monitoring program.

8.6 CONCLUSION

Archaea have historically been overlooked as a problem or a corrosive threat in oil and gas production. However, the last decade of research has propelled archaea and methanogens to the forefront of oil and gas MIC culprits. Despite the decades of work on SRB, methanogens possess the only fully understood corrosion mechanism for direct Fe^0 oxidation. The question remains if this is one of or the single explanation for methanogen corrosion. Considering Archaea's diversity and how little we know of all their novel methods to survive the past 3+ billion years, its likely many more direct or indirect corrosion mechanisms will be found.

One crucial perspective the industrially focused researcher or engineer can take from the medical field is that methanogens seem to indirectly participate in pathogenicity, supporting other microbes' growth, which may be directly involved in pathogenesis. This indirect role should not be minimized. Much is needed to elucidate the syntrophic associations essential for a healthy relation among microbes (including methanogens) and the host organism and uncover those associations leading to a disease state (Macario, et al. 2009).

This chapter provides a general overview of what we know about methanogens and discusses specific examples where methanogens were suspected in the MIC mechanism. Indeed, the oil and gas industry must fully appreciate the role Archaea and methanogens play in corrosion and other bioprocesses in oil reservoirs. Advances in metagenomics, transcriptomics, and proteomics have produced Archaea's valuable metabolic blueprints with fresh hypotheses on how metabolism, electron transport, and energy conservation may function. Despite many recent advances and discoveries, we find out there is much more to unpack in this enigmatic and under the appreciated domain of life.

REFERENCES

Adam, P.S., G. Borrel, C. Brochier-Armanet, and S. Gribaldo. 2017. "The growing tree of archaea: new perspectives on their diversity, evolution and ecology." *ISME J* 2407–2425.

Allers, T., and M. Mevarech. 2005. "Archaeal Genetics - The Third Way." *Nat. Rev. Gen.*

Amin Ali, O., E. Aragon, A. Fahs, S. Davidson, B. Ollivier, and A. Hirschler-Rea. 2020. "Iron corrosion induced by the hyperthermophilic sulfate-reducing archaeon Archaeoglobus fulgidus at 70°C." *Int. Biodeterior. Biodegrad.*

An, B.A., Shen, Y., Voordouw, J., Voordouw, G. 2019. "Halophilic methylotrophic methanogens may contribute to the high ammonium concentrations found in shale oil and shale gas reservoirs." *Frontiers Energy Res.*

Annie An, B., S. Kleinbub, O. Ozcan, and A. Koerdt. 2020. "Iron to gas: versatile multiport flow-column revealed extremely high corrosion potential by methanogen-induced microbiologically influenced corrosion (Mi-MIC)." *Fountiers Microbiol.*

Bapteste, E., C. Brochier, and Y. Boucher 2005. "Higher-level classification of the Archaea: evolution of methanogenesis and methanogens." *Archaea* 353–363.

Berghuis, B.A., F.B. Yu, F. Schulz, P.C. Blainey, T. Woyke, and S.R. Quake. 2019. "Hydrogenotrophic methanogenesis in archaeal phylum Verstraetearchaeota reveals the shared ancestry of all methanogens." *Proceedings of the National Academy of Sciences*, 5037–5044.

Bhattad, Ujwal H. 2012. "Preservation of methanogenic cultures to enhance anaerobic digestion." *Dissertations.* https://epublications.marquette.edu/dissertations_mu/209?utm_source=epublications.marquette.edu%2Fdissertations_mu%2F209&utm_medium=PDF&utm_campaign=PDFCoverPages.

Boucher, Y. et al. 2003. "Lateral gene transfer and the origins of prokaryotic groups." *Annu. Rev. Genet.* 283–328.

Bouwer, E. J., and P.L. McCarty. 1984. "Modeling of trace organics biotransformation in the subsurface." *Groundwater.*

Boyd, J.A., S.P. Jungbluth, A.O. Leu. 2019. "Divergent methyl-coenzyme M reductase genes in a deep-subseafloor Archaeoglobi." *ISME J* 1269–1279.

Bräuer, S.L., H. Cadillo-Quiroz, E. Yashiro, J.B. Yavitt, and S.H. Zinder. 2006. "Isolation of a novel acidiphilic methanogen from an acidic peat bog." *Nature* 192–194.

Brileya. 2014. "The class archaeoglobi." In *The Prokayrotes*, edited by Rosenberg, 15–23. Berlin: Springer.

Brown, J.R., and W.F. Doolittle. 1997. "Archaea and the prokaryote-to-eukaryote transistion." *Microbiol Mol. Biol Rev.* 456–502.

Buan, Nicole. 2018. "Methanogens: pushing the boundaries of biology." *Emerg. Top Life Sci.* 629–646.

Canfield, D.E. 2006. "Gas with an ancient history." *Nature* 426–427.

Costello, J.A. 1974. "Cathodic depolarization by sulphate-reducing bacteria." *S. Afr. J. Sci.* 202.

Dainels, L. 1987. "Bacterial Methanogenesis and Growth from CO2 with elemental iron as the sole Source of Electrons." *Science.*

Deppenmeier, U., et al. 2002. "The genome of Methanosarcina mazei: evidence for lateral gene transfer between bacteria and archaea." *J. Mol. Microbiol. Biotechnol.*

Deutzmann, J.S., M. Sahin, and A.M. Spormann. 2015. "Extracellular enzymes facilitate electron uptake in biocorrosion and bioelectrosynthesis." *mBio.*

Dlugokencky, E.J., E.G. Nisbet, R. Fisher and D. Lowry. 2011. "Global atmospheric methane: budget, changes and dangers." *Philos. Trans. A Math. Phys. Eng. Sci.* 2058–2072.

Dolfing, J. 2014. "Syntrophy in microbial fuel cells." *ISME J.*

Doolittle, W.F. 2003. "How big is the iceberg of which organellar genes in nuclear genomes are but the tip?" *Philos. Trans. R. Soc. Lond. B* 57–58.

Empadhinhas, N., and M.S. da Costa. 2008. "Osmoadaptation mechanisms in prokaryotes: distribution of compatible solutes." *Int. Microbiol.* 151–161.

Enning, D., H. Venzlaff, J. Garrelfs, H.T. Dinh, V. Meyer, and K. Mayrhofer. 2012. "Marine sulfate-reducing bacteria cause serious corrosion of iron under electroconductive biogenic mineral crust." *Environ. Microbiol.* 1772–1787.

Enning, D., and J. Garrelfs 2014. "Corrosion of iron by sulfate-reducing bacteria: new views of an old problem." *Appl. Environ. Microbiol.*

Enzmann, F., F. Mayer, M. Rother, and D. Holtmann. 2018. "Methanogens: biochemical background and biotechnological applications." *AMB Express (AMB Express.).*

Evans, P.N., et al. 2019. "An evolving view of methane metabolism in the Archaea." *Nature Rev Microbiol* 219–232.

Ferry, J.G. 1992. "Methane from acetate." *J. Bacteriol.* 5489–5495.

Fournier, G.P. 2008. "Evolution of acetoclastic methanogenesis in Methanosarcina via horizontal gene transfer from cellulolytic Clostridia." *J. Bacteriol.* 1124–1127.

Fournier, G.P., and J.P. Gogarten. 2008. "Evolution of acetoclastic methanogenesis in Methanosarcina via horizontal gene transfer from cellulolytic Clostridia." *J. Bacteriol.* 1124–1127.

Gaines, H.A. 1910. "Bacterial activity as a corrosion induced in the soil." *J. Ind. Eng. Chem.* 128–130.

Garrett, J.H., M.D. London, and H.K. Lewis. 1891. "The action of water on Lead. Being an Inquiry into the cause and mode of the action and its prevention." *Crookes, S.W.*

Hai-Yan, Tang, Dawn E. Holmes, Toshiyuki Ueki, Paola A. Palacios, and Derek R. Lovley. 2004. "Iron corrosion via direct metal-microbe electron transfer." *Appl. Environ. Sci.*

Head, I.M., M.D. Jones, and S.R. Lartner. 2003. "Biological activity in the deep subsurface and the origin of heavy oil." *Nature* 344–352.

Holler, T., G. Wegener, and H. Niemman. 2011. "Carbon and sulfur back flux during anaerobic microbial oxidation of methane and coupled sulfate reduction." *Proceedings of the National Academy of Sciences of the USA*, 1484.

Hug, L.A., B.J. Baker, K. Anantharaman, and C. Brown. 2016. "A new view of the tree of life." *Nat Microbiol* 16048.

Hungate. 1969. "Chapter IV: a roll tube method for cultivation of strict anaerobes." In *Methods in Microbiolgy*, edited by Hungate, 117–132. London: Elsevier.

Hungate. 1950. "The anaerobic mesophilic cellulolytic bacteria." *Bacteriol. Rev.*

Jones, D.M., et al. 2008. "Crude-oil biodegradation via methanogenesis in subsurface petroleum reservoirs." *Nature* 176–180.

Kato, S, I. Yumoto, and Y. Kamagata. 2015. "Isolation of acetogenic bacteria that induce biocorrosion by utilizing metallic iron as the sole electron donor." *Appl. Environ. Microbiol.* 67–73.

Kennedy, S. P., W. V. Ng, S. L. Salzberg, L. Hood & S. DasSarma 2001. "Understanding the adaptation of Halobacterium species NRC-1 to its extreme environment through computational analysis of its genome sequence." *Genome Res.* 1641–1650.

Koonin, E.V., K.S. Makarova, and L. Aravind 2001. "Horizontal gene transfer in prokaryotesL quantification and classification." *Annu. Rev. Mircobiol.* 709–742.

Lahme, S., J. Mand, J. Longwell, R. Smith, and D. Enning. 2020. "Severe corrosion of carbon steel in oil field produced water can be linked to methanogenic archaea containing a special type of [NiFe] hydrogenase." *bioRxi.*

Lang, K., et al. 2015. "New mode of energy metabolism in the seventh order of methangens as recealed by comparative genome analysis of 'Cadidatus Methanoplasma termitum'." *Appl. Environ. Microbiol.* 1338–1352.

Li, J, C.F. Wong, M.T. Wong, H. Huang, and F.C. Leung 2014. "Modularized evolution in archaeal methanogens phylogenetic forest." *Genome Biol. Evol.* 3344–3359.

López-García, P., and D. Moreira 2020. "The syntrophy hypothesis for the origin of eukaryotes revisited." *Nat. Microbiol.*

Lovey, D.R. 2003. "Extending the upper temperature limit for Life." *Science* 934.

Lurie-Weinberger, M.N., and U. Gophna. 2015. "Archaea in and on the human body: health implications and future directions." *PLoS Pathog.*

Macario, E., Conway de Macario, and A.J. Macario. 2009. "Methanogenic archaea in health and disease: a novel paradigm of microbial pathogenesis." *Int. J. Med. Microbiol.* 99–108.

Mand, J., H.S. Park, T.R. Jack, and G. Voordouw 2014. "The role of acetogens in microbially influenced corrosion of steel." *Front Microbiol.*

Meyerdierks, O. Glöckner, R. Amann, M. Krüger, F. Widdle, R. Böcher, R.K. Thauer, and S. Shima. 2003. "A conspicuous nickel protein in microbial mats that oxidize methane anaerobically." *Nature* 878–881.

Nelson-Sathi, S., et al. 2015. "Origins of major archaeal clades correspond to gene acquisitions from bacteria." *Nature* 77–80.

Nitschke, W., and M.J. Russell. 2012. "Redox bifurcations: mechanisms and importance to life now, and at its origin: a widespread means of energy conversion in biology unfolds…." *Bioessays.*

Nobu, M.K., T. Narihiro, K. Kuroda, R. Mei, and W. Liu. 2016. "Chasing the elusive Euryarchaeota class WSA2: genomes reveal a uniquely fastidious methyl-reducing methanogen." *ISME J.* 2478–2487.

Okoro, C., and J. Lin. 2017. "Persistence of halophylic methanogens and oil-degrading bacteria in an offshore oil-producing facility." *Geomicrobiology.*

Okoro, Chuma Conlette, Samuel, Olusegun, and Johnson Lin. 2016. "The effects of Tetrakis-hydroxymethyl phosphonium sulfate (THPS), nitrite and sodium chloride on methanogenesis and corrosion rates by methanogen populations of corroded pipelines." *Corros. Sci.*

Ondov, B.D., N.H. Bergman, and A.M. Phillippy. 2011. "Interactive metagenomic visualization in a Web browser. BMC Bioinformatics." *BMC Bioinf.*

Oren, A. 2008. "Microbial life at high salt concentrations: phylogenetic and metabolic diversity." *Saline Syst.*

Pflüger, W., and V. Müller 2003. "Lysine-2,3-aminomutase and beta-lysine acetyltransferase genes of methanogenic archaea are salt induced and are essential for the biosynthesis of Nepsilon-acetyl-beta-lysine and growth at high salinity." *Appl. Environ. Microbiol.* 6047–6055.

Postgate, J.R. 1963. "Versatile medium for the enumeration of sulfate-reducing bacteria." *ASM.* https://aem.asm.org/content/aem/11/3/265.full.pdf.

Rivera, M.C., R. Jain, J.E. Moore, and J.A. Lake. 1998. "Genomic evidence for two functionally distinct gene classes." *Natl. Acad. Sci. USA* 6239–6244.

Rotherman, D.H., G.P. Fournier, K.L. Fournier, E.J. French, and R.E. Summons. 2014. "Methanogenic burst in the end-Permian carbon cycle." *PNAS* 5462–5467.

Saiki R.K., et al. 1985. "Enzymatic amplification of beta-globin genomic sequences and restriction site analysis for diagnosis of sickle cell anemia." *Science* 1350–1354.

Scheller, G., R. Boecher, and R.K. Thauer. 2010. "The key nickel enzyme of methanogenesis catalyses the anaerobic oxidation of methane." *Nature* 606–608.

Schink, B. and Stams, J.M."Syntrophism amoung Prokaryotes." 2006. In *Prokaryotes*, 309–335.

Schmitt, E., et al. 2020. "Recent advances in archaeal translation initiation." *Front. Microbiol.*

Shima, S, and R.K. Thauer. 2005. "Methyl-coenzyme M reductase and the anaerobic oxidation of methane in methanotrophic Archaea." *Curr. Opin. Microbiol.* 643–648.

Söllinger, A., and T. Urich. 2019. "Methylotrophic methanogens everywhere - physiology and ecology of novel players in global methane cycling." *Biochem. Soc. Trans.*

Sorokin, D.Y., K.S. Makarova, B. Abbas, M. Ferrer, P.N. Golyshin, E.A. Galinski, S. Ciordia, M.C. Mena, A.Y. Merkel, and Y.I. Wolf. 2017. "Discovery of extremely halophilic, methyl-reducing euryarchaea provides insights into the evolutionary origin of m." *Nat. Microbiol.*

Sousa, F., et al. 2016. "Lokiarchaeon is hydrogen dependent." *Nat. Microbiol.*

Stams, A.J.M., and C.M. Plugge. 2009. "Electron transfer in syntrophic communities of anaerobic bacteria and archaea." *Nat. Rev. Microbiol.* 568–577.

Steinberg, Lisa M., and John M. Regan. 2009. "mcrA-targeted real-time quantitative PCR method to examine methanogen communities." *Appl. Enviorn. Microbiol.* 4435–4442.

Stephenson, M. and L.C.H. Strickland 1933. "Hydrogenase: the bacterial formation of methane by the reduction of one-carbon compounds by molecular hydrogen." *Biochem.*

Strąpoć, D. 2017. "Biogenic methane." In *Encyclopedia of Geochemistry. Encyclopedia of Earth Sciences Series*, edited by W. White. Springer.

Takai, K., et al. 2008. "Cell proliferation at 122 C and isotopically heavy CH4 production by a hyperthermophilic methanogen under high-pressure cultivation." *Proc. Natl. Acad. Sci. U.S.*

Thauer, R.K. 1998. "Biochemistry of methanogenesis: a tribute to Marjory Stephenson. 1998 Marjory Stephenson Prize Lecture." *Microbiology* 144.

Thauer, R.K., A.K. Kaster, H. Seedorf, W. Buckel. 2008a. "Methanogenic archaea: ecologically relevant differences in energy conservation." *Nat. Rev. Microbiol.* 579–591.

Thauer, R.K., A.K. Kaster, H. Seedorf, W. Buckel and R. Hedderich. 2008b. "Methanogenic archaea: ecologically relevant differences in energy conservation." *Nat. Rev. Microbiol.* 579–591.

Thierry, D., and W. Sand. 2002. "Microbially influenced corrosion." In *Corrosion Mechanisms in Theory and Practice*, edited by P. Marcus, 583–603. CRC Press.

Tidwell, T.J., Z.R. Broussard, R. DePaula, and V.V. Keassler. 2017. "Mitigation of severe pitting corrosion caused by MIC in a CDC biofilm reactor." *NACE Int.* 51317–59604.

Timmers, Peer H.A., et al. 2017. "Reverse methanogenesis and respiration in methanotrophic archaea." *Archaea.*

Tori, K., H. Tsurumaru, and S. Harayama. 2010. "Iron corrosion activity of anaerobic hydrogen-consuming microorganisms isolated from oil facilities." *J Biosci Bioeng.*

Tsurumaru, H., N. Ito, K. Mori, S. Wakai, T. Uchiyama, T. Lino, A. Hosoyama, H. Ataku, and S. Harayama. 2018. "An extracellular [NiFe] hydrogenase mediating iron corrosion is encoded in a genetically unstable genomic island in Methanococcus maripaludis." *Sci. Rep.*

Uchiyama, T., K. Ito, K. Mori, H. Tsurumaru, and S. Harayama. 2010. "Iron-Corroding Methanogen Isolated from a Crude-Oil Storage Tank." *Appl. Environ. Microbiol.* 1783–1788.

Ueno, Yuichiro. 2006. "Evidence from fluid inclusions for microbial methanogenesis in the early Archaean era." *Nature.*

Vanwonterghem, I., et al. 2016. "Methylotrophic methanogenesis discovered in the archaeal phylum Verstraetearchaeota." *Nat. Microbiol.* 16170.

Vemuri, R., E.M. Shankar, M. Chieppa, R. Eri, and K. Kavanagh 2020. "Beyond just bacteria: functional biomes in the gut ecosystem including virome, mycobiome, archaeome and helminths." *Microorganisms.*

von Wolzogen Kuhr C.A.H., and L.S. van der Vlugt. 1934. "The graphitization of cast iron as an electrobiochemical process in anaerobic soil." *Water.*

Kim, W., and W.B. Whitman. 2014. *Encyclopedia of Food Microbiology*, 2nd Edition). Cambridge: Elsevier.

Welte, C., and U. Deppenmeier. 2011. "Membrane-bound electron transport in Methanosaeta thermophila." *J. Bacteriol.*

Woese, C. R., and G. E. Fox. 1977. "Phylogenetic structure of the prokaryotic domain: the primary kingdoms." *Proc. Natl. Acad. Sci. USA* 5088–5090.

Wolfe. 1976. "New approach to cultivation of methanogenic bacteria: 2-mercaptoethanesulfonic acid (HS-CoM)-dependent growth of Methanobacterium ruminantium in a pressurized atmosphere." *Appl. Environ. Microbiol.*

Wolfe, R.S.. 1993. *An Historical Overview of Methanogenesis.* Boston, MA: Chapman & Hall.

Wongnate, Sliwa, B. Ginovska, Dayle Smith, M.W. Wold, and N. Lehnert. 2016. "The radical mechansim of biological methane synthesis by methyl-coenzyme M reductase." *Science* 953–958.

Zeikus, J.G. 1977. "The biology of methanogenic bacteria." *Bacteriological Rev.* 514–541.

9 Molecular Methods for Assessing Microbial Corrosion and Souring Potential in Oilfield Operations

Gloria N. Okpala, Rita Eresia-Eke, and Lisa M. Gieg

CONTENTS

9.1 Introduction 170
 9.1.1 Microbial Reservoir Souring and Its Control 170
 9.1.2 Microbiologically Influenced Corrosion (MIC) 172
9.2 MMM for Monitoring Oil Field-Associated Microbial Communities 173
 9.2.1 Sampling and Preservation for MMM 174
 9.2.2 Nucleic Acids Extraction 176
 9.2.3 PCR Amplification Using Universal Primers 177
 9.2.4 PCR Amplification of Group-Specific Genes 178
 9.2.5 RNA Analysis 184
 9.2.6 Metagenomics 185
 9.2.7 Multi-Omics Approaches 186
9.3 Case Studies Highlighting the Use of MMM Related to Souring and MIC 187
 9.3.1 Case Study 1: Using MMM to Pinpoint the Cause of Corrosion in a Water-Transporting Pipeline 188
 9.3.2 Case Study 2: Multi-Omics Analyses for Assessing MIC 189
 9.3.3 Case Study 3: Use of MMM to Determine the Effects of Temperature on Nitrate Treatment of Souring 190
 9.3.4 Case Study 4: Using MMM to Determine the Effects of Seawater Flooding to Produce Crude Oil from an Offshore Reservoir 191
 9.3.5 Case Study 5: Use of MMM to Reveal Dominant Taxa and Metabolisms Associated with Hydraulic Fracturing Fluids 192
9.4 Summary and Future Directions 194
References 194

9.1 INTRODUCTION

Microbial souring and microbiologically influenced corrosion (MIC) are detrimental processes associated with oilfield operations. The detrimental impacts of souring include an increase in the sulfide content of produced crude oil and gas due to the activity of sulfide-producing microorganisms. This phenomenon can result in reservoir plugging, a health and safety hazard for oil facility workers, and decreased value of the crude oil as increased production costs are needed to decrease the sulfide content (Hubert 2010; Gieg et al. 2011; Johnson et al. 2017). Oilfield souring due to the activity of sulfide-generating microorganisms can also lead to the corrosion of oil facility infrastructure, although MIC can be due to many different types of microorganisms. Energy-related companies collectively spend on the order of billions of dollars each year to treat and/or prevent these detrimental processes through such activities as applying chemical inhibitors or scavengers, microbial biocides, replace corroded production equipment, and/or site remediation (Hubert and Voordouw 2007; Koch et al. 2016). In the United States alone, it is estimated that the oil and gas industry suffers an annual loss of $3–7 billion (USD), a substantial portion of which is due to MIC (Vigneron et al. 2018). Having a better understanding of the microbial communities inhabiting various locations within oil and gas operations (e.g. within reservoirs and associated with crude oil and gas processing, transportation, and storage infrastructure) and the conditions under which they thrive can help operators better predict when detrimental microbial processes may present problems. With the emergence of robust and low-cost molecular microbiological methods (MMM) within the past two decades, identifying microbial communities in oilfield systems through the use of MMM has become an invaluable tool for helping to make operational predictions and decisions around microbial control. Several recent reviews on the topic of microbial souring and its control and MIC are available (Hubert 2010; Gieg et al. 2011; Enning and Garrelfs 2014; Sharma and Voordouw 2016; Johnson et al. 2017; Little et al. 2020) thus we only briefly overview these phenomena here. In addition, we summarize the various considerations needed for obtaining and processing oilfield samples for MMM analysis, overview the MMM approaches that have been used for assessing microorganisms in oilfields, and describe some examples wherein MMM have helped to better understand microbial communities and metabolisms that may lead to microbial souring or corrosion problems in oilfield operations.

9.1.1 Microbial Reservoir Souring and Its Control

To satisfy the increasing global energy demand, the oil and gas sector is continually seeking ways to secure new energy resources while boosting recovery and production of the existing ones. A substantial amount of the world's oil is produced by waterflooding, a secondary oil recovery process that typically allows the extraction of up to 40–60% of oil from subsurface reservoirs. Waterflooding operations are carried out through the injection of water to re-pressurize the reservoir, displacing the oil from reservoir rock pores and pushing it toward producer wells. While sources of water used for waterflooding can vary immensely, the presence of sulfate in a given water source can have a major impact on the potential of a reservoir to become sour. Increased sulfate content can stimulate the metabolism of sulfate-reducing microorganisms (SRM), a diverse group of anaerobic microorganisms that use sulfate as a terminal electron acceptor during anaerobic respiration, generating hydrogen sulfide, the causative agent of souring (Reinsel et al. 1996). The high sulfate content of seawater (28–30 mM) poses a major challenge, as its use for waterflooding in offshore oil recovery operations can increase reservoir sulfate content from 0–0.6 to 22–28 mM (Nilsen et al. 1996a; Gittel et al. 2009; Gittel et al. 2012). Other sources of water containing sulfate can also lead to souring of land-based crude oil reservoirs (Voordouw et al. 2009). Thus, reservoirs flooded with sulfate-containing water can potentially become sour, although the rate of emergence of bio-generated hydrogen sulfide differs from one reservoir to another (Eden et al. 1993; Zhang et al. 2012). Water injection in general can also change the physical and chemical characteristics and microbial composition of reservoirs as it mixes with the organic substrate rich formation water, creating an environment favorable for microbial growth (Eden et al. 1993; Magot 2005; Vigneron et al. 2017). Waterflooding can also increase the readily degradable oil organics (Gieg et al. 2011) to reservoir microorganisms and cause a temperature reduction to <45°C in the near injection wellbore region (NIWR) due to injection of cold seawater (Okpala et al. 2017).

While SRM are the most well-recognized agents of microbial souring, thiosulfate-reducers are becoming more frequently identified as important contributors to this phenomenon (Gieg et al. 2011; Piceno et al. 2014; Hu et al. 2016), particularly in hydraulic fracturing operations used to recover oil and gas from low porosity subsurface reservoirs (Liang et al. 2016; Mouser et al. 2016; Booker et al. 2017). Unlike SRM that utilize sulfate to produce sulfide, thiosulfate-reducing microorganisms can utilize thiosulfate as a terminal electron acceptor to either produce sulfite and sulfide (Fardeau et al. 1993; Smock et al. 1998; Leavitt et al. 2014), or to disproportionate it into sulfate and sulfide (Finster et al. 1998; Jackson and McInerney 2000). When this metabolism occurs, there can be a significant contribution to souring from the production of sulfide, even in the absence of sulfate (Booker et al. 2017). Also, the production of sulfate through disproportionation reactions can stimulate the growth of SRM to produce even more sulfide. Although the role of thiosulfate-reducing microorganisms in sulfidogenesis has not been as thoroughly investigated as SRM, these kinds of microorganisms are becoming of increasing concern in the petroleum industry (Mouser et al. 2016).

Control of souring is usually necessary for both environmental and economic reasons. Currently, there are two main strategies that are used for preventing or mitigating souring. One strategy is through the use of biocides to inhibit the growth of SRM and other microbes. This has a high success rate in above-ground facilities and but is less effective in the reservoir (Nemati et al. 2001). Biocide application in above-ground facilities below certain threshold values can result in the acclimatization of SRM and other microbes. Also, the continuous injection of biocides may lead to biocide-tolerant microbial communities (Vilcaez et al. 2007; Basafa and Hawboldt 2019), though this phenomenon remains poorly understood. Glutaraldehyde, benzalkonium chloride, and tetrakishydroxymethyl phosphonium sulfate (THPS) are non-oxidizing biocides commonly used to eliminate microbes in oil field operations

(McDonnell and Russell 1999; Videla and Herrera 2005; McGinley et al. 2009; Enzien and Yin 2011). A second strategy often used for controlling souring is through the injection of nitrate. Nitrate can permeate reservoir formations more readily than biocides to stimulate the growth of nitrate-reducing bacteria (NRB) that reduce nitrate to nitrite. The addition of nitrate can inhibit the growth/activity of SRM by (1) stimulating organotrophic NRB which, for bioenergetics reasons, can outcompete SRM for available carbon and energy sources (Hubert and Voordouw 2007; Gieg et al. 2011); (2) stimulate sulfide-oxidizing NRB which can reduce the accumulation of sulfide (Greene et al. 2003; Hulecki et al., 2009; Lahme et al. 2019); or (3) produce nitrite, which is an inhibitor of sulfite reductase that catalyzes the reduction of sulfite to sulfide (Gittel et al. 2012; Callbeck et al. 2013) in the dissimilatory sulfate-reducing pathway. There are some limitations to using nitrate. For example, the ratio of N/S in a system can lead to the production of sulfur and/or nitrite as corrosive by-products (Lahme et al. 2019). Further, some mesophilic sulfate-reducers are able to overcome nitrite inhibition by using a periplasmic nitrite reductase which reduces nitrite to ammonium (Greene et al. 2003; Haveman et al. 2004). Nonetheless, nitrate treatment to control microbial souring has been successfully applied both onshore (Voordouw et al. 2009) and offshore (Larsen et al. 2004). As an alternate to nitrate, perchlorate has also been more recently shown to be an effective microbial souring inhibitor (Gregoire et al. 2014; Liebensteiner et al. 2014), though additional research on its efficacy in oil reservoirs is needed. It must be noted that the successful mitigation of H_2S production in oil reservoirs using nitrate depends largely on reservoir temperature and the appropriate microbial community (Tsesmetzis et al. 2016).

9.1.2 Microbiologically Influenced Corrosion (MIC)

Microbial souring and corrosion often go hand-in-hand in the oil and gas industry, as biologically produced sulfide can react with ferrous ions released from carbon steel, yielding the formation of corrosive iron sulfides as end-products. To date, MIC due to the action of SRM is the best understood mechanism, well described in previous reviews (Zhang et al. 2011; Enning and Garrelfs 2014; Lv and Du 2018). However, MIC of metal infrastructure can involve many different types of microorganisms and potentially different mechanisms, many of which are not well characterized or defined (Liang et al. 2014; Kato et al. 2015; Sharma and Voordouw, 2016; Lv and Du 2018; Blackwood 2018; Jia et al. 2019). MIC is a complex process in which different microorganisms act in sequence or collectively to influence different electrochemical reactions, secreting products (such as proteins and metabolic intermediates) that can have secondary impacts on surfaces, particularly in a localized manner (Procopio 2019). Microorganisms can trigger the onset of corrosion by altering the pH of the surrounding environment, secreting corrosive metabolites (e.g. organic acids, sulfide), and engaging in direct oxidation-reduction reactions mediated by extracellular enzymes (Little and Lee 2007; Procopio 2019). However, MIC mechanisms are far from being understood. This is largely due to the complexity of processes—chemical, physical, and microbiological—occurring concomitantly, making it very challenging to underpin a singular mechanism as the main culprit

(Kip and van Veen 2015; Little et al. 2020). Microorganisms involved in the corrosion of metal infrastructure are usually embedded in biofilms—surface-attached communities interacting differently on a metal surface (Bryers 2008; Little and Lee 2007; Davidova et al. 2012; Wang and Melchers, 2017, Vigneron et al. 2018). Biofilms have complex architectures, with altering nutrient and redox potential gradients from the surface of attachment to the outside, thus stimulating the growth and activity of different species in the biofilm (Eckert 2015). Due to their changing nature, biofilms are not easily identified or controlled. The production of extracellular polymeric substances (EPS) not only allows biofilm microbes to attach to each other or to surfaces but confers protection against several external perturbations such as changes in temperature, chemicals (e.g. biocides), pH, flow, etc. occurring around the biofilm environment (Beech and Gaylarde 1999; Lejeune 2003). The EPS produced by biofilms forms complex microenvironments on metal surfaces that can also change the properties of deposits such as sand and corrosion products to be more electroactive and aggressive (Suarez et al. 2019). The presence of multiple species in a biofilm does not necessarily imply high functional diversity within the biofilm, because multiple species may perform similar functions resulting in some biochemical or metabolic overlap, referred to as functional redundancy (Kotu et al. 2019). However, the presence of multispecies biofilms on metal surfaces does imply that MIC may be caused by more than one mechanism and corrosion of this sort is often quite severe in nature compared to those caused by single species cultures (Lee et al. 2013; Vigneron et al. 2016; Suarez et al. 2019). Understanding the activity of microorganisms within a surface-attached community is essential for understanding how they influence the corrosion process. Linking microorganisms and/or their metabolic activities to actual corrosion (metal loss, or electrochemical measurements) is critical for attributing MIC as a corrosive mechanism (Eckert and Skovhus 2018). The use of MMM in recent years has helped to expand our understanding of microorganisms and how their activities link to metal corrosion.

9.2 MMM FOR MONITORING OIL FIELD-ASSOCIATED MICROBIAL COMMUNITIES

Until relatively recently, culturing was the primary method used to identify microorganisms inhabiting petroleum-rich environments, many of which showed that thermophilic and hyperthermophilic anaerobic microbes were abundant in oil reservoirs (e.g. Beeder et al. 1994; Mueller and Nielsen 1996; Nilsen et al. 1996a, 1996b; Orphan et al. 2000). Frequently isolated microorganisms have included sulfate-reducers and fermenters, methanogens, acetogens, and manganese-, nitrate-, and iron-reducers (Grassia et al. 1996; Greene et al., 1997; Slobodkin et al. 1999; Magot et al. 2000). Culture-dependent methods rely on the ability of the microbes of interest to grow in a laboratory-prepared medium under specific conditions of pH, salinity, pressure, redox conditions, and temperature that mimic the source environment (Röling et al. 2003). This approach allows for the isolation of microorganisms that can be studied in detail to better understand specific metabolic functions of interest. However, only a small fraction (<1%) of all microbial populations can be cultivated (Staley and Konopka, 1985), which greatly limits our understanding of the microbial diversity present in oil field environments (Grassia et al. 1996; Kaster et al. 2009; Konopka 2009; Vigneron et al. 2018).

To circumvent this limitation, culture-independent approaches are now more widely used to characterize microbial communities in many biomes, including subsurface petroliferous reservoirs and across oilfield operations. MMM based on sequencing 16S rRNA and/or functional (catabolic) genes are now routinely used to determine microbial community structure and to estimate the numbers of general or specific bacterial and archaeal communities inhabiting oilfields (Figure 9.1) (Gittel et al. 2009; Wang et al. 2014; Zhou et al. 2020), often in conjunction with laboratory enrichment techniques (Nilsen et al. 1996a; Orphan et al. 2000; Bonch-Osmolovskaya et al. 2003; Gittel et al. 2009). Following more recent advancements in high-throughput sequencing technologies, researchers now apply omics approaches such as metagenomics, transcriptomics, proteomics, and metabolomics to better understand the physiology and functions that specific groups of microbes may perform within a community (Figure 9.1).

9.2.1 Sampling and Preservation for MMM

One of the major limitations in studying oil reservoirs and characterizing microbial communities therein is having access to appropriate samples devoid of contamination (Hu et al. 2016). By far, the most common type of oilfield samples available for identifying microorganisms inhabiting oilfields are produced water samples collected from the production wellhead or following oil-water separation units, as they are comparatively inexpensive and easy to obtain. A downside to analyzing produced water samples is that this type of sample may only be representative of the planktonic microorganisms (water phase-associated) rather than the surface-attached microorganisms which are presumably the most abundant in subsurface rock-bearing environments such as petroleum reservoirs (Wentzel et al. 2013). To obtain surface-attached microorganisms, drilling and coring are required, which is largely cost-prohibitive. Obtaining produced water samples also alters the pressure of the fluids, wherein fluids are under higher pressures in the reservoir but drop to ambient pressure as they flow to the surface; this pressure differential can lyse cells (Kotlar et al. 2011). For some studies, pressurized water samples have been collected for microbial community analysis using special samplers in order to minimize such cell lysis (Kotlar et al. 2011). Samples have also been retrieved directly from oil deposits under high pressure and high temperature conditions using an oil phase single-phase reservoir sampler (Yamane et al. 2011). Following collection, the samples were slowly depressurized to minimize lysis of microbial cells prior to activity determinations and reservoir community analysis. This unique sampling procedure also prevents contamination of the samples by biofilms formed on the walls of wells and pipelines, but it is a very expensive procedure (Wentzel et al. 2013). Despite the limitations of using depressurized produced water samples to assess oilfield microbial communities, these kinds of samples have revealed the presence of SRM (or other sulfide-producing microorganisms), microbial souring activity, or corrosive microbial communities in dozens of oilfield surveys using cultivation or MMM techniques (Gieg et al. 2011; Vigneron et al. 2018). Given that MIC is surface phenomenon, samples attached to pipe

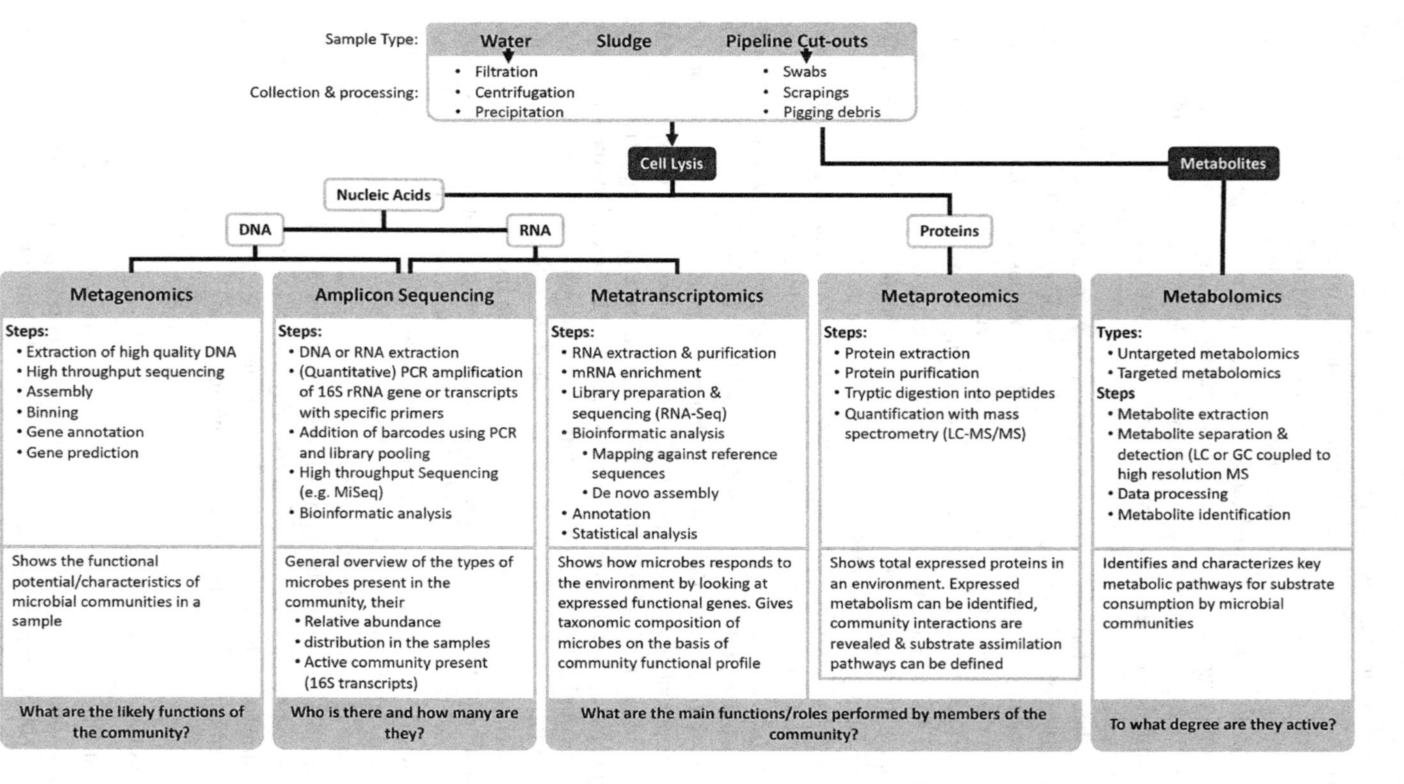

FIGURE 9.1 A summary of different MMM approaches that can be used to analyze oil field samples in order to identify and/or characterize microbial communities and activities involved in souring or MIC.

surfaces (including corrosion coupons) are the most ideal for identifying potentially corrosive microorganisms, especially from locations where leaks have actually occurred (Eckert 2015). However, produced water samples (as above), pipeline cleaning pig samples (that collect surface attached material from a long pipe segment), and tank bottom sludge samples have also been widely used for identifying potentially corrosive organisms and activities. Regardless of the type of sample collected, all samples should be collected with sterile implements and into sterile collection vessels. Industry standards are available that describe proper sampling and handling of oilfield samples for microbiological analyses (as summarized in Skovhus et al. 2017).

Samples collected for MMM (nucleic acid analysis) must be processed immediately, or preserved effectively as soon as possible after sampling. This is because the microbial community composition of an oilfield sample has been shown to change within a matter of hours (Kilbane 2014, De Paula et al., 2018; Rachel and Gieg 2020), and/or DNA degradation can occur within the sample which can limit the authenticity of results (Sharma and Huang 2019a; Kumar et al. 2020). While some field kits are emerging in the market to immediately process field samples for MMM (Sharma and Huang 2019b), these are not yet widely used and samples typically require processing following transport to laboratories. Filtration of produced water samples using a 0.2 or 0.45 µm filter immediately after collection followed by freezing the filters (e.g. at −20 or −80°C) or adding a chemical preservative (alcohols such as ethanol or isopropanol, or commercially available nucleic acids preservatives such as DNAzol®, RNAlater™, DNA/RNA Shield®, to name a few) have been reported to be good approaches for maintaining microbial community integrity for MMM analysis (Laramie et al. 2015; Oldham et al. 2019, Sharma and Huang 2019b, Salgar-Chaparro and Machuca 2019). In some cases, however, in-field filtration may not be possible, and samples may require several days for transport to a laboratory for analysis. In these cases, a chemical preservative should be used to maintain microbial community integrity (Rachel and Gieg 2020), which can include an alcohol, a commercially available nucleic acid preservative (e.g. as above), or a homemade buffer solution such as Longmire's buffer, DMSO-based solutions, or cetyltrimethyl ammonium bromide (CTAB) (May et al. 2011; Stein et al. 2013; Renshaw et al. 2014; Tatangelo et al. 2014; Tuorto et al. 2015; Menke et al. 2017). Flash freezing of samples on-site using liquid nitrogen is also an alternative way of preserving samples (Menke et al. 2017), though this option can be costly, hazardous, and unrealistic as spontaneous combustion and explosion risk may occur (Doorenweerd and Beentjes 2012). A summary of the advantages and limitations of several types of preservation methods have been recently provided (Sharma and Huang 2019b).

9.2.2 Nucleic Acids Extraction

Extraction of DNA (or RNA) is an important step when identifying microbial communities in oilfield samples (Figure 9.1). Hence, it is necessary that good quantity and quality DNA material is obtained. DNA extracted from oilfield samples is seldom clean enough for downstream processes such as polymerase chain reaction (PCR) amplification due to the co-precipitation of humic acids, proteins, polyphenolic compounds, complex organic matter, hydrocarbons, and other production chemicals. These PCR inhibitors can exert their effects by interacting with DNA or blocking the action of DNA polymerase during PCR amplification. For produced water samples containing an oil phase, some studies have suggested separating the oil phase from the aqueous phase by heating the produced water sample from 50°C to 60°C for 15–30 minutes and then separating the phases using sterile separation funnels (Wang et al. 2012; Gao et al. 2016). Furthermore, washing of soil, sediment, or core samples to remove contaminating hydrocarbon components with a TRIS buffer solution containing different strengths of EDTA has been shown to be effective in improving DNA recovery from these samples (Fortin et al. 2004; Gales et al. 2016).

There are many DNA extraction procedures that are in use today for obtaining DNA suitable for sequencing. Steps involved in DNA extraction include cell lysis (to release nucleic acids from microbial cells), removal of inhibitors and cell debris that can interfere with downstream applications (e.g., PCR), and nucleic acid mobilization, purification, and elution into a suitable buffer (Figure 9.1) (Oldham et al. 2012). Cell lysis can be achieved by using either enzymes to disrupt cell wall and cell membrane components or chemicals or chaotropic agents that solubilize lipid membranes (Lever et al. 2015). However, total genomic DNA (gDNA) isolation using commercial kits that use a bead-beating step for cell lysis has been reported to give more consistent DNA yields (Cruaud et al. 2014; Yuan et al. 2015). Nonetheless, no commercial kit universally suits all samples. Overall, DNA extraction protocols may be modified or adjusted to improve DNA recovery from different samples (Morono et al. 2014).

9.2.3 PCR Amplification Using Universal Primers

In oil production facilities or systems, the identification and characterization of microbial community composition relies heavily on sequencing the 16S rRNA gene, a universal marker gene for prokaryotes that has both conserved and variable regions. Sequencing and phylogenetic analysis of 16S rRNA genes have become the main approaches for studying microbial communities without the need for culturing (Figure 9.1) (Tremblay et al. 2015; Thijs et al. 2017), including those found in oil-field environments (Dahle et al. 2008; Gittel et al. 2009; van der Kraan et al. 2010). This process relies on the use of primers that amplify targeted regions on genomic DNA and ensures coverage of relevant microbial groups with detailed representation of microbial community members (Thijs et al. 2017; Hugerth and Andersson 2017). The choice of using the 16S rRNA gene stems from the fact that its conserved (slowly-evolving) regions are common across all bacteria and archaea and amenable for designing broad-spectrum primers for PCR, while the variable (rapidly-evolving) regions are shared among few numbers of species. Primer selection is therefore crucial for 16S rRNA gene sequencing as selecting the right primer pairs increases the success and coverage of targeted and relevant microorganisms during 16S rRNA sequencing (Fredriksson et al. 2013, Sambo et al. 2018).

Criteria for selecting 16S rRNA gene PCR primers for amplicon sequencing include sequencing depth, coverage of the taxa of interest (in this case Bacteria and Archaea), comparability of results with other studies, and phylogenetic resolution of the sequenced PCR products (Parada et al. 2016). The use of suboptimal primers results in the biased overestimation or underestimation of certain taxa, which in turn can lead to erroneous biological conclusions (Klindworth et al. 2013; Guo et al. 2016; Walters et al. 2016; Thijs et al. 2017). Furthermore, even with the nine hypervariable regions in 16S rRNA gene that enable species identification, no single variable region can identify all microbial community members (Chakravorty et al. 2007). At the same time, the issue of microbial taxa having different 16S rRNA gene copies in their genomes leads to overestimation of taxa abundance (Louca et al. 2018). Another limitation of using this MMM approach is that metabolic potential and/or function of identified taxa can only be inferred based on previous knowledge, and not actually identified (Martiny et al. 2013; Mohan et al. 2014; Aßhauer et al. 2015; Salgar-Chaparro and Machuca, 2019).

Despite known limitations, the use of 16S rRNA gene sequencing is a common and cost-effective MMM for identifying microbial communities in oilfield environments that could potentially be involved in souring or MIC and has enabled the identification of microbial communities present in different oil reservoirs around the world (e.g. Wentzel et al. 2013; Okoro et al. 2014; Gao et al. 2016; Li et al. 2017; Okpala et al. 2017). A compilation of commonly used primer pairs for targeting the 16S rRNA gene for PCR amplification in oilfield systems and other environments is provided in Table 9.1. While universal primers can be used in acquiring pertinent data on microbial composition and relative abundances of taxonomic groups, they do not give quantitative information on absolute numbers of the microorganisms and relative changes in microbial abundance when control strategies are being designed or implemented. Therefore, using primers specific for quantification is also needed; some primer sets recently reported for this purpose are also given in Table 9.1.

9.2.4 PCR Amplification of Group-Specific Genes

Although sequencing of the 16S rRNA gene enables the phylogenetic identification and characterization of different taxonomic groups in a sample, the role or function of members within the community is at best inferred (McGrath et al. 2010). Analysis of catabolic genes more specific to particular types of microorganisms is essential when studying how different environmental factors impact the diversity and numbers of different taxonomic groups. For example, in oil and gas fields where souring is a concern, specifically identifying and/or quantifying SRM can be a useful MMM approach (Johnson et al. 2017). In many cases, microbial groups such as SRM comprise a subset of the entire community detected or identified with 16S rRNA gene analysis and may be overshadowed by more abundant taxa within the community. Thus, the use of genes specific to individual microbial groups can allow for a better resolution of these members within whole microbial communities. A more group-specific, targeted PCR approach could further highlight community diversity and the role that environmental and experimental changes may have on that specific group (Imhoff 2016), such as the effects of nitrate treatment to control souring (Callbeck et al. 2013).

To successfully apply a catabolic gene approach, suitable primers that are specific to metabolic pathways under study must be designed. Specific primer sets are available for the detection and quantification of genes unique to different microbial groups in the oil industry. The functional genes of importance with respect to reservoir

TABLE 9.1
Primers Commonly Used for Amplifying the 16S rRNA Gene

16S rRNA Gene Primer Pair	Region Covered	Primer Sequence	Comment	References
515f/926r	V4-V5	515F: 5'- GTGNCAGCMGCCGCGGTAA-3' 926R: 5'-CCGYCAATTYMTTTRAGTTT-3'	Reduced bias and can detect environmentally important taxa including archaea. Shown to have high accuracy when tested against a mock community; Universal	Quince et al. 2011; Parada et al. 2016; Lomans et al. 2016
515f/806r	V4	515F: 5'-GTGCCAGCMGCCGCGGTAA-3' 806R; 5'-GGACTACHVGGGTWTCTAAT-3'	Underestimates Crenarchaeota, Thaumarchaeota, environmental important archaea and SAR11 clade. Overestimates *Gammaproteobacteria* and some other archaea	Caporaso et al. 2011; Walters et al. 2016
Modified 515F/806R	V4	515FB: 5'-GTGYCAGCMGCCGCGGTAA-3' 806RB; 5'-GGACTACNVGGGTWTCTAAT-3'	Modification was done to remove biases against Crenarchaeota/Thaumarchaeota and SAR11 clade	Parada et al. 2016; Apprill et al. 2015
27f /338r	V1-V2	27F: 5' AGAGTTTGATCCTGGCTCAG-3' 338R: 5' GCTGCCTCCCGTAGGAGT-3'	Identifies bacteria and some archaea	Klindworth et al. 2013; dos Santos et al. 2020
926F/1392R	V6-V8	926F: 5'-AAACTYAAAKGAATWGRCGG-3' 1392R: 5'-ACTTTCGTTCTTGATYRA-3'	Biased against Verrucomicrobia	Lane 1985; An et al. 2013; Guo et al. 2016
27F/533R	V1-V3	27F: 5'-AGAGTTTGATCCTGGCTCAG-3' 533R: 5' -AGA GTT TGA TCC TGG CTC AG-3'	Bacterial 16S rRNA gene	Gao et al. 2015
344F/915R	V3-V5	344F: 5' -ACG GGG YGC AGC AGG CGC GA-3' 915R: 5' -GTG CTC CCC CGC CAA TTC CT-3'	Archaeal 16S rRNA gene	Gao et al. 2015; Liang et al. 2018
341F/806R	V3-V4	341: 5'-CCTAYGGGRBGCASCAG-3' 806R: 5'- GGACTACNNGGGTATCTAAT-3'	Bacterial and archaeal	Lee et al. 2015
341f/785r	V3-V4	341F: 5'-CCTACGGGNGGCWGCAG-3' 785R: 5'- GACTACHVGGGTATCTAATCC-3'	Mainly bacteria	Klindworth et al. 2013; Fortunato et al. 2012
8F/805R		8F: 5'AGAGTTTGATYMTGGCTCAG-3' 805R: 5'GACTACCAGGGTATCTAATCC-3'	Bacterial	Savage et al. 2010; Li et al. 2017

(Continued)

TABLE 9.1 (Continued)
Primers Commonly Used for Amplifying the 16S rRNA Gene

16S rRNA Gene Primer Pair	Region Covered	Primer Sequence	Comment	References
109F/912R		109F: 5'-ACKGCTCAGTAACACGT-3' 912R: CTCCCCCGCCAATTCCTTTA	Archaea	Nazina et al. 2006
338F/806R	V3-V4	338F: 5′-ACTCCTACGGGAGG-CAGCAG-3′ 806R: 5′-GGACTACHVGGGTWTCTAAT-3′	Bacteria	Ren et al. 2014
515F/907R	V4-V5	515F: 5′-GTGCCAGCMGCCGCGG-3′ 907R: 5′-CCGTCAATTCMTTTRAGTTT-3′	Archaea	Klindworth et al. 2013
S-D-Arch-0008-b-S-18/ S-D-Arch-0519-a-A-19		S-D-Arch-0008-b-S-18: 5'-TCYGGTTGATCCTGSCGG-3' S-D-Arch-0519-a-A-19: 5'-GGTDTTACCGCGGCKGCTG-3'		
S-D-Bact-0516-a-S-18/ S-D-Bact-0907-a-A-20		S-D-Bact-0516-a-S-18: 5'-TGCCAGCAGCCGCGGTAA-3' S-D-Bact-0907-a-A-20: 5'- CCGTCAATTCMTTTGAGTTT-3'	Bacteria; can also be used for qPCR	Klindworth et al. 2013; Lomans et al. 2016
BACT1369F/ BACT1492R	qPCR qPCR	BACT1369F: 5'- CGGTGAATACGTTCYCGG -3' BACT1492R: 5'- GGWTACCTTGTTACGACTT -3'	Bacteria Bacteria	Suzuki et al. 2000; Vigneron et al. 2016
8F/338R		8F: 5'-AGAGTTTTGATYMTGGCTC-3' 338R: 5'-GCTGCCTCCCGTAGGAGT-3'		Vigneron et al. 2016
ARC787F/ ARC1059R	qPCR qPCR	ARC787F: 5'- ATTAGATACCCSBGTAGTCC -3' ARC1059R: 5'- GCCATGCACCWCCTCT -3'	Archaea Archaea	Yu et al. 2005; Vigneron et al. 2016
A806f/A958r	qPCR	A806f: 5'-ATTAGATACCCSBGTAGTCC-3' A958r: 5'-YCCGGCGTTGAMTCCAATT-3'	Archaea	Delong et al. 2002; Vigneron et al. 2016
A519F/A906R		A519F: 5'-CAGCMGCCGCGGTAA-3' A906R: 5'-CCCGCCAATTCCTTTAAGTTTC-3'		Klindworth et al. 2013; Lomans et al. 2016

souring are the *dsrAB* and *aprAB* genes that are exclusive to SRM. The *aprAB* gene encodes adenosine 5′-phosphosulfate reductase which is responsible for the reduction of adenosine 5′-phosphosulfate (APS) to sulfite, while the *dsrAB* gene encodes dissimilatory (bi)sulfite reductase, which catalyzes the last step in the sulfate reduction pathway, catalyzing the conversion of sulfite to sulfide. Müller et al. (2015) conducted a survey of SRM genes identified from multiple biomes and found that *dsrAB* sequences were highly diverse, spanning multiple phyla. Thus, a single primer pair cannot capture the diversity of SRM. However, as an outcome of the survey, Müller et al. (2015) did propose the use of nine different primer sets that can be used to provide the best coverage for SRM capture in a given environmental sample. Several studies that used 16S rRNA amplicon sequencing have shown that thiosulfate-reducers such as *Halanaerobium* are frequent contributors of biogenic sulfide in oilfields and shale formations (Vigneron et al. 2016, 2017; Bonifay et al. 2017; Booker et al. 2017), thus detecting and/or quantifying the population of thiosulfate-reducers is of interest in some operations such as in hydraulically fractured fields (Liang et al. 2016). Recent research that examined the genome of *Halanaerobium* using metagenomics (Booker et al. 2017; Lipus et al., 2017) identified the rhodanese-like protein, rdlA, which catalyzes the reduction of thiosulfate to sulfite, the anaerobic sulfite reductase asrABC complex, which reduces sulfite to sulfide, and thiosulfate reductase (phs) as key proteins that potentially function during thiosulfate reduction by this organism. Ravot et al. (2005) reported the successful amplification of *rdlA* in *H. congolense* and four other thiosulfate-reducers using degenerate primers; additional primers targeting *Halanaerobium* spp. retrieved from hydraulic fracturing operations (e.g., Liang et al. 2014; Lipus et al. 2017; Booker et al. 2017) should be designed to help track such thiosulfate-reducing genes in produced water samples. Genes involved in nitrate reduction can be used to track the efficacy of souring control using nitrate treatment (Fida et al. 2016; Table 9.2).

Microbial taxa involved in MIC include sulfidogens along with many other groups, such as fermentative organisms, acetogens, methanogens, sulfide-oxidizers, iron-oxidizers, iron-reducers, and nitrate-reducers (Sharma and Voordouw 2016; Vigneron et al. 2016). Primer sets quantifying several of these types of microorganisms can be used to determine the abundances of these organisms in corrosive systems; examples of such qPCR primers are listed in Table 9.2. While members of such groups may be involved in MIC, their mere presence or high numbers in a system does not necessarily indicate that they are involved in corrosion; a gene or set of genes linked to a corrosion mechanism would be the best biomarker for indicating MIC in an oilfield setting. A few recent studies have sought such "MIC" gene targets. Using *Desulfovibrio ferrophilus* strain IS5, a SRM known to directly use electrons from carbon steel as an electron donor, Deng et al. (2018) showed that outer membrane cytochrome genes were upregulated under starvation conditions, suggesting their roles in extracellular electron transfer reactions. However, a link to such expression in a metal corrosion scenario is yet to be confirmed. A series of studies using different strains of *Methanococcus maripaludis* showed that iron corrosion by methanogens can be due to the expression and activity of an extracellular [NiFe] hydrogenase. Deutzmann et al. (2015) initially proposed that an extracellular hydrogenase functioned during iron corrosion, which was later confirmed by Tsurumaru et al. (2018).

TABLE 9.2
Examples of qPCR Primer Sets Used to Enumerate Sulfate-Reducers, Nitrate-Reducers, and Methanogens

Gene Specific Primer Pair	Target	Primer Sequence	References
Sulfate-reducers			
aprA	Adenosine-5-phosphosulfate reductase	apR-1FWF: 5'- TGGCAGATCATGATYMAYGG-3' aprA-5-RV': 5'-GCGCCAACNGGDCCRTA-3'	Meyer and Kuever 2007; Aoki et al. 2015
aprA	Adenosine-5-phosphosulfate reductase	RH1-aps-F: 5'-CGCGAAGACCTKATCTTCGAC-3' RH2-aps-R: 5'-ATCATGATCTGCCAgCGgCCGGA-3'	Ben-Dov et al. 2007; Duncan et al. 2017
dsrA	Dissimilatory sulfite reductase	DSR-1Fdeg: 5'-ACSCAYTGGAARCACG-3' PJdsr853Rdeg; 5'-CGGTGMAGYTCRTCCTG-3'	Quillet et al., 2012; Perez-Jimenez and Kerkhof 2005
dsrA	Dissimilatory sulfite reductase	Dsr1f: 5'-ACSCACTGGAAGCACGGCGG-3' DsrR: 5'-GTGGMRCCGTGCAKRTTGG-3'	Spence et al. 2008, Korenblum et al. 2013
dsrAB	Dissimilatory sulfite reductase AB	DSR-1Fdeg: 5'-ACSCAYTGGAARCACG-3' DSR-4Rdeg: 5'-GTGTARCAGTTDCCRCA-3'	Klein et al. 2001
dsrB	Dissimilatory sulfite reductase B	Dsr2060F: 5'-CAACATCGTYCAYACCCAGGG-3' Dsr4R: 5'-GTGTAGCAGTTACCGCA-3'	Geets et al., 2006, Wagner et al., 1998; Priha et al. 2013; Tian et al. 2017
dsrB	Dissimilatory sulfite reductase B	**DSR1728Fmix** DSR1728FmixA: 5'- CAYACCCAGGGNTGG-3' DSR1728FmixB: 5'- CAYACBCAAGGNTGG-3' DSR1728FmixC: 5'-CATACDCAGGGHTGG-3' DSR1728FmixD: 5'- CACACDCAGGGNTGG-3' DSR1728FmixE: 5'- CATACHCAGGGNTAY-3' **Dsr4Rmix** DSR4Ra: 5'-GTGTAACAGTTTCCACA-3' DSR4Rb: 5'-GTGTAACAGTTACCGCA-3' DSR4Rc: 5'-GTGTAGCAGTTKCCGCA-3' DSR4Rd: 5'-GTGTAGCAGTTACCACA-3' DSR4Re: 5'-GTGTAACAGTTACCACA-3' DSR4Rf: 5'-GTATAGCARTTGCCGCA-3' DSR4Rg: 5'-GTGAAGCAGTTGCCGCA-3'	Vigneron et al. 2017; Müller et al. 2015

(*Continued*)

TABLE 9.2 (Continued)
Examples of qPCR Primer Sets Used to Enumerate Sulfate-Reducers, Nitrate-Reducers, and Methanogens

Gene Specific Primer Pair	Target	Primer Sequence	References
Nitrate-reducers			
napA	Periplasmic nitrate reductase	V17m/napA-4r: 5'- TGGACVATGGGYTTYAAYC-3' napA4r: 5' - ACYTCRCGHGCVGTRCCRCA-3'	Bru et al. 2007
narG	Membrane bound nitrate reductase	narG-f: 5' - TCGCCSATYCCGGCSATGTC -3' narG-r: 5' - GAGTTGTACCAGTCRGCSGAYTCSG -3'	Flanagan et al. 1999; Bru et al. 2007
nirK	Nitrite reductase (copper- and heme d1)	GnirK2F: 5'-GGK GTV TTT ATG TAC CAT TGC-3' GnirK2R: 5'-SCC GCT YGC CGG AAG CAT CAC-3'	Verbaendert et al. 2014; Fida et al. 2016
nirS	Nitrite reductase (heme c and heme d)	nirS2F: 5'-TACCACCC(C/G)GA(A/G)CCGCGCGT-3' nirS3R: 5'-GCCGCCGTC(A/G)TG(A/C/G)AGGAA-3'	Braker et al. 1998; Fida et al. 2016
Methanogens			
mcrA	Methyl CoM reductase	mcrF: 5'-GGTGGTGTMGGATTCACACARTAYGCWACAGC-3'	Luton et al. 2002; Morris et al. 2013
mcrA	Methyl CoM reductase	mcrR: 5'-TTCATTGCRTAGTTWGGRTAGTT-3'	Nunoura et al. 2008
mcrA	Methyl CoM reductase	ME3MF: 5'-ATGTCNGGTGGHGTMGGSTTYAC-3' ME2r': 5'- TCATBGCRTAGTTDGGRTAGT-3' METH-f: 5'-RTRYTMTWYGACCARATMTG-3' METH-r: 5'-YTGDGAWCCWCCRAAGTG-3'	Colwell et al. 2008

This latter group isolated *M. maripaludis* strain OS7 that was corrosive to iron and obtained a mutant of this strain that was not corrosive. Genetic analyses showed that the mutant was devoid of an "MIC island" containing a [NiFe] hydrogenase along with other genes involved in its extracellular transport, providing further support that this enzyme was actually functioning to accelerate iron corrosion (Tsurumaru et al. 2018). Building on this, Lahme et al. (2020) designed qPCR primers targeting this hydrogenase (designated *micH*) and found this gene to be present only in highly corrosive biofilms in the lab and in oilfield settings, thus offering a robust gene-based biomarker for MIC assessments.

9.2.5 RNA Analysis

Because DNA-based sequencing techniques currently in use cannot differentiate between active and inactive bacterial species, RNA-based transcriptomic approaches have now been reported to gain additional information about active members in microbial communities related to souring or MIC. Some transcriptome analysis has been described for different environments including in oilfields (Li et al. 2017; Liu et al. 2018; Salgar-Chaparro and Machuca 2019). Only a few studies have combined DNA-based 16S rRNA gene sequencing with transcriptomics in oily environments due to the difficulty in obtaining high-quality RNA from oil reservoir samples (Zhou et al. 2020). Also, tools like PICRUSt (Langille et al. 2013) and Tax4Fun (Aßhauer et al. 2015) that can be used to predict the functional profile of prokaryotic communities based on the 16S rRNA gene data and reference genomes available in public databases or metagenome-assisted genome binning have only been recently developed (Salgar-Chaparro and Machuca 2019).

Zhou et al. (2020) showed that sequencing of the 16S rRNA gene and 16S rRNA transcripts retrieved from water samples in two production wells from a high temperature offshore oilfield in China revealed differences in the active microbial communities. One of the key findings from the study was that active microorganisms were in some cases very low in abundance or not present at all in the 16S rRNA gene dataset. Secondly, the dominant members in the genomic community were substantially reduced in the active community. For instance, the abundance of the thermophilic sulfate-reducers *Thermodesulforhabdus*, *Thermodesulfobacterium*, and *Thermodesulfovibrio* were low in the DNA-based community analysis, whereas they represented a high percentage of the RNA-based community analysis. Other active members identified in the RNA-based community analysis were mesophilic and thermophilic nitrate-reducers and sulfur-oxidizing bacteria. The major metabolic functions detected in the active community were sulfur, nitrogen, and carbon metabolism, correlating with the physicochemical data and active microorganisms identified. The use of both DNA- and RNA-based sequencing allowed for a more exhaustive view of the microbial communities and their activities in the oilfield. A similar approach was used by Salgar-Chaparro and Machuca (2019) to characterize corrosive microbial communities in biocide-treated and untreated sampling locations in an oil production facility. These authors also observed that using DNA-based sequencing data alone could lead to an underestimation of active microorganisms in samples. Using a transcriptomic approach, functional capabilities of active members of the community can be predicted which can help enhance the understanding of processes that take place in oilfields.

9.2.6 Metagenomics

The development of robust high-throughput sequencing technology platforms has made it more economically feasible for prokaryotic community surveys to be performed using metagenomics. This MMM approach (also known as shotgun metagenomics) is used to analyze the total microbial genomic content from an environmental sample (Riesenfeld et al. 2004; Thomas et al. 2012), and not just a single gene (such as the 16S rRNA or *dsrAB* genes). For this MMM approach, as outlined in Figure 9.1, genomic DNA is first extracted from environmental samples and the DNA sheared into low-sized fragments. Then, these fragments are independently sequenced using high-throughput sequencing technologies (such as Illumina®, PacBio®, or Nanopore® sequencing platforms). Individual microbial genomes are then assembled (MAGs, or metagenome-assembled genomes) and annotated/interpreted to provide valuable insights into potential community diversity and metabolic functioning. Some recent publications overview the many different computational (bioinformatics) programs that can be used to quality control, assemble, and annotate metagenomics data (Niu et al. 2018; Breitwieser et al. 2019). A metagenomics approach is often desired for identifying microbial community composition as well because it eliminates bias introduced in the PCR amplification step of 16S rRNA amplicon sequencing. The use of metagenomics sequencing along with robust binning algorithms has also yielded new reference genomes from uncultured taxa (Hug et al. 2016; Castelle and Banfield 2018). Thus, using these reference genomes, relevant information on metabolic potential of communities can be deduced without having to rely on cultivated organisms to obtain this information (Hug et al. 2016).

Limitations of shotgun metagenomics include the requirement for high quality and quantity DNA as starting material (up to a few micrograms DNA) which is also free of any environmental contamination (Procopio 2019) and interpretation of generated sequencing data which requires robust bioinformatic pipelines and know-how for data analysis and interpretation (Zhou et al. 2015). Sequencing costs are also higher than for 16S rRNA gene sequencing. However, this MMM approach can yield enormous amounts of information that can help to better predict microbial processes that may occur in oilfield environments. A metagenomics approach has now been reported in several studies aimed at characterizing microbial communities and potential metabolic functions in oilfields to better understand oilfield microbiota functionality in general (such as hydrocarbon biodegradation potential), to better predict environmental effects (such as waterflooding or temperature) that may lead to or inhibit souring, and to gain deeper insight into corrosive microorganisms associated with oilfield infrastructure (e.g. An et al. 2013, 2016; Hu et al. 2016; Vigneron et al. 2017; Bonifay et al. 2017; Wang et al. 2019).

9.2.7 Multi-Omics Approaches

The application of diverse high-throughput molecular techniques to provide a better understanding into the genomic architecture, functional, and metabolic potential along with activity of complex microbial communities is known as multi-omics. A general overview of the multi-omics approaches discussed in this section is given in Figure 9.1. Over the last decade, there has been an increase in the use of omics approaches to better understand microbial ecosystem functioning. Multi-omics approaches can target all levels of information flow—from DNA (metagenomics) to RNA (metatranscriptomics) to protein (metaproteomics) to biochemical pathway metabolites (metabolomics). As described above, 16S rRNA gene sequencing has given us a means of evaluating and estimating the diversity and composition of microbial communities across many biomes, including oilfield systems, in a simple and cost-effective manner. Metagenomics, on the other hand, has been used to provide a comprehensive snapshot of members of a microbiome and can shed light on potential genes and biochemical pathways in microbial communities. However, metagenomics does not differentiate between active and inactive contributors to processes occurring in an environment (Shakya et al. 2019). RNA analysis of genes expressed within a community to uncover functional potential can help to give a more thorough representation of the active members of a community. Metatranscriptomics can provide insight into the genes expressed by an entire community in each sample at a given moment under a specific environmental condition (Urich et al. 2008, Shi et al. 2009). This MMM approach can allow for a better understanding of the active biochemical functions taking place in complex microbial communities, and how environmental perturbations can alter those functions (Aliaga Goltsman et al. 2014). One of the major challenges for the use of metatranscriptomics is the susceptibility of environmental RNA to degradation. Secondly, since total RNA consists of 95% rRNA and 5% mRNA, an mRNA enrichment step is done which leads to reduction of RNA and causes loss of low abundant transcripts. In addition, mRNA has a very short lifespan of about 1.3 minutes at 37°C (Arraiano et al., 2010). However, when combined with metagenomics, metatranscriptomic analysis can be a powerful tool for improving our understanding of microbial community functioning.

Metaproteomics is a technique that measures, identifies, and characterizes the total protein complement of a microbial community in an environment using mass spectrometry methods (Mosier et al. 2015; Beale et al. 2016) and can reveal which proteins are actually being made by a microbial community under given conditions. Metaproteomics aids in obtaining real data for microbial activity as the method measures the total expressed proteins in an environment and can thus expose the dynamics of complex microbial communities (Mosier et al. 2015). The first step in a metaproteomics workflow is the extraction and purification of proteins. This is followed by conversion of the good quality proteins into peptides via a tryptic (enzymatic) digestion step and finally, quantification with mass spectrometry. Commonly used quantification techniques include two-dimensional polyacrylamide gels coupled with mass spectrometry (2D nano-LC/MS-MS), liquid chromatography coupled with mass spectrometry (LCMS/MS), and matrix-assisted laser desorption/ionization (MALDI) coupled with time-of-flight mass spectrometer (TOF-MS) (Zhou et al. 2015;

Beale et al. 2016). Metaproteomics have been applied in the study of the adaptation of biofilms in nutrient-deficient environments (Mosier et al. 2015). Bell et al. (2018) recently combined metagenomics and metaproteomics approaches to show that the cycling of incoming sulfur compounds in deep groundwater can result in biocorrosion occurring within the system even with low microbial diversity.

Metabolomics, which identifies the intermediates or end-products of either all or a subset of biochemical pathways being carried out by a microbial community, is another powerful omics approach that can be used to better understand the functioning of a given microbial community. Metabolites can be directly identified in a sample through the use of chromatographic separation methods like liquid chromatography (LC) or gas chromatography (GC) coupled with low- or high-resolution mass spectrometry (Duncan et al. 2009; Beale et al. 2014; Lenhart et al. 2014; Bonifay et al. 2017). Since metabolites are products of metabolic pathways of active community members, the expression of these leading to the production of these metabolites in a community can be monitored (Eckert and Skovhus 2018; Kotu et al. 2019). For example, identifying metabolites in an oilfield sample that are diagnostic of hydrocarbon metabolism can help to pinpoint carbon sources that are driving processes in oil reservoirs, while amino acids, fatty acids, fatty alcohols, and carboxylic acids have been identified and found to be associated with corrosion in different MIC studies (Beale et al. 2014; Lenhart et al. 2014; Bonifay et al. 2017).

Overall, combining different multi-omic tools (metagenomics, metatranscriptomics, metaproteomics, and metabolomics) can resolve microbial community structure, complex interactions between microbes and their environment, and other community dynamics. One big hurdle akin to multi-omics is the inherent requirement of advanced computational methods for processing and interpreting data generated (Beale et al. 2016; Procopio 2019). These must be taken into consideration when planning to include multi-omics in souring or MIC studies.

9.3 CASE STUDIES HIGHLIGHTING THE USE OF MMM RELATED TO SOURING AND MIC

Within the past 15 years, hundreds of studies have reported the use of MMM of some type in order to characterize and better understand microbial communities and functions in hydrocarbon-associated natural and industrial ecosystems, including reservoirs and oilfield processing systems. As overviewed above, 16S rRNA gene sequencing and qPCR targeting whole microbial communities or specific microbial groups or functions are now widely used MMM approaches for identifying and/or quantifying microbial communities in oilfield operations. Though currently lesser used, RNA-based approaches, metagenomics, and multi-omics approaches can provide enormous amounts of metabolic information thus will likely see additional application for the characterization of oilfield systems. The MMM toolbox can thus be a highly valuable part of any kind of oilfield system analysis and can also be used to help pinpoint problems and find solutions related to detrimental processes such as souring and MIC. There are numerous excellent examples in the literature describing how MMM approaches have helped to better understand and potentially solve an oilfield operational issue; providing an exhaustive list of all reported cases is beyond the scope of this chapter. Instead, we highlight only a few of these examples wherein MMM approaches, often in conjunction with laboratory physiology studies, were used to help pinpoint operational problems such as microbial souring or corrosion and/or help to devise solutions.

9.3.1 Case Study 1: Using MMM to Pinpoint the Cause of Corrosion in a Water-Transporting Pipeline

Pipelines transporting brackish subsurface water to steam generation plants used for steam-assisted gravity drainage (SAGD) for bitumen recovery in Alberta's oil sands operations can be prone to corrosion. To minimize the rate and severity of corrosion occurring in these pipelines, the water samples are degassed and treated with the oxygen scavenger sodium bisulfite (SBS) to mitigate the occurrence of potential oxygen-mediated corrosion. Even with this control measure in place, a pipeline transporting water experienced frequent failures, and it was of interest to the operator to determine the underlying reasons, including whether microorganisms could be involved (Park et al. 2011). An investigation into the microbial community composition of fluid samples transported through the pipeline as well as cut-outs from the failed pipeline sections at leak sites was thus undertaken using an MMM approach (Park et al. 2011). Samples were collected upstream and downstream of the SBS injection site and the effect of SBS on microbial community composition was determined. Sequencing of the 16S rRNA gene of the water samples revealed *Pseudomonas* and members of *Methanobacteriaceae* (*Methanolobus* and *Methanocalculus*) as the most abundant microbial taxa in the brackish source water samples. The case was different for the upstream and downstream pipeline sessile communities. The biofilm microbial community composition of pipeline cut-out upstream of SBS injection harbored mainly *Methanobacteriaceae*, *Coriobacteriaceae*, and *Desulfuromonas*. However, biofilm samples downstream of the SBS injection site consisted of mainly of the SRM *Desulfocapsa*, *Desulfurivibrio*, and *Desulfomicrobium*. Notably, *Desulfocapsa* can disproportionate bisulfite to sulfide and sulfate, thus the addition of bisulfite as an oxygen scavenger was likely stimulating the activity of *Desulfocapsa*, leading to the formation of sulfate that could stimulate other SRM along with the generation of corrosive products (Park et al. 2011). The identification of methanogenic taxa in both upstream and downstream pipe sessile community additionally indicated their possible involvement in the observed corrosion occurring in the water transporting pipeline system.

In a follow-up study, An et al. (2016) used a metagenomics approach to provide a more in-depth understanding of the effect of the bisulfite injection on the microbial community. The upstream pipe sessile metagenome was dominated by the archaea *Methanobacteriales* and *Methanosarcinales*, while the SRM *Desulfovibrionales* (*Desulfomicrobium*), *Desulfobacterales* (*Desulfocapsa*), and *Desulfuromonadales* formed the most abundant bacterial taxa in the metagenome. Following bisulfite injection, the downstream pipe sessile metagenome revealed *Methanobacteriales* as the dominant archaeal member, while *Desulfomicrobium* and *Desulfocapsa* were the main SRM observed. However, the use of 16S rRNA amplicon sequencing alone of the same brackish water samples had shown very low relative abundances of these microbial taxa (Park et al. 2011; An et al. 2016). Furthermore, metagenomics analysis revealed that members of the *Epsilonproteobacteria*, namely *Sulfuricurvum*, *Sulfurovum*, and *Arcobacter*, were also potentially involved in the pipeline corrosion. An increase in the abundance of the *hynL* gene for [NiFe]-hydrogenase, which is used to oxidize H_2 for reduction of sulfur, was affiliated with *Sulfuricurvum* and *Sulfurovum* genomes in the samples collected after the bisulfite injection point (An et al., 2016). These observations, together with low organic carbon content of the brackish water transported through this failed pipeline, indicated that *Epsilonproteobacteria* contributed to corrosion of the pipe. In these two related studies, MMM helped to identify that microorganisms with specific metabolisms were contributing to the corrosion occurring in this water-transporting pipeline.

9.3.2 Case Study 2: Multi-Omics Analyses for Assessing MIC

To date, a couple of studies have used a multi-omics approach to better understand MIC in oilfield operations. Lenhart et al. (2014) demonstrated that sessile (biofilm) microbial communities within a closed pipeline system are established from the planktonic communities in the bulk phase fluid transported through the pipeline. To show this, corroded sections recovered from a high-temperature oil transporting pipeline (referred hereafter as cookies) from a production facility in the Alaskan North slope oilfield were sampled and evaluated by 16S rRNA gene sequencing. In addition, metabolites retrieved from surface deposits were also analyzed and compared to those extracted from the bulk fluid. The microbial community was dominated by thermophilic sulfidogenic anaerobes—either sulfate- or thiosulfate-reducers such as *Thermodesulfobacterium, Desulfanicium, Thermoacetogenium,* and *Thermoanaerobacter.* In addition to these core sulfidogens, other organisms such as *Thermovirga lienii*, *Anaerobaculum thermoterrenum*, and *Anaerobaculum hydrogeniformans* OS1 capable of reducing elemental sulfur were identified. Comparison of the cookie biofilm/sessile community in this study and the microbial community from PIG samples and produced water (PW) from the same oil production facility sampled earlier (Duncan et al., 2009) showed that the dominant taxa in the PW (*Thermacetogenium phaeum* and *T. commune*) and the PIG community (*Thermoanaerobacter* and *Thermacetogenium*) were similar. The observation that similar microbial taxa were found in the cookie biofilm, PW, and PIG samples from the same operation led to the conclusion that the evaluation of a planktonic bacterial community could provide an understanding of MIC occurring within a pipeline. Another highlight of this study was the application of metabolomics to profile hydrocarbon degradation metabolites present in the cookie surface deposits. In this study, high-pressure LC coupled with quadrupole time-of-flight mass spectrometry (HPLC-Q-TOF-MS) was used to analyze the acid wash samples from cookies, while GC/MS was used to measure metabolites in PW samples transported in the pipelines (Duncan et al. 2009). The metabolic profiles of the sessile metabolites detected from residual surface deposits and corrosion products on the cookies were different from those in the PW samples from the pipelines. The metabolic products, e.g., alkyl succinates, of anaerobic hydrocarbon degradation were found in PW (planktonic) samples, whereas sessile metabolites had saturated and aromatic hydrocarbons as the major products (Duncan et al. 2009). The conclusion drawn from these studies that combined 16S RNA gene sequencing and metabolomics was that planktonic communities degraded *n*-alkanes to fatty acids that were used by the biofilm community to reduce sulfate and thiosulfate which ultimately resulted in the production of H_2S in the biofilm matrix.

In a separate study, Bonifay et al. (2017) examined two pipelines in the Norwegian sector of two adjacent North Sea oil production system, one of which exhibited severe corrosion, using metabolomics and metagenomics analyses to ascertain the contribution of MIC. The two North Sea oil production systems were treated with nitrate, and the pipelines showed similar chemical and physical attributes. Metabolic analysis of pigging debris from the corroded pipeline revealed metabolites produced from the anaerobic degradation of hydrocarbons such as alkyl- and benzyl succinates, whereas the non-corroded pipeline had more of aerobic degradation metabolites such as methylbenzylalcohol and dimethylcathechols (Bonifay et al. 2017). Metagenomic data showed a high abundance of *Pseudomonas stutzeri*, a nitrate reducer in the non-corroded pipeline, and the binned genome contained all the genes for dissimilatory nitrate reduction (Bonifay et al. 2017). The highly corroded pipeline metagenome, on the other hand, revealed haloalkalophilic bacteria such as *Desulfonatronospira thiodismutans* that can directly reduce sulfate, sulfite, and thiosulfate with either H_2 or simple organics as electron donors, or ferment sulfite and thiosulfate. Overall, the metabolites and the microbial community compositions correlated, and favored MIC as an important corrosion mechanism contributing to the corroded pipeline (Bonifay et al. 2017).

9.3.3 Case Study 3: Use of MMM to Determine the Effects of Temperature on Nitrate Treatment of Souring

One of the mechanisms supporting the injection of nitrate into oil reservoirs for souring control is the inhibition of sulfide production in SRM by nitrite (Voordouw et al. 2009). However, in most low-temperature reservoirs, nitrite production is transient as it is further reduced to N_2 via the denitrification pathway (Agrawal et al. 2012). This pathway involves nitrite reductase encoded by the *nirS* (heme c- and heme d1-containing nitrite reductase) and *nirK* (copper- and heme d1-containing nitrite reductase) gene, or via the dissimilatory nitrate reduction to ammonium (DNRA) pathway (Zumft 1997; Fida et al. 2016). However, at higher temperatures, nitrite had been reported to accumulate in some oilfield-derived tNRB enrichments (Reinsel et al. 1996; Agrawal et al. 2014). Using the MMM approaches of 16S rRNA sequencing and qPCR to enumerate nitrite reduction genes, Fida et al. (2016) investigated the temperature limit of nitrite reduction by oilfield-derived thermophilic nitrate reducers, and the effect of temperature on the nitrite reductase genes. Produced water samples from a low temperature oil field that were continuously treated with nitrate (2 mM) to prevent souring were used for establishing laboratory microcosms at 40 to 70°C. 16S rRNA gene sequencing data revealed that temperature caused a shift in microbial community composition. At 50–70°C, *Geobacillus* and *Petrobacter* were the dominant nitrate-reducers, while *Pseudomonas* and *Thauera* were most abundant at lower temperatures (40–45°C). *Thauera*, *Petrobacter*, and *Geobacillus* strains were isolated from the enrichments and the abundance of the nitrite reductase genes *nirS* and *nirK* were quantified using qPCR at different temperatures. At lower temperatures of 40–45°C, the *nirS* gene was abundant in *Thauera* but was in comparatively low abundance in the thermophilic *Petrobacter* and *Geobacillus* spp. However, at higher incubation temperatures, where nitrite accumulated, the *nirS* and *nirK* genes could not be amplified/quantified from *Petrobacter* and *Geobacillus* spp. A further probe into the distribution of nitrite reductase genes revealed that the *nirS* gene was not active at high temperatures. Thus, using MMM in combination with culturing, this study suggested that nitrite accumulation at high temperatures is effective, due to the fact that nitrite reductase is not active at higher temperatures (Fida et al. 2016).

Okpala et al. (2017) also investigated the impact of temperature gradients on the success of nitrate treatment for souring control in a high temperature oilfield. To this end, multiple microbial enrichments were established to determine the temperature (40–70°C) dependence of nitrate and sulfate reduction of samples collected from a high-temperature oilfield (*in situ* temperature of about 95°C). Sequencing of the 16S rRNA amplicons for water samples from the oilfield (PW and injection water (IW)) indicated that PW samples were dominated by thermophiles including *Thermoanaerobacter, Methanothermococcus, Methermicoccus, Thermococcus*, and *Archaeoglobus*, while the IW samples were mainly comprised of aerobic mesophilic marine bacteria (Okpala et al. 2017). Enrichment of sulfate- and nitrate-reducers from the PW at different temperatures revealed the activity of thermophilic SRM only at 55–65°C (dominated by *Desulfotomaculum*) and no SRM activity at lower temperatures. Nitrate-reducing activity was only detected in the IW at 50°C (*Marinobacter*) and 60°C (*Geobacillus*). Similar to Fida et al. (2016), nitrite accumulated in the thermophilic NRB enrichments and pure isolates. The mapping of NRB and SRM in the NIWR of this oilfield using MMM demonstrated that successive thermal gradient zones (mesophilic, thermophilic, and abiotic zones), observed in NIWR of high temperature reservoirs as a result of the injection of cold seawater for oil recovery, impacts the types of microbes enriched and, potentially, the efficacy of souring control strategies. The study suggested that reinjection of hot PW will eliminate the mesophilic zone in the NIWR which would in turn result in efficient control of souring using a nitrate injection strategy (Okpala et al. 2017).

9.3.4 Case Study 4: Using MMM to Determine the Effects of Seawater Flooding to Produce Crude Oil from an Offshore Reservoir

Vigneron et al. (2017) investigated the impact of long-term seawater flooding over the operational lifespan of 32 offshore reservoirs and the shifts in microbial community composition and thus potential metabolic functionality across the oilfield. The field under study was the Halfdan oil field located 2.1 km below the seafloor with a reservoir temperature of 80°C. The injected seawater at this field is deoxygenated and about 100–150 mg/L of nitrate is dosed daily to prevent souring and corrosion issues. In addition to the produced water samples collected from each well, two formation waters, which were chosen based on their location in the oil field; a) FWA bounded on the North, and b) FWB bounded by the south, east, and west. FWA was characterized by lower salinity and low metal ion concentration, while FWB had higher salinity and increased metal ion and sulfate concentrations. Each sample was subject to both physicochemical and metagenomics analysis. The interactions between reservoir conditions (pre- and post (current) waterflooding) and microbial community shifts were also evaluated. The results showed that the higher the % of injected water reaching the producing well, the lower the core temperature (around 42°C), and the higher the sulfate, nitrate, and nitrite concentrations. However, well fluids impacted by a lower % of injected water contained higher concentrations of ammonium and VFA (such as acetate). How did these observations affect microbial community composition? With high seawater breakthrough, archaeal abundances decreased significantly and were dominated by *Archaeoglobus* and *Thermococcales*. In contrast, the bacterial community was in high relative abundance in wells with seawater breakthrough, with *Deferribacteres* observed in wells with 56 to 91% of injected water fractions. Other frequently identified bacterial lineages included *Thermotogales, Clostridia, Flexistipes, Archaeoglobales*, and *Deltaproteobacteria*. The abundance of *Deferribacteres*, some members of which are nitrate-reducers, correlated strongly with the well that had high nitrite breakthrough. For wells having greater than or equal to a 10% water cut, *Clostridia* (*Firmicutes*), *Pelobacter* (*Deltaproteobacteria*), and (*Synergistes*) were predominant, while wells fed from high salinity formations and with a <10% water cut were dominated by *Petrotoga* (*Thermotogae*) and *Desulfotomaculum* (*Firmicutes*). A closer look at the metabolic profile to ascertain the prominent metabolic potential in each of the wells indicated that wells with a high % of injected seawater, irrespective of the salinity or fueling formation water, had high potential for nitrate reduction and sulfide oxidation, which supported the presence of the *Epsilonproteobacteria* and *Deferribacteraceae*. Other genes found in low percentages included those of denitrification and thiosulfate, sulfate, and polysulfur reduction. In wells with low seawater breakthrough, hydrocarbon degradation potential (*Deltaproteobacteria, Firmicutes, Archaeoglobales, Synergistes*), acetogenesis (*Firmicutes, Synergistes, Thermotogales*, and *Deltaproteobacteria*), and polysaccharide degradation (*Firmicutes, Thermotogales*, and Candidate division OP9 lineages) was seen. The following conclusions were made based on the physicochemical and microbial data. Firstly, for formation water, the lesser the % of injected seawater, the higher proportion of indigenous microbes that were seen within the community. Secondly, seawater injection is the major cause of souring in oil reservoirs and as more seawater breaks through the threat is raised further as temperature in the reservoir decreases further and the microbial community composition continues to change. Thirdly, testing and monitoring of a single well in a field will not give adequate information as to what is happening in other wells or parts of the reservoir. Overall, the study showed that adequate monitoring of oil field microbes using MMM can aid in the planning and implementation of preventive or mitigation strategies that will be effective in curbing souring and corrosion issues that may arise during the operational life of the reservoir.

9.3.5 Case Study 5: Use of MMM to Reveal Dominant Taxa and Metabolisms Associated with Hydraulic Fracturing Fluids

There are thousands of wells in deep shale formations which have been hydraulically fractured for gas and oil recovery (Mouser et al. 2016). To further understand the microbial metabolism and populations within these environments leading to detrimental effects such as souring and corrosion, several recent reports describe the use of MMM to not only reveal microbial community compositions but also their metabolic potential and functions such as carbon and energy metabolism, nutrient acquisition, sporulation and dormancy, stress responses, and syntrophic interactions that support microbial metabolisms of halophilic microbial communities that develop in these manufactured ecosystems (Mohan et al. 2014; Mouser et al. 2016).

Mouser et al. (2016) overviewed a number of studies using MMM approaches, including metagenomics, to understand microbial functioning in these subsurface environments. For example, Mohan et al. (2014) used a metagenome approach to determine microbial community compositions and potential functions from a Marcellus shale natural gas well. The authors analyzed the metagenome of hydraulic fracturing source water along with produced water (flowback water) after day 1 and day 9. The microorganisms changed notably, with members of the order *Rhodobacterales* dominating in both the source water and day one produced water but decreasing in the day 9 produced water. There was also an increase in members of the order *Thermoanaerobacterales* in the day 1 produced water, but this decreased in relative abundance by day 9. However, *Gammaproteobacteria*, which constituted a minor fraction of the source water community, were present at over 50% relative abundance in the day 9 produced water. Shifts were also found in functional categories of metagenomes. There was an increase in carbohydrate metabolism genes in day 9 produced water, which suggested the ability of the microbial community to use available carbon sources from the shale formation or those added in fracturing fluids. Stress response genes (such as for heat shock or acid, periplasmic, or osmotic stress) were primarily associated with *Vibrionales* and *Altermonadales*, while genes involved in sulfur metabolism were associated with *Vibrionales* and *Bacteriodales* in the day 9 produced water. This kind of shotgun metagenomic analysis showed the emergence of distinct bacterial taxa and functionality as the result of changes to the environment, providing information to operators that may guide decision-making with respect to biocide control treatment options, and source water and wastewater management.

Lipus et al. (2017) further examined produced water from 42 Marcellus shale gas wells using 16S rRNA gene and quantitative PCR (qPCR) to identify microbial taxa and abundance. Members of the order *Halanaerobiales* were found in comparatively high abundance; *Halanaerobium* was present in 40 out of 42 wells with an abundance of 99.1% in one single well. Shotgun metagenomics sequencing and binning of a *Halanaerobium* draft genome (MDAL1) retrieved from produced water found it to be closely related to an oil field isolate, *H. congolense* capable of thiosulfate reduction, acid production, and biofilm formation—all characteristics that may contribute to souring and MIC. It was interesting to note that classical sulfate reduction genes (such as *dsrAB*, *aps*) were not identified in the produced well fluids.

Booker et al. (2017) analyzed produced fluids from a different black shale field (Utica shale) and also discovered the predominance of *Halanaerobium* catalyzing thiosulfate-dependent sulfidogenesis. In subsurface systems, because thiosulfate is frequently measured at low concentrations, the hypothesis is that it is rapidly converted through disproportionation to sulfate and sulfide or to sulfite. Laboratory tests prepared from well fluids demonstrated thiosulfate conversion to sulfide. From the proteogenomics and laboratory growth experiments carried out in the study, the presence of rhodanese-like proteins and anaerobic sulfite reductase (*asr*) complexes which can convert thiosulfate to sulfide were discovered from the analysis of the *Halanaerobium* genome and genomes reconstructed from metagenomic data sets. It was discovered that rhodanese enzymes were present in several phyla (e.g. *Firmicutes*, *Proteobacteria*) while both rhodanese genes and *asr* complexes were found within *Clostridia* such as *Halanaerobium*, confirming the potential contribution of this taxon to souring in hydraulic fracturing operations. Laboratory tests, field sample geochemistry, and MMM approaches in such studies collectively demonstrated the potential importance of thiosulfate-reducers such as *Halanaerobium* in contributing to souring and MIC in hydraulic fracturing operations (Booker et al. 2017).

9.4 SUMMARY AND FUTURE DIRECTIONS

Molecular microbiological methods are becoming widely applied in the oil and gas industry and can be used to help determine whether microbial reservoir souring or MIC may be a problem in an oilfield operation. Methods such as 16S rRNA gene sequencing for identifying microbial community members (from which metabolic function can only be inferred), and PCR or qPCR for targeting and quantifying total numbers of microorganisms or specific groups of microorganisms in given oilfield operations are most commonly used (Figure 9.1). The more recent use of metagenomics has expanded the knowledge base past mere microbial identification to additionally identify potential metabolic functions associated with a whole microbial community, yielding new insights into such environments as waterflooded oil reservoirs, hydraulic fracturing operations, and corroding pipelines. However, whether the genes identified via metagenomics are actually expressed under the environmental conditions defining an oilfield operation requires the use of additional omics approaches, such as metatranscriptomics and/or metabolomics, ideally coupled with empirical testing (Figure 9.1). To date, these latter omics approaches have only been used in a couple of cases; thus, there is a clear need for their more widespread use to more deeply understand the microbiota inhabiting many oilfield systems in order to prevent or mitigate detrimental souring or corrosion issues in a rapid and cost-effective manner. Along these lines, there is also a great need for the development of diagnostic biomarkers that can be designed as an outcome of MMM data, such as genes or metabolites that can serve as leading indicators for MIC. With such tools in hand, a better view of the microbial communities and their functions will allow for improved monitoring and mitigation strategies for microbial souring and corrosion, and help to establish best practices for the use and interpretation of MMM methods.

REFERENCES

Aßhauer, K. P., B. Wemheuer, R. Daniel, and P. Meinicke. 2015. Tax4Fun: predicting functional profiles from metagenomic 16S rRNA data. *Bioinformatics* 31: 2882–2884.

Agrawal, A., H.-S. Park, S. Nathoo, et al. 2012. Toluene depletion in produced oil contributes to souring control in a field subjected to nitrate injection. *Environmental Science & Technology* 46: 1285–1292.

Agrawal, A., D. An, A. Cavallaro, and G. Voordouw. 2014. Souring in low-temperature surface facilities of two high-temperature Argentinian oil fields. *Applied Microbiology & Biotechnology* 98: 8017–8029.

Aliaga Goltsman, D.S., L.R. Comolli, B.C. Thomas, and J.F. Banfield. 2014. Community transcriptomics reveals unexpected high microbial diversity in acidophilic biofilm communities. *ISME Journal* 9: 1014–1023.

An, D., S.M. Caffrey, J. Soh, et al. 2013. Metagenomics of hydrocarbon resource environments indicates aerobic taxa and genes to be unexpectedly common. *Environmental Science & Technology* 47: 10708–10717.

An, D., X. Dong, A. An, H.-S. Park, M. Strous, and G. Voordouw. 2016. Metagenomic analysis indicates *Epsilonproteobacteria* as a potential cause of microbial corrosion in pipelines injected with bisulfite. *Frontiers in Microbiology* 7: 28.

Aoki, M., R. Kakiuchi, T. Yamaguchi, K. Takai, F. Inagaki, and H. Imachi. 2015. Phylogenetic diversity of *aprA* genes in subseafloor sediments on the northwestern pacific margin off Japan. *Microbes & Environment* 30: 276–280.

Apprill, A., S. McNally, R. Parsons, and L. Weber. 2015. Minor revision to V4 region SSU rRNA 806R gene primer greatly increases detection of SAR11 bacterioplankton. *Aquatic Microbial Ecology* 75: 129–137.

Arraiano, C. M., J. M. Andrade, S. Domingues, et al. 2010. The critical role of RNA processing and degradation in the control of gene expression. *FEMS Microbiology Reviews* 34: 883–923.

Basafa, M., and K. Hawboldt. 2019. Reservoir souring: sulfur chemistry in offshore oil and gas reservoir fluids. *Journal of Petroleum Exploration and Production Technology* 9: 1105–1118.

Beale, D. J., P. D. Morrison, and C. Key. 2014. Metabolic profiling of biofilm bacteria known to cause microbial influenced corrosion. *Water Science and Technology* 69: 1–8.

Beale, D. J., A. V. Karpe, S. Jadhav, T. H. Muster, and E. A. Palombo. 2016. Omics-based approaches and their use in the assessment of microbial-influenced corrosion of metals. *Corrosion Reviews* 34: 1–15.

Beech, I.B., and C. C. Gaylarde. 1999. Recent advances in the study of biocorrosion - an overview. *Revista de Microbiologia* 30: 117–190.

Beeder, J., R. K. Nilsen, J. T. Rosnes, T. Torsvik, and T. Lien. 1994. *Archaeoglobus fulgidus* isolated from hot North Sea oil field waters. *Applied and Environmental Microbiology* 60: 1227–1231.

Bell, E., T. Lamminmäki, J. Alneberg, et al. 2018. Biogeochemical cycling by a low-diversity microbial community in deep groundwater. *Frontiers in Microbiology* 9: 2129.

Ben-Dov, E., A. Brenner, and A. Kushmaro. 2007. Quantification of sulfate-reducing bacteria in industrial wastewater, by real-time polymerase chain reaction (PCR) using *dsrA* and *apsA* genes. *Microbial Ecology* 54: 439–451.

Blackwood, D. 2018. An electrochemist perspective of microbiologically influenced corrosion. *Corrosion and Materials Degradation* 1: 59–76.

Bonch-Osmolovskaya, E. A., M. L. Miroshnichenko, A. V. Lebedinsky, et al. 2003. Radioisotopic, culture-based, and oligonucleotide microchip analyses of thermophilic microbial communities in a continental high-temperature petroleum reservoir. *Applied and Environmental Microbiology* 69: 6143–6151.

Bonifay V., B. Wawrik, J. Sunner, et al. 2017. Metabolomic and metagenomic analysis of two crude oil production pipelines experiencing differential rates of corrosion. *Frontiers in Microbiology* 8: 99.

Booker, A. E., M. A. Borton, R. A. Daly, et al. 2017. Sulfide generation by dominant *Halanaerobium* microorganisms in hydraulically fractured shales. *mSphere* 2: e00257–e00217.

Braker, G., A. Fesefeldt, and K.-P. Witzel. 1998. Development of PCR primer systems for amplification of nitrite reductase genes (*nirK* and *nirS*) to detect denitrifying bacteria in environmental samples. *Applied and Environmental Microbiology* 64: 3769–3775.

Breitwieser, F. P., J. Lu, and S. L. Salzberg. 2019. A review of methods and databases for metagenomic classification and assembly. *Briefings in Bioinformatics* 20: 1125–1136.

Bru, D., A. Sarr, and L. Philippot. 2007. Relative abundance of proteobacterial membrane-bound and periplasmic nitrate reductases in selected environments. *Applied and Environmental Microbiology* 73: 5971–5974.

Bryers, J. D. 2008. Medical biofilms. *Biotechnology and Bioengineering* 100: 1–18.

Callbeck, C. C., A. Agrawal, and G. Voordouw. 2013. Acetate production from oil under sulfate-reducing conditions in bioreactors injected with sulfate and nitrate. *Applied and Environmental Microbiology* 79: 5059–5068.

Caporaso, J. G., C. L. Lauber, and W. A. Walters. 2011. Global patterns of 16S rRNA diversity at a depth of millions of sequences per sample. *Proceedings of the National Academy of Sciences* 108: 4516–4522.

Castelle, C. J., and J. F. Banfield. 2018. Major new microbial groups expand diversity and alter our understanding of the tree of life. *Cell* 172: 1181–1197.

Chakravorty, S., D. Helb, M. Burday, N. Connell, and D. Alland. 2007. A detailed analysis of 16S ribosomal RNA gene segments for the diagnosis of pathogenic bacteria. *Journal of Microbiological Methods* 69: 330–339.

Colwell, F.S., S. Boyd, M.E. Delwiche, D.W. Reed, T.J. Phelps, and D.T. Newby. 2008. Estimates of biogenic methane production rates in deep marine sediments at Hydrate Ridge, Cascadia Margin. *Applied and Environmental Microbiology* 74: 3444–3452.

Cruaud, P., A. Vigneron, C. Lucchetti-Miganeh, P. E. Ciron, A. Godfroy, and M. A. Cambon-Bonavita. 2014. Influence of DNA extraction method, 16S rRNA targeted hypervariable regions, and sample origin on microbial diversity detected by 454 pyrosequencing in marine chemosynthetic ecosystems. *Applied and Environmental Microbiology* 80: 4626–4639.

Dahle, H., F. Garshol, M. Madsen, and N. K. Birkeland. 2008. Microbial community structure analysis of produced water from a high-temperature North Sea Oil-Field. *Antonie van Leeuwenhoek* 93: 37–49.

Davidova I. A., K. E. Duncan, B. M. Perez-Ibarra, and J. M. Suflita. 2012. Involvement of thermophilic archaea in the biocorrosion of oil pipelines. *Environmental Microbiology* 14: 1762–1771.

Deng, X., N. Dohmae, K.H. Nealson, K. Hashimoto, and A. Okamoto. 2018. Multi-heme cytochromes provide a pathway for survival in energy-limited environments. *Science Advances* 4: eaao5682.

De Paula, R., C. St. Peter, A. Richardson, et al. 2018. DNA sequencing of oilfield samples: impact of protocol choices on the microbiological conclusions. NACE Corrosion 2018, NACE International, Paper #11662.

Deutzmann, J.S., M. Sahin, and A.M. Spormann. 2015. Extracellular enzymes facilitate electron uptake in biocorrosion and bioelectrosynthesis. *mBio* 6: e00496–e00415.

Doorenweerd, C., and K. Beentjes. 2012. Extensive guidelines for preserving specimen or tissue for later DNA work. Version 1.

dos Santos, C. J., D. Romaskevis, G. Lopes, et al. 2020. Diversity of sulfate-reducing prokaryotes in petroleum production water and oil samples. *International Biodeterioration and Biodegradation* 151: 104966.

Duncan, K.E., I.A. Davidova, H.S. Nunn, et al. 2017. Design features of offshore oil production platforms influence their susceptibility to biocorrosion. *Applied Microbiology and Biotechnology* 101: 6517–6529.

Duncan, K. E., L. M. Gieg, V. A. Parisi, et al. 2009. Biocorrosive thermophilic microbial communities in Alaskan North Slope oil facilities. *Environmental Science & Technology* 43: 7977–7984.

Eckert, R. B. 2015. Emphasis on biofilms can improve mitigation of microbiological influenced corrosion in oil and as industry. *Corrosion Engineering, Science and Technology* 50 (3): 163–168.
Eckert, R. B., and T. L. Skovhus. 2018. Advances in the application of molecular microbiological methods in the oil and gas industry and links to microbiologically influenced corrosion. *International Biodeterioration and Biodegradation* 126: 169–176.
Eden, B., P. J. Laycock, and M. Fielder. 1993. Oilfield reservoir souring, health and safety report – *Offshore Report* 92: OTH 385.
Enning, D., and J. Garrelfs. 2014. Corrosion of iron by sulfate-reducing bacteria: new views of an old problem. *Applied and Environmental Microbiology* 80: 1226–1236.
Enzien, M., and B. Yin. 2011. *New biocide formulations for oil and gas injection waters with improved environmental footprint.* In *Offshore Technology Conference*, 2–5 May, 0TC 21794: 1–7.
Fardeau, M.-L., J.-L. Cayol, M. Magot, and B. Ollivier. 1993. H_2 oxidation in the presence of thiosulfate, by a *Thermoanaerobacter* strain isolated from an oil-producing well. *FEMS Microbiology Letters* 113: 327–332.
Fida, T.T., C. Chen, G. Okpala, and G. Voordouw. 2016. Implications of limited thermophilicity of nitrite reduction for control of sulfide production in oil reservoirs. *Applied and Environmental Microbiology* 82: 4190–4199.
Finster, K., W. Liesack, and B. Thamdrup. 1998. Elemental sulfur and thiosulfate disproportionation by *Desulfocapsa sulfoexigens* sp. nov., a new anaerobic bacterium isolated from marine surface sediment. *Applied and Environmental Microbiology* 64: 119–125.
Flanagan, D. A., L. G. Gregory, J. P. Carter, A. Karakas-Sen, D. J. Richardson, and S. Spiro. 1999. Detection of genes for periplasmic nitrate reductase in nitrate respiring bacteria and in community DNA. *FEMS Microbiology Letters* 177: 263–270.
Fortin, N., D. Beaumier, K. Lee, and C. W. Greer. 2004. Soil washing improves the recovery of total community DNA from polluted and high organic content sediments. *Journal of Microbiological Methods* 56: 181–191.
Fortunato, C. S., L. Herfort, P. Zuber, A. M. Baptista, and B. C. Crump. 2012. Spatial variability overwhelms seasonal patterns in bacterioplankton communities across a river to ocean gradient. *ISME Journal* 6: 554–563.
Fredriksson, N. J., M. Hermansson, and B. M. Wilén. 2013. The choice of PCR primers has great impact on assessments of bacterial community diversity and dynamics in a wastewater treatment plant. *PLoS One* 8(10): e76431.
Gales, G., N. Tsesmetzis, I. Neria, et al. 2016. Preservation of ancestral cretaceous microflora recovered from a hypersaline oil reservoir. *Scientific Reports* 6: 22960.
Gao, P. K., G. Q. Li, H. M. Tian, Y. S. Wang, H. W. Sun, and T. Ma. 2015. Differences in microbial community composition between injection and production water samples of water flooding petroleum reservoirs. *Biogeosciences* 12: 3403–3414.
Gao, P., H. Tian, Y. Wang, et al. 2016. Spatial isolation and environmental factors drive distinct bacterial and archaeal communities in different types of petroleum reservoirs in China. *Scientific Reports*, 6: 20174.
Geets, J., B. Borremans, L. Diels, et al. 2006. DsrB gene-based DGGE for community and diversity surveys of sulfate-reducing bacteria. *Journal of Microbiological Methods* 66: 194–205.
Gieg, L. M., T. R. Jack, and J. M. Foght. 2011. Biological souring and mitigation in oil reservoirs. *Applied Microbiology and Biotechnology* 92: 263–282.
Gittel, A., M.V. W. Kofoed, K. B. Sørensen, K. Ingvorsen, and A. Schramm. 2012. Succession of *Deferribacteres* and *Epsilonproteobacteria* through a nitrate-treated high-temperature oil production facility. *Systematic and Applied Microbiology* 35: 165–174.
Gittel, A., K. B. Sørensen, T. L. Skovhus, K. Ingvorsen, and A. Schramm. 2009. Prokaryotic community structure and sulfate reducer activity in water from high-temperature oil reservoirs with and without nitrate treatment. *Applied and Environmental Microbiology* 75: 7086–7096.

Grassia, G. S., K. M. McLean, P. Glénat, A. Bauld, and A. J. Sheehy. 1996. A systematic survey for thermophilic fermentative bacteria and archaea in high temperature petroleum reservoirs. *FEMS Microbiology Ecology* 21: 47–58.

Greene, A. C., B. K. C. Patel, and A. J. Sheehy. 1997. *Deferribacter themzophilus* gen. nov., sp. nov., a novel thermophilic manganese- and iron-reducing bacterium isolated from a petroleum reservoir. *Strain* 47: 505–509.

Greene, E. A., C. Hubert, M. Nemati, G. E. Jenneman, and G. Voordouw. 2003. Nitrite reductase activity of sulphate-reducing bacteria prevents their inhibition by nitrate-reducing, sulphide- oxidizing bacteria. *Environmental Microbiology* 5: 607–617.

Gregoire, P., A. Engelbrektson, C. G. Hubbard, et al. 2014. Control of sulfidogenesis through bio-oxidation of H_2S coupled to (per)chlorate reduction. *Environmental Microbiology Reports* 6: 558–564.

Grigoryan, A., S. L. Cornish, B. Buziak, et al. 2008. Competitive oxidation of volatile fatty acids by sulfate- and nitrate-reducing bacteria from an oil field in Argentina. *Applied and Environmental Microbiology* 74: 4324–4335.

Guo, J., J. R. Cole, Q. Zhang, C. T. Brown, and J. M. Tiedje. 2016. Microbial community analysis with ribosomal gene fragments from shotgun metagenomes. *Applied and Environmental Microbiology* 82:157–166.

Haveman, S. A., A. Greene, C. P. Stilwell, J. K. Voordouw, and G. Voordouw. 2004. Physiological and gene expression analysis of inhibition of *Desulfovibrio vulgaris* Hildenborough by nitrite. *Journal of Bacteriology* 186: 7944–7950.

Hu, P., L. Tom, A. Singh, et al. 2016. Genome-resolved metagenomic analysis reveals roles for candidate phyla and other microbial community members in biogeochemical transformations in oil reservoirs. *MBio* 7 (1): 1–12.

Hubert, C., and G. Voordouw. 2007. Oil field souring control by nitrate-reducing *Sulfurospirillum* spp. that outcompete sulfate-reducing bacteria for organic electron donors. *Applied and Environmental Microbiology* 73: 2644–2652.

Hubert, C. 2010. Microbial ecology of oil reservoir souring and its control by nitrate injection. In *Handbook of Hydrocarbon and Lipid Microbiology*, Ed. K. N. Timmis, 2753–2766. Berlin, Heidelberg: Springer.

Hug, L. A., B. J. Baker, K. Anantharaman, et al. 2016. A new view of the tree of life. *Nature Microbiology* 11: 16048.

Hugerth, L. W., and A. F. Andersson. 2017. Analysing microbial community composition through amplicon sequencing: from sampling to hypothesis testing. *Frontiers in Microbiology* 8: 1561.

Hulecki, J. C., J. M. Foght, M. Gray, and P. M. Fedorak. 2009. Sulfide persistence in oilfield waters amended with nitrate and acetate. *Journal of Industrial Microbiology & Biotechnology* 36: 1499–1511.

Imhoff, J. F. 2016. New dimensions in microbial ecology-functional genes in studies to unravel the biodiversity and role of functional microbial groups in the environment. *Microorganisms* 4: 19.

Jackson, B. E., and M. J. McInerney. 2000. Thiosulfate disproportionation by *Desulfotomaculum thermobenzoicum*. *Applied and Environmental Microbiology* 66: 3650–3653.

Jia, R., T. Unsal, D. Xu, Y. Lekbach, and T. Gu. 2019. Microbiologically influenced corrosion and current mitigation strategies: a state of the art review. *International Biodeterioration and Biodegradation* 137: 42–58.

Johnson, R. J., B. D. Folwell, A. Wirekoh, M. Frenzel, and T. L. Skovhus. 2017. Reservoir souring – latest developments for application and mitigation. *Journal of Biotechnology* 256: 57–67.

Kaster, K. M., K. Bonaunet, H. Berland, G. Kjeilen-Eilertsen, and O. G. Brakstad. 2009. Characterisation of culture-independent and -dependent microbial communities in a high-temperature offshore chalk petroleum reservoir. *Antonie van Leeuwenhoek* 96: 423–439.

Kato, S., I. Yumoto, and Y. Kamagata. 2015. Isolation of acetogenic bacteria that induce biocorrosion by utilizing metallic iron as the sole electron donor. *Applied and Environmental Microbiology* 81: 67–73.

Kilbane, J. 2014. Effect of sample storage conditions on oilfield microbiological samples. NACE Corrosion 2014, NACE International, Paper #3788.

Klein, M., M. Friedrich, A. J. Roger, et al. 2001. Multiple lateral transfers of dissimilatory sulfite reductase genes between major lineages of sulfate-reducing prokaryotes. *Journal of Bacteriology* 183: 6028–6035.

Klindworth, A., E. Pruesse, T. Schweer, et al. 2013. Evaluation of general 16S ribosomal RNA gene PCR primers for classical and next-generation sequencing-based diversity studies. *Nucleic Acids Research* 41: e1.

Kip, N., and J. A. Van Veen. 2015. The dual role of microbes in corrosion. *ISME Journal* 9: 542–551.

Koch, G., J. Varney, N. Thopson, O. Moghissi, M. Gould, and J. Payer. 2016. International measures of prevention, application, and economics of corrosion technologies study (IMPACT). *NACE International*, Houston, TX, A-19.

Konopka, A. 2009. What is microbial community ecology? *ISME Journal* 3: 1223–1230.

Kotlar, H. K., A. Lewin, J. Johansen, et al. 2011. High coverage sequencing of DNA from microorganisms living in an oil reservoir 2.5 kilometres subsurface. *Environmental Microbiology Reports* 3: 674–681.

Kotu, S. P., M. S. Mannan, and A. Jayaraman. 2019. Emerging molecular techniques for studying microbial community composition and function in microbiologically influenced corrosion. *International Biodeterioration and Biodegradation* 144:v104722.

Korenblum, E., F. de Vasconcelos Goulart, I. de Almeida Rodrigues, et al. 2013. Antimicrobial action and anti-corrosion effect against sulfate reducing bacteria by lemongrass (*Cymbopogon citratus*) essential oil and its major component, *AMB Express* 3: 44.

Kumar, G., J. E. Eble, and M. R. Gaither. 2020. A practical guide to sample preservation and pre-PCR processing of aquatic environmental DNA. *Molecular Ecology Resources* 20: 29–39.

Lahme, S., J. Mand, J. Longwell, R. Smith, and D. Enning. 2020. Severe corrosion of carbon steel in oil field produced water can be linked to methanogenic archaea containing a special type of [NiFe] hydrogenase. *Applied and Environmental Microbiology*. doi: 10.1128/AEM.01819-20.

Lahme, S., D. Enning, C.M. Callbeck, et al. 2019. Metabolites of an oil field sulfide-oxidizing, nitrate-reducing *Sulfurimonas* sp. cause severe corrosion. *Applied and Environmental Microbiology* 85: e01891–e01818.

Lane, D. J. 1985. Rapid determination of 16S ribosomal RNA sequences for phylogenetic analyses. *Proceedings of the National Academy of Science USA* 82: 6955–6959.

Langille M., J. Zaneveld, J. Caporaso, et al. 2013. Predictive functional profiling of microbial communities using 16S rRNA marker gene sequences. *Nature Biotechnology* 31: 814–821.

Laramie, M. B., D. S. Pilliod, and C. S. Goldberg. 2015. Characterizing the distribution of an endangered salmonid using environmental DNA analysis. *Biological Conservation* 183: 29–37.

Larsen, J., M. H. Rod, and S. Zwolle. 2004. Prevention of reservoir souring in the Halfdan Field by nitrate injection. NACE Corrosion 2004, NACE International, Paper # 04761.

Leavitt, W. D., R. Cummins, M. L. Schmidt, et al. 2014. Multiple sulfur isotope signatures of sulfite and thiosulfate reduction by the model dissimilatory sulfate-reducer, *Desulfovibrio alaskensis* str. G20. *Frontiers in Microbiology* 5: 591.

Lee, C., B. Brown, D. Jones, et al. 2015. Field applications for on-site DNA extraction and qPCR. NACE Corrosion 2015, NACE International, Paper # 2015-5686.

Lee, J. S., J. M. McBeth, R. I. Ray, B. J. Little, and D. Emerson. 2013. Iron cycling at corroding carbon steel surfaces. *Biofouling* 29: 1–10.

Lejeune, P. 2003. Contamination of abiotic surfaces: what a colonizing bacterium sees and how to blur it. *Trends in Microbiology* 11: 179–184.

Lenhart, T. R., K. E. Duncan, I. B. Beech, et al. 2014. Identification and characterization of microbial biofilm communities associated with corroded oil pipeline surfaces. *Biofouling* 30: 823–835.

Lever, M. A., A. Torti, P. Eickenbusch, A. B. Michaud, T. Šantl-Temkiv, and B. B. Jørgensen. 2015. A modular method for the extraction of DNA and RNA, and the separation of DNA pools from diverse environmental sample types. *Frontiers in Microbiology* 6: 476.

Li, X.-X., J.-F. Liu, L. Zhou, et al. 2017. Diversity and composition of sulfate-reducing microbial communities based on genomic DNA and RNA transcription in production water of high temperature and corrosive oil reservoir. *Frontiers in Microbiology* 8: 1011.

Liang, R., I.A. Davidova, C.R. Marks, et al. 2016. Metabolic capability of a predominant *Halanaerobium* sp. in hydraulically-fractured gas wells and its implication in pipeline corrosion. *Frontiers in Microbiology* 7: 988.

Liang, R., R. S. Grizzle, K. E. Duncan, M. J. McInerney, and J. M. Suflita. 2014. Roles of thermophilic thiosulfate-reducing bacteria and methanogenic archaea in the biocorrosion of oil pipelines. *Frontiers in Microbiology* 5: 89.

Liang, B., K. Zhang, L. Y. Wang, et al. 2018. Different diversity and distribution of archaeal community in the aqueous and oil phases of production fluid from high-temperature petroleum reservoirs. *Frontiers in Microbiology* 9: 841.

Liebensteiner, M. G., N. Tsesmetzis, A. J. M. Stams, and B. P. Lomans. 2014. Microbial redox processes in deep subsurface environments and the potential application of (per)chlorate in oil reservoirs. *Frontiers in Microbiology* 5: 428.

Little, B. J., D.J. Blackwood, J. Hinks, et al. 2020. Microbially influenced corrosion - any progress? *Corrosion Science* 170: 108641.

Little, B., and J. S. Lee. 2007. *Microbiologically Influenced Corrosion*, 1st ed. New York: Wiley-Interscience.

Lipus, D., A. Vikram, D. Ross, et al. 2017. Predominance and metabolic potential of *Halanaerobium* spp. in produced water from hydraulically fractured Marcellus shale wells. *Applied and Environmental Microbiology* 83: 8.

Liu, Y., D. D. Galzerani, S. M. Mbadinga, et al. 2018. Metabolic capability and in situ activity of microorganisms in an oil reservoir. *Microbiome* 6: 5.

Lomans, B. P., R. De Paula, B. Geissler, C. A. T. Kuijvenhoven, and N. Tsesmetzis. 2016. Proposal of improved biomonitoring standard for purpose of microbiologically influenced corrosion risk assessment. SPE Paper #179919-MS.

Louca, S., M. Doebeli, and L. W. Parfrey. 2018. Correcting for 16S rRNA gene copy numbers in microbiome surveys remains an unsolved problem. *Microbiome* 6: 41.

Luton, P. E., J. M. Wayne, R. J. Sharp, and P. W. Riley. 2002. The *mcrA* gene as an alternative to 16S rRNA in the phylogenetic analysis of methanogen populations in landfill. *Microbiology* 148: 3521–3530.

Lv, M., and M. Du. 2018. A review: microbiologically influenced corrosion and the effect of cathodic polarization on typical bacteria. *Reviews in Environmental Science and Bio/Technology* 17: 431–446.

Magot, M., B. Ollivier, and B. K. C. Patel. 2000. Microbiology of petroleum reservoirs. *Antonie van Leeuwenhoek* 77: 103–116.

Magot, M. 2005. Indigenous microbial communities in oil fields, In *Petroleum Microbiology*, Eds. B. Ollivier and M. Magot, pp. 21–34. Washington, DC: ASM Press.

Martiny, A. C., K. Treseder, and G. Pusch. 2013. Phylogenetic conservatism of functional traits in microorganisms. *ISME Journal* 7: 830–838.

May L. A., J. L. Higgins, and C. M. Woodley. 2011. Saline-saturated DMSO-EDTA as a storage medium for microbial DNA analysis from coral mucus swab samples. *NOAA Technical Memorandum NOS NCCOS* 127 and CRCP 15: 14.

McDonnell, G., and D. Russell. 1999. Antiseptics and disinfectants: activity, action, and resistance. *Clinical Microbiology Reviews* 12: 147–179.

McGrath, K. C., R. Mondav, R. Sintrajaya, B. Slattery, S. Schmidt, and P. M. Schenk. 2010. Development of an environmental functional gene microarray for soil microbial communities. *Applied and Environmental Microbiology* 76: 7161–7170.

McGinley, H. R., M. V. Enzien, G. Hancock, S. Gonsior, and M. Miksztal. 2009. Glutaraldehyde: an understanding of of its ecotoxicity profile and environmental chemistry. NACE Corrosion 2009, NACE International, Paper #09405.

Menke S., M. A. F. Gillingham, K. Wilhelm, and S. Sommer. 2017. Home-made cost-effective preservation buffer is a better alternative to commercial preservation methods for microbiome research. *Frontiers in Microbiology* 8: 102.

Meyer B., and J. Kuever. 2007. Molecular analysis of the diversity of sulfate-reducing and sulfur-oxidizing prokaryotes in the environment, using *aprA* as functional marker gene. *Applied and Environmental Microbiology* 73: 7664–7679.

Mohan A. M., K. J. Bibby, D. Lipus, R. W. Hammack, and K. B. Gregory. 2014. The functional potential of microbial communities in hydraulic fracturing source water and produced water from natural gas extraction characterized by metagenomic sequencing. *PLoS One* 9: e107682.

Morono, Y., T. Terada, T. Hoshino, and F. Inagaki. 2014. Hot-alkaline DNA extraction method for deep-subseafloor archaeal communities. *Applied and Environmental Microbiology* 80: 1985–1994.

Morris, R., A. Schauer-Gimenez, U. Bhattad, et al. 2013. Methyl coenzyme M reductase (*mcrA*) gene abundance correlates with activity measurements of methanogenic H_2/CO_2-enriched biomass. *Microbial Biotechnology* 7: 77–84.

Mosier, A., Z. Li, B. Thomas, H. L. Hettich, C. Pan, and J. F. Banfield. 2015. Elevated temperature alters proteomic responses of individual organisms within a biofilm community. *ISME Journal* 9: 180–194.

Mouser P.J., M. Borton, T. H. Darrah, A. Hartsock, and K. C. Wrighton. 2016. Hydraulic fracturing offers view of microbial life in the deep terrestrial subsurface. *FEMS Microbiology Ecology* 92: fiw166.

Mueller, R. F., and P. H. Nielsen. 1996. Characterization of thermophilic consortia from two souring oil reservoirs. *Applied and Environmental Microbiology* 62: 3083–3087.

Müller, A.L., Kjeldsen, K.U., T. Rattei, M. Pester, and A. Loy. 2015. Phylogenetic and environmental diversity of DsrAB-type dissimilatory (bi)sulfite reductases. *ISME Journal* 9: 1152–1165.

Nazina, T. N., N. M. Shestakova, A. A. Grigor'yan, et al. 2006. Phylogenetic diversity and activity of anaerobic microorganisms of high-temperature horizons of the Dagang oil field (P.R. China). *Microbiology* 75: 55–65.

Nemati, M., G. E. Jenneman, and G. Voordouw. 2001. Mechanistic study of microbial control of hydrogen sulfide production in oil reservoirs. *Biotechnology and Bioengineering* 74: 424–434.

Nilsen, R. K., T. Terje, and T. Lien. 1996a. *Desulfotomaculum Thermocisternum* sp. nov., a sulfate reducer isolated from a hot North Sea oil reservoir. *International Journal of Systematic Bacteriology* 46: 397–402.

Nilsen, R. K., J. Beeder, T. Thorstenson, and T. Torsvik. 1996b. Distribution of thermophilic marine sulfate reducers in north sea oil field waters and oil reservoirs. *Applied and Environmental Microbiology* 62: 1793–1798.

Niu, S., S. Yang, S. McDermaid, et al. 2018. Bioinformatics tools for quantitative and functional metagenome and metatranscriptome data analysis in microbes, *Briefings in Bioinformatics* 19: 1415–1429.

Nunoura, T., H. Oida, J. Miyazaki, A. Miyashita, H. Imachi, and K. Takai. 2008. Quantification of *mcrA* by fluorescent PCR in methanogenic and methanotrophic microbial communities. *FEMS Microbiology Ecology* 64: 240–247.

Oldham, A. L., V. Sandifer, and K. E. Duncan. 2019. Effects of sample preservation on marine microbial diversity analysis. *Journal of Microbiological Methods* 158: 6–13.
Oldham, A. L., H. S. Drilling, B. W. Stamps, B. S. Stevenson, and K. E. Duncan. 2012. Automated DNA extraction platforms offer solutions to challenges of assessing microbial biofouling in oil production facilities. *AMB Express* 2: 60.
Okoro, C., S. Smith, L. Chiejina, et al. 2014. Comparison of microbial communities involved in souring and corrosion in offshore and onshore oil production facilities in Nigeria. *Journal of Industrial Microbiology and Biotechnology* 41: 665–678.
Okpala, G. N., C. Chen, T. Fida, and G. Voordouw. 2017. Effect of thermophilic nitrate reduction on sulfide production in high temperature oil reservoir samples. *Frontiers in Microbiology* 8: 1573.
Orphan, V. J., L. T. Taylor, D. Hafenbradl, and E. F. Delong. 2000. Culture-dependent and culture-independent characterization of microbial assemblages associated with high-temperature petroleum reservoirs. *Applied and Environmental Microbiology* 66: 700–711.
Parada, A. E., D. M. Needham, and A. M. Fuhrman. 2016. Every base matters: assessing small subunit rRNA primers for marine microbiomes with mock communities, time series and global field samples. *Environmental Microbiology* 18: 1403–1414.
Park, H.-S., Chatterjee, I., X. Dong, et al. 2011. Effect of sodium bisulfite injection on the microbial community composition in a brackish-water-transporting pipeline. *Applied and Environmental Microbiology* 77: 6908–6917.
Perez-Jimenez, J. R., and L. J. Kerkhof. 2005. Phylogeography of sulfate reducing bacteria among disturbed sediments, disclosed by analysis of the dissimilatory sulfite reductase genes (*dsrAB*). *Applied and Environmental Microbiology* 71: 1004–1011.
Piceno, Y. M., F. C. Reid, L. M. Tom, et al. 2014. Temperature and injection water source influence microbial community structure in four Alaskan North Slope hydrocarbon reservoirs. *Frontiers in Microbiology* 5: 13.
Priha, O., M. Nyyssönen, M. Bomberg, et al. 2013. Application of denaturing high-performance liquid chromatography for monitoring sulfate-reducing bacteria in oil fields. *Applied and Environmental Microbiology* 79: 5186–5196.
Procopio, L. 2019. The role of biofilms in the corrosion of steel in marine environments. *World Journal of Microbial Biotechnology* 35: 73.
Ravot, G., L. Casalot, O. Bernard, G. Loison, and M. Magot. 2005. *rdlA*, a new gene encoding a rhodanese-like protein in *Halanaerobium congolense* and other thiosulfate-reducing anaerobes. *Research in Microbiology* 156: 1031–1038.
Quillet, L., L. Besaury, M. Popova, S. Paissé, J. Deloffre, and B. Ouddane. 2012. Abundance, diversity and activity of sulfate-reducing prokaryotes in heavy metal-contaminated sediment from a salt marsh in the Medway Estuary (UK). *Biotechnology* 14: 363–381.
Quince, C., A. Lanzen, R. J. Davenport, and P. J. Turnbaugh. 2011. Removing noise from pyrosequenced amplicons. *BMC Bioinformatics* 12: 38.
Rachel, N.M., and L.M. Gieg. 2020. Preserving microbial community integrity in oilfield produced water. *Frontiers in Microbiology*, 11: 581387.
Reinsel, M. A., J. T. Sears, P. S. Stewart, and M. J. Mclnerney. 1996. Control of microbial souring by nitrate, nitrite or glutaraldehyde injection in a sandstone column. *Journal of Industrial Microbiology* 17: 128–136.
Ren, G., H. Zhang, X. Lin, J. Zhu, and Z. Jia. 2014. Response of phyllosphere bacterial communities to elevated CO_2 during rice growing season. *Applied Microbiology and Biotechnology* 98: 9459–9471.
Riesenfeld, C. S., P. D. Schloss, and J. Handelsman. 2004. Metagenomics: genomic analysis of microbial communities. *Annual Review of Genetics* 38: 525–552.
Renshaw, M. A., B. P. Olds, C. L. Jerde, M. M. Mcveigh, D. M. Lodge. 2014. The room temperature preservation of filtered environmental DNA samples and assimilation into a phenol–chloroform–isoamyl alcohol DNA extraction. *Molecular Ecology Resources* 15: 1.

Röling, W. F., I. M. Head, and S. R. Larter. 2003. The microbiology of hydrocarbon degradation in subsurface petroleum reservoirs: perspectives and prospects. *Research in Microbiology* 154: 321–328.

Salgar-Chaparro, S. J., and L. L. Machuca. 2019. Complementary DNA/RNA-based profiling: characterization of corrosive microbial communities and their functional profiles in an oil production facility. *Frontiers in Microbiology* 10: 2587.

Sambo, F., F. Finotello, E. Lavezzo, et al. 2018. Optimizing PCR primers targeting the bacterial 16S ribosomal RNA gene. *BMC Bioinformatics* 19: 343.

Savage, K. N., L. R. Krumholz, L. M. Gieg, et al. 2010. Biodegradation of low-molecular-weight alkanes under mesophilic, sulfate-reducing conditions: metabolic intermediates and community patterns. *FEMS Microbiology Ecology* 72: 485–495.

Shakya, M., C. C. Lo, and P. S. G. Chain. 2019. Advances and challenges in metatranscriptomic analysis. *Frontiers in Genetics* 10: 904.

Sharma, N., and W. Huang. 2019a. Expanding industry access to molecular microbiological methods: development of an off-the-shelf laboratory workflow for qPCR and NGS analysis. NACE Corrosion 2019, NACE International, Paper #13033.

Sharma, N., and W. Huang. 2019b. Rapid in-field collection and ambient temperature preservation of corrosion-related microbial samples for downstream molecular analysis. In *Oilfield Microbiology*, Eds. T. L. Skovhus, and C. Whitby. Boca Raton, FL: CRC Press.

Sharma, M., and G. Voordouw. 2016. MIC detection and assessment - A holistic approach. In: *Microbiologically Influenced Corrosion in the Upstream Oil and Gas Industry*, Eds. T.L. Skovhus, D. Enning, and J.S. Lee, 177–212. Boca Raton, FL: CRC Press.

Shi, Y., G. W. Tyson, and E. F. DeLong. 2009. Metatranscriptomics reveals unique microbial small RNAs in the ocean's water column. *Nature* 459: 266–269.

Skovhus, T. L., R. B. Eckert, and E. Rodrigues. 2017. Management and control of microbiologically influenced corrosion (MIC) in the oil and gas industry - overview and a North Sea case study. *Journal of Biotechnology* 256: 31–45.

Slobodkin, A. I., D. G. Zavarzina, and T. G. Sokolova. 1999. Dissimilatory reduction of inorganic electron acceptors by thermophilic anaerobic prokaryotes. *Mikrobiologiia* 68: 522–542.

Smock, A. M., M. E. Böttcher, and H. Cypionka. 1998. Fractionation of sulfur isotopes during thiosulfate reduction by *Desulfovibrio desulfuricans*. *Archives of Microbiology* 169: 460–463.

Spence, C., T. R. Whitehead, and M. A. Cotta. 2008. Development and comparison of SYBR Green quantitative real-time PCR assays for detection and enumeration of sulfate-reducing bacteria in stored swine manure. *Journal of Applied Microbiology* 105: 2143–2152.

Staley, J., and A. Konopka. 1985. Measurement of in situ activities of nonphotosynthetic microorganisms in aquatic and terrestrial habitats. *Annual Review of Microbiology* 39: 321–3346.

Stein E. D., B. P. White, R. D. Mazor, P. E. Miller, and E. M. Pilgrim. 2013. Evaluating ethanol-based sample preservation to facilitate use of dna barcoding in routine freshwater biomonitoring programs using benthic macroinvertebrates. *PLoS One* 8(1): e51273.

Suarez, C., M. Piculell, O. Modin, et al. 2019. Thickness determines microbial community structure and function in nitrifying biofilms via deterministic assembly. *Scientific Reports* 9: 5110.

Suzuki, M.T., L. T. Taylor, and E. F. DeLong. 2000. Quantitative analysis of small-subunit rRNA genes in mixed microbial populations via 5'-nuclease assays. *Applied and Environmental Microbiology* 66: 4605–4714.

Tatangelo, V., A. Franzetti, I. Gandolfi, G. Bestetti, and R. Ambrosini. 2014. Effect of preservation method on the assessment of bacterial community structure in soil and water samples. *FEMS Microbiology Letters* 356:32–38.

Tremblay, J., K. Singh, A. Fern, et al. 2015. Primer and platform effects on 16S rRNA tag sequencing. *Frontiers in Microbiology* 6: 771.

Thijs, S., D. P. De Beeck, B. Beckers, et al. 2017. Comparative evaluation of four bacteria-specific primer pairs for 16S rRNA gene surveys. *Frontiers in Microbiology* 8: 494.

Thomas, T., J. Gilbert, and F. Meyer. 2012. Metagenomics - a guide from sampling to data analysis. *Microbial Informatics and Experimentation* 21: 3.

Tian, H., P. Gao, Z. Chen, et al. 2017. Compositions and abundances of sulfate-reducing and sulfur-oxidizing microorganisms in water-flooded petroleum reservoirs with different temperatures in China. *Frontiers in Microbiology* 8: 143.

Tsesmetzis, N., M. J. Maguire, I. M. Head, and B. P. Lomans. 2016. Protocols for investigating the microbial communities of oil and gas reservoirs. In *Hydrocarbon and Lipid Microbiology Protocols*, Eds T. McGenity, K.N. Timmis, B. Nogales Fernández. Berlin Heidelberg: Springer-Verlag.

Tsurumaru, H., N. Ito, K. Mori, et al. 2018. An extracellular [NiFe] hydrogenase mediating iron corrosion is encoded in a genetically unstable genomic island in *Methanococcus maripaludis*. *Scientific Reports* 8: 15149.

Tuorto, S. J., C. M. Brown, K. D. Bidle, L. R. McGuinness, and L. J. Kerkhof. 2015. BioDry: an inexpensive, low-power method to preserve aquatic microbial biomass at room temperature. *PLoS One* 10: e0144686.

Urich, T., A. Lanzén, J. Qi, D. H. Huson, C. Schleper, and S. C. Schuster. 2008. Simultaneous assessment of soil microbial community structure and function through analysis of the meta-transcriptome. *PLoS One* 3: e2527.

van der Kraan, G. M., J. Bruining, B. P. Lomans, M. C. M. van Loosdrecht, and G. Muyzer. 2010. Microbial diversity of an oil–water processing site and its associated oil field: the possible role of microorganisms as information carriers from oil-associated environments, *FEMS Microbiology Ecology* 71: 428–443.

Verbaendert, I., S. Hoefman, P. Boeckx, N. Boon, and P. De Vos. 2014. Primers for overlooked *nirK*, *qnorB*, and *nosZ* genes of thermophilic Gram-positive denitrifiers. *FEMS Microbiology Ecology* 89: 162–180.

Videla, H. A., and L. K. Herrera. 2005. Microbiologically influenced corrosion: looking to the future. *International Microbiology* 8: 169–180.

Vigneron, A., I. M. Head, and N. Tsesmetzis. 2018. Damage to offshore production facilities by corrosive microbial biofilms. *Applied Microbiology and Biotechnology* 102: 2525–2533.

Vigneron, A., E. B. Alsop, B. P. Lomans, N. C. Kyrpides, I. M. Head, and N. Tsesmetzis. 2017. Succession in the petroleum reservoir microbiome through an oil field production life cycle. *ISME Journal* 11: 2141–2154.

Vigneron, A., E. B. Alsop, B. Chambers, B. P. Lomans, I. M. Head, and N. Tsesmetzis. 2016. Complementary microorganisms in highly corrosive biofilms from an offshore oil production facility. *Applied and Environmental Microbiology* 82: 2545–2554.

Vilcaez, J., M. Sanzo, S. Koichi, and I. Chihiro. 2007. Numerical evaluation of biocide treatment against sulfate reducing bacteria in oilfield water pipelines. *Journal of the Japan Petroleum Institute* 50: 208–217.

Voordouw, G., A. A. Grigoryan, A. Lambo, et al. 2009. Sulfide remediation by pulsed injection of nitrate into a low temperature canadian heavy oil reservoir. *Environmental Science & Technology* 43: 9512–9518.

Wagner, M., A. J. Roger, J. L. Flax, G. A. Brusseau, and D. A. Stahl. 1998. Phylogeny of dissimilatory sulfite reductases supports an early origin of sulfate respiration. *Journal of Bacteriology* 180: 2975–2982.

Walters, W., E. R. Hyde, D. Berg-Lyons, et al. 2016. Improved bacterial 16S rRNA gene (V4 and V4-5) and fungal internal transcribed spacer marker gene primers for microbial community surveys. *MSystems* 1: e00009–e00015.

Wang, L. Y., R. Y. Duan, J. F. Liu, S. Z. Yang, J. D. Gu, and B. Z. Mu. 2012. Molecular analysis of the microbial community structures in water-flooding petroleum reservoirs with different temperatures. *Biogeosciences* 9: 5177–5203.

Wang, L., W. Ke, X. Sun, et al. 2014. Comparison of bacterial community in aqueous and oil phases of water-flooded petroleum reservoirs using pyrosequencing and clone library approaches. *Applied Microbiology & Biotechnology* 98: 4209–4221.

Wang, I., and R. E. Melchers. 2017. Long-term under-deposit pitting corrosion of carbon steel pipes. *Ocean Engineering* 133: 231–243.

Wang, X., X. Li, L. Yu, et al. 2019. Characterizing the microbiome in petroleum reservoir flooded by different water sources. *Science of the Total Environment* 653: 872–885.

Wentzel, A., A. Lewin, F. J. Cervantes, S. Valla, and H. K. Kotlar. 2013. Deep subsurface oil reservoirs as poly-extreme habitats for microbial life. A current review. In *Polyextremophiles: Life under Multiple Forms of Stress*, Eds. J. Seckback, O. Aharon, and S. Helga, 27:439–466. Dordrecht: Springer.

Yamane, K., H. Yoshiyuki, O. Hiroshi, and F. Kazuhiro. 2011. Microbial diversity with dominance of 16S rRNA gene sequences with high GC contents at 74 and 98°C subsurface crude oil deposits in Japan. *FEMS Microbiology Ecology* 76: 220–235.

Yu, Y., C. Lee, J. Kim, and S. Hwang. 2005. Group-specific primer and probe sets to detect methanogenic communities using quantitative real-time polymerase chain reaction. *Biotechnology and Bioengineering* 89: 670–679.

Yuan, J., M. Li, and S. Lin. 2015. An improved DNA extraction method for efficient and quantitative recovery of phytoplankton diversity in natural assemblages. *PLoS One* 10: e0133060.

Zhang, C., F. Wen, and Y. Cao. 2011. Progress in research of corrosion and protection by sulfate-reducing bacteria. *Procedia Environmental Sciences*, 10 (Part B): 1177–1182.

Zhang, F., S. Yue-Hui, S. Chai, et al. 2012. Microbial diversity in long-term water-flooded oil reservoirs with different in situ temperatures in China. *Scientific Reports* 2: 1–10.

Zhou, J., H. Zhili, Y. Yunfeng, et al. 2015. High-throughput metagenomic technologies for complex microbial community analysis: open and closed formats. *MBio* 6(1): e02288.

Zhou, L., Y. Lu, D. Wang, et al. 2020. Microbial community composition and diversity in production water of a high-temperature offshore oil reservoir assessed by DNA- and RNA-based analyses. *International Biodeterioration and Biodegradation* 151: 104970.

Zumft, W. G. 1997. Cell biology and molecular basis of denitrification. *Microbiology Molecular Biology Reviews* 61: 533–616.

10 Microbial Reservoir Souring

Communities Relating to the Initiation, Propagation, and Remediation of Souring

Matthew Streets and Leanne Walker

CONTENTS

10.1 Introduction 207
10.1.1 Overview of Souring 207
10.1.2 Sulfide-Generating Microorganisms 208
10.1.3 Mitigation Strategies 209
10.1.4 Analytical Methods 209
10.2 Initiation and Propagation 210
10.2.1 Overview 210
10.2.2 Laboratory Experiments 211
10.3 Microbiological Nitrate Reduction 214
10.3.1 Overview 214
10.3.2 Field Observations 216
10.3.3 Laboratory Experiments 217
10.4 Conclusion 221
Acknowledgments 222
References 222

10.1 INTRODUCTION

10.1.1 Overview of Souring

Microbiological sulfide generation was recognized in the late 1980s as the leading cause of reservoir souring, which is defined as the downhole generation of hydrogen sulfide during secondary oil recovery operations, and the subsequent production of "sour" (i.e. sulfide containing) fluids at topsides facilities (Eden et al. 1993). The introduction of sulfate and a complex microflora via seawater injection during secondary recovery, which has the primary aims of pressure maintenance and some additional oil production via "sweep," results in the establishment of a downhole microbiological community that is able to enzymatically reduce sulfate and generate sulfide at potentially deleterious concentrations to production and export facilities (Gieg et al. 2011). In our own studies, we have observed aqueous sulfide concentrations in excess of 31mM (1,000mg/L).

Reservoir souring results in significant additional expenditure for the operation of oil and gas fields due to health and safety implications, sulfide stress cracking of metallurgy, increased production chemical demands (e.g. biocide, sulfide scavenger, and nitrate formulations) and also the diminished value of sour crude oil (Crook et al. 2015). It is therefore essential for oil and gas operators to be aware of the key microorganisms related to reservoir souring, as well as determining the most effective treatment regimes in order to mitigate deleterious microbiological growth downhole.

Within this chapter, we will be reviewing the results from pressurized, sand-packed bioreactor experiments, some of which were sponsored by operators and service companies within the Oil and Gas Industry. These results can be used to determine the microflora which may be present within a reservoir formation (i.e. at simulated pressure and temperature (PT) conditions) under secondary recovery conditions, as well as highlight any shifts in the downhole microbiological community due to the introduction of various souring prevention and souring remediation chemical treatment strategies.

10.1.2 Sulfide-Generating Microorganisms

Sulfide-generating microorganisms are introduced into the downhole environment via the injection of seawater during secondary recovery. These microorganisms can utilize various sulfur-containing ionic and non-ionic species, for example sulfate, thiosulfate, and elemental sulfur, and generate hydrogen sulfide as a by-product.

Sulfate-reducing microorganisms (SRM) are the main contributing group of microorganisms associated with reservoir souring and are able to utilize the sulfate in the injected seawater alongside metabolizable carbon sources found downhole (e.g. volatile fatty acids (VFAs) or hydrocarbon components of the crude oil) within an anaerobic respiratory process resulting in the final production of hydrogen sulfide as a waste product.

Various genera of SRM have been isolated from oil and gas production fluids throughout the years, and current offshore monitoring techniques for both souring and microbiologically influenced corrosion (MIC) include the assessment of SRM abundance in injection and production facilities.

Common SRM genera comprise both mesophilic and thermophilic microorganisms; however, the greater proportion of downhole sulfide is assumed to be produced by the faster growing mesophilic SRM, located close to the near-wellbore of the injector that is introduced into the formation by the commencement of seawater injection (Vigneron et al. 2017). Sulfate-reducing microorganisms are able to colonize the water-flooded zone between an injector-producer (I/P) pair and are postulated to be located mainly along fracture faces (in-house research, Rawwater Engineering Company Limited). However, the microorganisms are also able to penetrate the formation itself in reservoirs with certain geological compositions (e.g. unconsolidated sand and highly permeable sandstone reservoirs).

This chapter will present the results from various pressurized laboratory simulation studies whereby the effects of pressure, temperature, flow rates, and injection chemistries on the establishment of a sulfide-producing microbiological consortium are analyzed. These environmental factors are known to be important selective pressures on the injected marine microbiome and result in the selection and proliferation of microorganisms downhole.

10.1.3 Mitigation Strategies

Typical mitigation strategies applied in the field to combat microbiological reservoir souring can be classified into two groups: biocides and competitive exclusion treatments. Both groups involve chemical treatment of injection water or the topsides water injection facilities to prevent and control SRM activity.

Biocides are broad-spectrum organic chemical treatments that have wide-ranging modes of action to kill microbiological species. Common biocides utilized within the Oil and Gas Industry include glutaraldehyde and tetrakis(hydroxymethyl)phosphonium sulfate (THPS) (Xue and Voordouw 2015).

Competitive exclusion treatments include the addition of nitrate or potentially perchlorate ions into the injection water, with the aim of promoting the growth of nitrate- or perchlorate-reducing microorganisms (NRM and PRM) within the downhole formation. As both nitrate and perchlorate reduction are more energetically favorable to sulfate reduction, the injection of these treatments results in the promotion of NRM/PRM growth which is then able to out-compete the SRM for surface area and metabolizable carbon (Wu et al. 2018).

10.1.4 Analytical Methods

Traditional methods utilized within the Oil and Gas Industry for determining the presence and abundance of SRM typically involve the use of serial dilution selective media vials to calculate the most probable number (MPN) of SRMs within a given sample (National Association of Corrosion Engineers 2004). This technique has been widely accepted as an industry standard for determining the abundance of SRM. The results from MPN analysis are used in the calculations for generating potential risk ratings for operations; however, there are well known limitations with this method.

MPN analysis can be readily performed offshore by those working on the platform and can provide results after only 28 days of incubation. However, one of the key limitations of this method is that not all SRM can be cultured within Modified Postgate B media and would therefore be excluded from any subsequent risk analysis and yet could be playing a significant role in reservoir souring.

Various publications have compared the strengths and weaknesses of traditional microbiological techniques with newer molecular microbiological methods (MMM), such as qPCR and 16S rRNA metagenomic analysis, for use in the Oil and Gas Industry (Juhler, 2012, Johnson et al., 2017). However, MMM are becoming more prevalent within the industry to profile microbiological community shifts overtime in order to evaluate the potential for production-associated problems and the development of microbiological control strategies (Folarin et al. 2013, Vigneron et al. 2017).

The qPCR method of analysis utilises specific gene primers to target and amplify the signal for target genes in an initial sample. For example, a primer targeting the *dsrA* gene is commonly used for determining the presence and abundance of SRMs within a sample. Due to ongoing development in MMM, the costs and turnaround times associated with the analysis of samples via qPCR have decreased significantly over the years. However, this method of analysis is not able to differentiate between viable and non-viable cells within a sample and therefore can often result in the over-representation of microorganisms within specific metabolic groups. qPCR techniques are also not able to determine specific genera that may be present within a sample, as the gene primers utilised are commonly related to target genes associated with specific metabolic processes.

The 16S rRNA metagenomic analysis technique involves DNA amplification via targeting 16S rRNA gene sequences with universal primers to isolate bacterial and archaeal DNA within the sample. The amplified sequences are then processed through an analytical pipeline (commonly Quantitative Insights Into Microbial Ecology, or 'QIIME') and are clustered into operational taxonomic units (OTUs) for further processing and sequence analysis. The end product of this analytical method is that all prokaryotic 16S rRNA gene sequences within the samples can be identified and compared against gene libraries, resulting in the identification and estimated abundance of specific prokaryotic classes, families, and genera (Klindworth et al. 2013, Bonifay et al. 2017).

Analyses of produced water samples from North Sea oil fields using 16S rRNA metagenomic analysis have commonly identified bacteria within the *Firmicutes*, *Bacteroidetes*, and δ-proteobacteria classifications. However, variations between produced fluid samples have also been observed, indicating that differences in reservoir operational parameters (for example reservoir pressure, temperature, injection chemistry, and oxygen control) can also have a significant effect on the communities within produced fluids (Dahle et al. 2008, Stevenson et al. 2011, Vigneron et al. 2017).

Produced fluids and core samples from the pressurized bioreactor experiments described in this chapter have been analyzed with a combination of both qPCR and 16S rRNA metagenomic analysis, in order to demonstrate how microbiological communities associated with reservoir souring shift overtime during the development of a competent sulfide-generating biomass, and also how microbiological communities respond to various remediation treatments.

10.2 INITIATION AND PROPAGATION

10.2.1 Overview

As outlined previously, the initiation and propagation of downhole microbiological souring is facilitated by the commencement of seawater injection into the formation during secondary recovery processes. The seawater includes sulfate ions which, along with other compounds such as thiosulfate, can be enzymatically reduced by specific prokaryotes downhole. This results in the production of hydrogen sulfide species as a waste product which are subsequently produced at the topsides facilities upon the breakthrough of injection fluids at the production platform (Eden et al. 1993).

Seawater contains a diverse microbiological community of both aerobic and anaerobic species (Fuhrman et al. 2015, Ribicic et al. 2018). The downhole environment of the oil reservoir enacts a strong selection pressure which shifts the relative abundance of specific microorganisms toward those which are able to proliferate under these extreme conditions. It has been well documented that the formation PT conditions and the strong reducing environment of the downhole formation all combine to result in a population shift away from a seawater community profile toward those commonly observed in production fluids (Vigneron et al. 2017).

In the following pressurized bioreactor studies, the initiation and propagation of a sulfidogenic microbiological community was investigated under various PT conditions to replicate the downhole environment in the field. This resulted in the selection and proliferation of specific microorganisms that were significantly distinct from one another depending on growth conditions.

10.2.2 Laboratory Experiments

Historically, most laboratory-based experiments have been performed under quasi-environmental field conditions, only considering the impact of temperature. More recently, it has been shown that the impact of studying oilfield microbiology under truly simulated field conditions (environmental PT and pH combinations) is vital when translating data from laboratory studies to the field.

Streets et al. (2014) described a laboratory experiment in which a diverse microbiological community was exposed to different simulated field conditions of a Ghanaian oilfield. The study was conducted in order to evaluate the variation in microbiological, hydrocarbon-mediated sulfide production under different PT conditions, which could feedback into an existing souring forecast of the field. The experiments were performed in three sand-packed columns which were saturated with crude oil from the Ghanaian asset and flooded with anoxic (deoxygenated) synthetic seawater to remove the mobile oil phase, leaving only a residual oil phase saturating the core matrix.

A low-pressure, low-temperature (LP/LT) bioreactor that was 0.75m in length with an internal diameter (ID) of 0.025m was operated under daily batch injection (35mL per day over a 30-minute injection cycle) at 7.0MPa, 297K (1,000psig, 24°C / 75°F) in order to determine the souring propensity of the crude oil under environmental conditions known to support significant SRM activity. Two high-pressure bioreactors (2m length, 0.008m ID) were operated under monthly batch injection (150mL per injection cycle over a 30-minute injection cycle) at simulated, downhole field conditions: a high-pressure, low-temperature (HP/LT) bioreactor was operated at 48.4MPa, 297K (7,000psig, 24°C / 75°F) to mimic the near wellbore of an injector within the Ghanaian asset, and a high-pressure, high-temperature (HP/HT) bioreactor was operated at 48.4MPa, 368K (7,000psig, 95°C / 203°F) which simulated a deeper proportion of the reservoir between the I/P pair. All influent water consisted of anoxic synthetic seawater (90% v/v) and anoxic local (Irish Sea) seawater (10% v/v) containing viable microorganisms with an influent sulfate concentration of

28mM, naturally buffered at pH 8.2 (termed “anoxic 90:10”). The anoxic condition was achieved through nitrogen gas sparging to a dissolved oxygen concentration of less than 100ppbv prior to treatment with oxygen scavenger. The influent waters contained no significant sulfide concentrations (<0.02mM).

The inoculum source for all three bioreactors was taken from an archived, oil-saturated, sour bioreactor which had historically operated at 7.0MPa, 297K (1,000psig, 20°C / 203°F), known to contain sulfate-reducers, nitrate-reducers, hydrocarbon utilizers, acid producers, and general heterotrophs. The influent of the two HP bioreactors was also supplemented with additional mixed inoculum from asset produced water samples prior to injection to ensure continuous microbiological challenging of the high-pressure systems.

From traditional MPN analysis, viable SRM (in excess of 10^3 cells per mL) were detected in the anoxic effluent waters of the LP/LT bioreactor after 14 days of operation, and total sulfide concentrations in excess of the 0.02mM limit of quantification (LoQ) were measured from the LP/LT bioreactor after 60 days of operation, peaking at 0.46mM after 180 days of injection. No significant total sulfide concentrations were measured from either HP bioreactor after 270 days of operation. However, viable SRM were detected in the anoxic effluent waters of the HP/LT bioreactor after 120 days of operation. Throughout the 360-day study, no viable SRM were detected in effluent waters from the HP/HT bioreactor.

Following confirmation of persistent viable SRM in the bioreactor effluent water, the operating pressure of the HP/LT bioreactor was decreased from 48.4MPa to 7.0MPa for the final 90 days of the experiment. In addition, the influent for the HP/LT column was switched to non-supplemented baseline water, similar to the LP/LT bioreactor. The subsequent measured production of sulfide (0.42mM) from the HP/LT column was attributed to the previously dormant microbiological communities within the pressurized column, confirmed via MPN analysis from the 3-log increase in the effluent MPN/mL. The experimental results from the three pressurized bioreactors indicated that the elevated operating pressure of 48.4MPa (7,000psig) was sufficiently high to suppress significant microbiological hydrocarbon-mediated sulfide production.

In another study, Streets et al. (2015) described how a mesophilic sulfide-producing community would perform optimally at its preconditioned operating pressure.

The laboratory pressure range study consisted of six oil-saturated, sand-packed bioreactors, all 0.75m in length with an ID of 0.025m. The saturating crude oil was known to support significant microbiological sulfide production under mesophilic conditions. The columns all received anoxic 90:10 under weekly batch injection (150mL per injection cycle over a 30-minute injection cycle) with an influent sulfate concentration of 28mM. The influent water was also supplemented with a mixed VFA concentration of 0.64mM (56:4:4 ratio for acetate, propionate, and butyrate, respectively) which, from stoichiometric calculations, could support 0.72mM sulfide production. As per the Ghanaian oilfield asset assessment study, the inoculum source was taken from an archived, oil-saturated, sour bioreactor which had historically operated at 7.0MPa, 297K (1,000psig, 20°C), known to contain sulfate-reducers, nitrate-reducers, hydrocarbon utilizers, acid producers, and general heterotrophs.

Duplicate pressurized bioreactors were operated under mesophilic conditions (297K) at one of three pressures: 0.1MPa, 7.0MPa, and 20.8MPa. Influent and effluent samples were captured at 0.1MPa and concentration measurements were made for VFAs and total sulfide at each injection cycle.

Throughout the study, all six bioreactors demonstrated VFA utilization. As described in Figure 10.1, significant sulfide production above that which would be supported by the influent mixed VFA concentration alone was measured, indicating microbiological hydrocarbon-mediated sulfide production. After 370 days of operation, the average total mass of sulfide produced from the 7.0MPa columns (8.1mg) was 1.6 times greater than the 0.1MPa columns (5.1mg), and 2.3 times greater than the 20.8MPa columns (3.5mg).

The water chemistry results confirmed that microbiological, VFA-, and hydrocarbon-mediated sulfide production in the sand-packed columns was not inhibited between 0.1MPa and 20.8MPa.

During the first 110 days of the bioreactor study, a bottle test study was conducted in parallel in order to compare the activity of the microbiological communities in the two different incubation methods. The bottles contained sand, anoxic 90:10, mixed VFAs, and crude oil and were inoculated using effluent from the archived, oil-saturated, sour bioreactor. From multiplex sequencing of the V4-V5 regions of the 16S rRNA genes (Ion Torrent Personal Genome Machine (PGM) and analysis using QIIME), significant microbiological community differences were observed between the water extracted from the bottle test study, and the effluent and core samples from all six bioreactors (de Rezende et al., 2016).

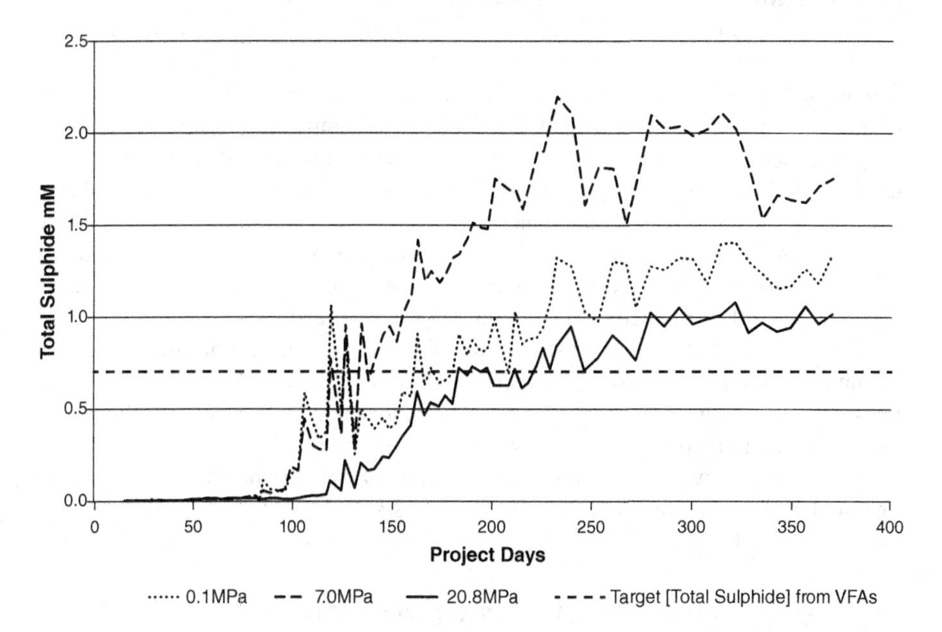

FIGURE 10.1 Average sulfide production profiles from the duplicate bioreactor sets. Effluent water samples were taken during each injection cycle for both sulfide and VFA determination. Samples for total sulfide concentration were collected directly onto zinc acetate crystals and measured as described by Moest (1975), and VFA determination was conducted via ion chromatography analysis.

From Principle Coordinate Analysis (PCoA), it was observed that the incubation method (bottle test or bioreactor effluent) was the most influential factor in separating the communities, having a greater effect than either incubation time or operating pressure. The dataset also described further 16S rRNA gene profiling comparing the microbiological communities from the effluent water samples with the communities associated with the core material from the injector and producer ends of the bioreactors. The distances between the communities were calculated by weighted UniFrac, which considered the phylogenetic distances and relative abundances. From alpha diversity analysis, it was demonstrated that microbiological communities associated with each bioreactor section were significantly different from one another, with favorable growth at the injector end.

In longer-term studies which complement the literature, mesophilic communities established under strict anoxic, residual oil conditions shift from the utilization of more soluble hydrocarbon components such as toluene (Ommedal and Torsvik 2007), to the degradation of longer-chain hydrocarbons, such as polyaromatic hydrocarbon, PAH (Terrisse et al. 2017).

One set of internal bioreactors was operated at 7.0MPa, 303K (1,000psig, 30°C) with the residual crude oil acting as the sole carbon energy source for microbiological growth. Microbiological sulfide production was observed over a period of 2,000 days, and the community shift in the systems was documented from a dominance of *Desulfotignum* in the initial 200 days, to marine obligate hydrocarbonoclastic microorganisms (MOHCM) genera, such as *Cycloclasticus* and *Alcanivorax*.

Thermophilic community analysis is also considered to be important in understanding its contribution to sour gas production in higher temperature sections of the downhole Thermal Viability Shell (TVS), defined by Eden et al. (1993) as that portion of a water-flooded reservoir which, by virtue of its temperature limits alone, could support either mesophilic or thermophilic SRM. From unpublished internal studies, it has been demonstrated that pressure plays a vital role in shaping the nature and productivity of the biofilm. The dominance of *Thermovirga*, *Thermotoga*, *Petrotoga*, and *Marinobacter* in pressurized bioreactors at 7.0MPa, 333K (1,000psig, 60°C) under VFA and residual oil conditions reflects field-based observations made in the North Sea and the Gulf of Mexico (Miranda-Tello et al. 2004; Dahle et al. 2008).

These pressurized bioreactor studies highlight the complexity of microbiological community development over long periods of time, the requirement to consider an optimized sampling strategy in order to obtain representative samples, and the importance of simulating the ecophysiology of oilfield microorganisms in the laboratory to create a realistic downhole ecosystem. At the time of writing, the oldest continuously operated bioreactors in our laboratories were six years old, and shut-in reactors fifteen years old.

10.3 MICROBIOLOGICAL NITRATE REDUCTION

10.3.1 Overview

Harnessing microbiological nitrate reduction to control the production of biogenically soured fluids has been historically and routinely used in the wastewater industry (Heukelekian 1943). Nitrate dosing of water diminishes biogenic sour gas production via a microbiological competitive exclusion mechanism in the presence of nitrate-reducing microorganisms (NRM). Alternatively, thermodynamics indicate that nitrate reduction is energetically more favourable to an organism than sulfate reduction (Lovley and Chapelle 1995), and a number of SRM can use nitrate as a terminal electron acceptor (Marietou 2016).

Nitrate-reducing microorganisms typically fall into one of two main categories: chemolithotrophs and chemoorganotrophs. Chemolithotrophs utilize inorganic compounds as electron donors, for instance thiosulfate (Chung et al. 2014) and sulfide (Gevertz et al. 2000). Chemoorganotrophs on the other hand couple nitrate reduction with the oxidation of organic compounds. The free energy yield comparison between SRM and NRM of one such organic compound, toluene, in anoxic environments is described below (Heider et al. 1998):

$$\begin{aligned} \text{SRM}: C_7H_8 + 4.5SO_4^{2-} + 3H_2O \rightarrow 7HCO_3^- + 2.5H^+ \\ + 4.5HS^- \; \Delta G = -205 \text{ kJ} \left(\text{mol of toluene}\right)^{-1} \end{aligned}$$

$$\begin{aligned} \text{NRM}: C_7H_8 + 7.2NO_3^- + 0.2H^+ \rightarrow 7HCO_3^- + 3.6N_2 \\ + 0.6H_2O \; \Delta G = -3,554 \text{ kJ} \left(\text{mol of toluene}\right)^{-1} \end{aligned}$$

Note it has been observed that these two terminal electron acceptors can be reduced in parallel by some microorganisms (Dalsgaard and Bak 1994). In addition to this contrast in energy gained from the degradation of the same hydrocarbon molecule, the presence of nitrate in a pre-soured system can facilitate the proliferation of nitrate-reducing, sulfide-oxidizing microorganisms (NRSOM) which further decrease the pre-existing dissolved sour gas (Dunsmore et al. 2006). Other mechanisms such as significant changes to the environmental redox potential and the accumulation of nitrite (a specific SRM inhibitor) can diminish and even suppress significant biogenic souring under specific environmental conditions.

Assimilatory nitrate reduction describes the utilization of nitrate for biosynthesis, facilitating biomass growth through macromolecule incorporation as part of complex biofilm formation (Glaser et al. 1995). In anaerobic respiration, the coupling of the oxidation of organic compounds and the reduction of nitrate as a terminal electron acceptor can result in one of two distinct dissimilatory pathways: dissimilatory nitrate reduction to ammonium (DNRA) and denitrification. A third pathway which does not require an additional carbon source is termed anaerobic ammonium oxidation, or "anammox." This process describes ammonium oxidation and nitrite reduction resulting in the formation of dinitrogen (Zhou et al. 2017). These pathways are summarized in Figure 10.2.

DNRA and denitrification can take place concurrently, with both pathways competing for available nitrate and electron donors. However, it has been observed that these two dissimilatory pathways are typically dominant under different environmental scenarios; DNRA is favored in the presence of excess carbon, whereas denitrification and the accumulation of nitrite is often observed under carbon-limited conditions (van den Berg et al., 2016).

DNRA can occur under both microaerophilic and strict anaerobic conditions, both of which are well documented (Maier 2009). This two-step process is managed by

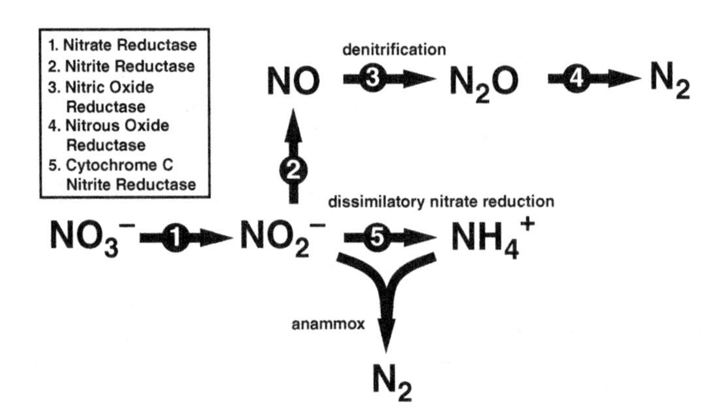

FIGURE 10.2 Schematic of microbiological nitrate reduction via denitrification, dissimilatory nitrate reduction to ammonium (DNRA) and anaerobic ammonium oxidation (anammox).

two separate enzymes: periplasmic nitrate reductase and cytochrome c nitrite reductase. The initial reduction step of nitrate to nitrite (catalyzed by periplasmic nitrate reductase) generates the energy in the DNRA process, whereas the subsequent reduction of nitrite to ammonium is mediated by cytochrome c nitrite reductase.

In an anoxic environment, denitrification is the most energetically favorable respiration process (Schürmann et al. 2003) which sees the microbiological reduction of nitrate to the nitrogen molecule. Complete dissimilatory nitrate reduction provides more energy per mole of nitrate reduced than DNRA, hence its favorability over DNRA in a low-carbon environment.

10.3.2 Field Observations

For over 20 years, nitrate injection has been implemented as a microbiological reservoir souring control strategy in the oil and gas industry (Jenneman et al. 1997). Since its introduction into North Sea assets in the late 1990s, oil companies have documented the use of nitrate-based chemistries in decreasing both downhole sour gas generation and subsequent topsides production (Voordouw and Telang 1999, Larsen 2004, Bødtker et al. 2008).

The decrease in sour gas production topsides as a result of nitrate dosing is ascribed to NRM outcompeting SRM for available carbon for growth and respiration in the downhole formation. Other proposed mechanisms include the proliferation of nitrate-reducing, sulfate-reducing microorganisms (NRSRM), microbiological oxidation of sulfide by nitrate-reducing, sulfide-oxidizing microorganisms (NRSOM) and the generation of toxic intermediate compounds such a nitrite which inhibit SRM activity. Subsequent bacterial enumeration in the production facilities of assets treated with nitrate injection has demonstrated an increased proportion of NRM relative to SRM (Eckford and Fedorak 2002).

The above sour fluid management technique does not therefore control biofouling and, under certain environmental conditions, can contribute to microbiological corrosion (Sanders and Sturman 2005).

Traditional application of nitrate has been through continuous treatment of the injection water. One such example is the BP Foinaven asset, notably the floating production, storage, and offloading (FPSO) vessel. Following reservoir modeling and souring forecasting evaluation of select I/P pairs, it was demonstrated that, upon breakthrough, a continuous injection dose of 0.31mM active nitrate concentration significantly diminished the sulfide concentration (>90%) of the produced water over a three-year period (Vance and Thrasher 2005).

The BP Schiehallion field, a neighboring North Sea asset, has also demonstrated successful application of nitrate treatment for reservoir souring control (Dunsmore et al. 2006). In this particular study, it was determined that one dominant, culturable NRM genus, *Marinobacter*, appeared to be heavily linked to successful nitrate dosing across multiple oilfield assets.

High concentration batch or pulsed nitrate injection has also been shown to further suppress downhole sulfide generation in conjunction with low, continuous nitrate dosing. The success of this particular treatment scenario is hypothesized to be a result of the disruption of the different microbiological zones; namely the active NRM zone tending to be closer to the near wellbore of the injection and SRM activity being located deeper into the formation (Voordouw et al. 2009).

Estimating the ever-changing nitrate demand within an oilfield throughout asset lifetime has proven to be demanding. One of the most notable assets to have received prolonged nitrate treatment is the Gullfaks field, based in the Norwegian section of the North Sea. Upon early breakthrough of injection water, souring was observed topsides. Subsequent nitrate injection was described to have diminished the production of sulfide species (Sunde et al. 2004). However, a secondary increase in sour fluid production was recorded many years later, suggesting that the initial reported nitrate effect could also be explained by the dynamic maturation of the oilfield and an ever-increasing nitrate concentration demand (Mitchell et al. 2017).

10.3.3 Laboratory Experiments

Numerous studies have investigated the applicability of microbiological nitrate reduction to inhibit sulfide production (Reinsel et al. 1996; Myhr et al. 2002). Recent studies have used modern bioinformatic techniques to gain further insight into the microbiome of the downhole reservoir (Vigneron et al. 2017).

In our recent study (Streets et al. 2019), we described a laboratory experiment in which mesophilic microbiological sulfate reduction and nitrate reduction were compared under different shut-in periods at simulated, downhole field conditions. The pressurized reservoir simulation systems consisted of 12 sand-packed columns that were 0.25m in length with an internal diameter of 0.025m, and operated at 7.0MPa, 303K (1,000psig, 30°C). The columns were saturated with crude oil and flushed with anoxic 90:10 to remove the mobile oil phase, leaving only a residual oil phase saturating the core matrix, which was the carbon source for microbiological activity, i.e. no additional VFA. Baseline influent water had a sulfate concentration of 28mM, and six of the columns were supplemented with additional nitrate (ranging from 0.3mM to 19.4mM), with higher nitrate concentrations used during the extended shut-in periods (ranging from a 1-day shut-in to 64-days shut-in).

The 12 bioreactors operated in four triplicate groups: daily injection of baseline influent water in the absence of nitrate (Bioreactor Group (BG) 1), daily injection of baseline water in the presence of nitrate (BG2), ever-increasing shut-in periods between injection cycles with baseline influent water in the absence of nitrate (BG3), and ever-increasing shut-in periods between injection cycles with baseline influent water in the presence of nitrate (BG4).

During the 430-day experiment, influent and effluent samples were captured at 0.1MPa and analyzed for concentrations of nitrate, nitrite, VFAs, and total sulfide.

In the absence of nitrate, significant effluent total sulfide concentrations were measured at each sampling timepoint from BG1 and BG3, ranging from 0.06mM to a maximum of 13.1mM following a 64-day shut-in period (subsequently normalized to 0.2mM/day on day 405 in Figure 10.3). Figure 10.3 demonstrates that similar rate profiles in the absence of nitrate were observed throughout from BG1 and BG3, with statistically greater average cumulative masses of sulfide being generated under extended shut-in periods (+39.6%, $P<0.1$).

In the presence of excess nitrate (BG2 and BG4), no significant total sulfide concentrations were measured (<0.02mM). Nitrate utilization was observed at all sampling timepoints, peaking at an average of 10.14mM following a 64-day shut-in period (subsequently normalized to 0.16mM/day on day 341 in Figure 10.4). As outlined in Figure 10.4, nitrate utilization in BG2 was relatively stable between days 65 and 430. Conversely, the utilization profile for BG4 emulated the souring profiles for BG1 and BG3. In total, the nitrate demand for BG4 was statistically significantly greater than BG3 (+85.2%, $P<0.1$).

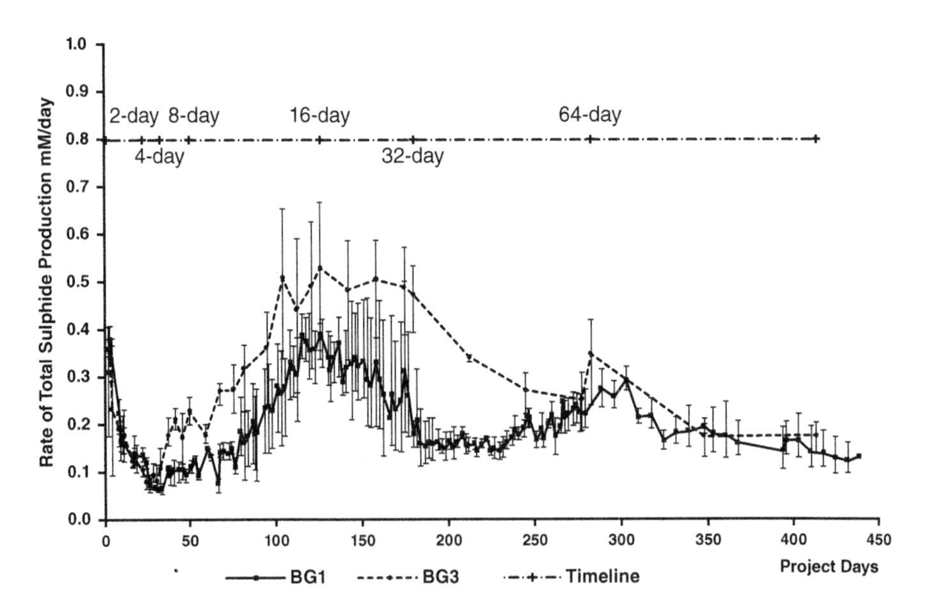

FIGURE 10.3 Normalized profiles of sulfide production rate for BG1 and BG3. Effluent water samples were taken during each injection cycle for sulfide determination throughout the 430-day experiment. Note that the timeline indicates the implementation of increased shut-in periods for BG3.

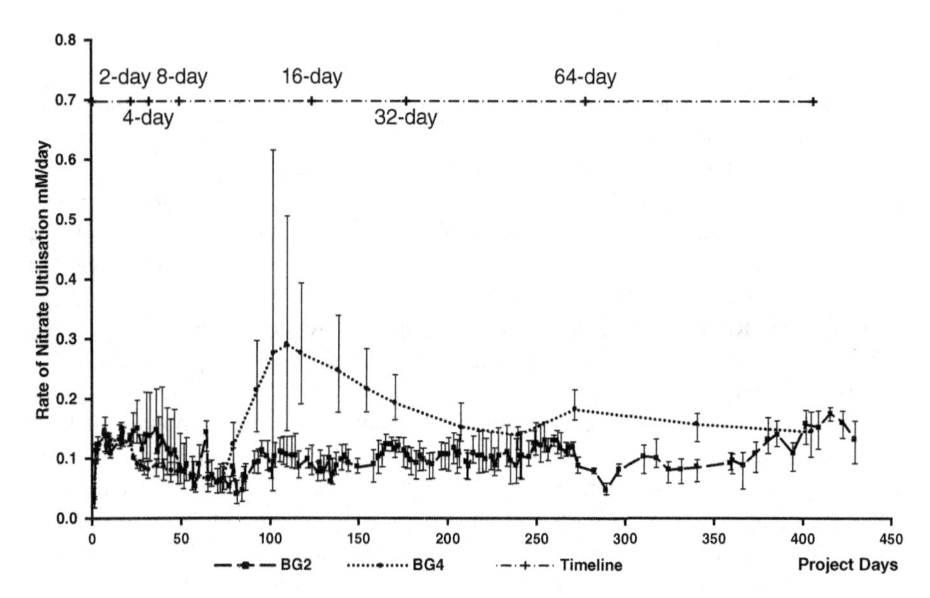

FIGURE 10.4 Normalized profiles of nitrate utilization rate for BG2 and BG4. Effluent water samples were taken during each injection cycle for nitrate determination via ion chromatography analysis throughout the 430-day experiment. Note that the timeline indicates the implementation of increased shut-in periods for BG4.

The results from BG3 and BG4 suggested an increased average rate of sulfide production and nitrate reduction respectively under ever-increasing shut-in periods, and that inhibition from soluble hydrocarbon components or toxic biogenic by-products did not limit activity. Throughout the study, no significant concentrations of VFA or nitrite were measured in the produced fluids from the bioreactors.

Comparative microbiological compositional analysis was conducted on the late planktonic communities of each bioreactor. Microbiological enumeration of effluent samples was conducted by qPCR. Rather than the traditional analysis which requires two separate primer sets to quantify the presence of total bacteria and archaea, a single primer set was used (Earth Microbiome Project, 515F-806R). This primer set was optimized for oilfield microbiology and was included in a proposed standard for oilfield MMM.

At day 340, higher bacterial counts per mL were observed from the shut-in bioreactors BG3 and BG4 (1.2×10^6 and 2.4×10^6 respectively) when compared with BG1 and BG2 (1.2×10^5 and 7.7×10^4 respectively). However, the extended shut-in periods appeared to result in an overall decrease in the diversity of the microbiological communities irrespective of the presence of nitrate.

Using a combination of qPCR and 16S rRNA metagenomic analysis, microbiological classes were identified, generating quantitative data output. Metabolic classes of all planktonic microorganisms identified in effluent water samples were assigned from a comprehensive database, populated with literature-based known species and their associated metabolisms. As biofilm analysis was not undertaken in this particular study, the precise relationship between the planktonic and sessile communities was unknown. However, from other work, it was quite clear that in combination with observations made from the effluent water chemistry dataset, a relationship existed and thus, the planktonic communities were represented within the sessile biofilm in the pressurised sand-packs.

In BG2 and BG4, the presence of nitrate was correlated with a decreased sulfide-producing capability (*dsrAB* gene), and an increased acid-producing (*fhs* gene) and nitrate-reducing and nitrite-reducing capability (*narG* and *nirK* genes) of the microbiological communities compared with BG1 and BG3. A significant increase in the sulfide-oxidizing capability (*sqr* gene) was observed from BG4.

From 16S rRNA gene sequencing, a significant difference was observed between the most abundant genus of BG1 and BG3 (*Desulfotignum*) compared with those bioreactors which received nitrate injection, BG2 and BG4 (*Marinobacter*).

Figures 10.5 and 10.6 illustrate that the dominant sulfidogenic genus from all 12 pressurized bioreactors was *Desulfotignum.* However, the comparison between relative abundances was significantly different between BG1 (31.2%) and BG3 (25.3%), compared with BG2 (5.6%) and BG4 (7.3%). In the presence of nitrate, there were significant increases in the relative abundance of known NRM and NRSOM, namely *Marinobacter* and *Desulfotomaculum*, respectively.

When comparing laboratory experiments and field samples, it is apparent that remediation through nitrate dosing significantly alters the downhole microbiome, often limiting the souring propensity in the oilfield reservoir. However, the potential for increased microbiological corrosion in and around the injection facilities must always be considered prior to implementation.

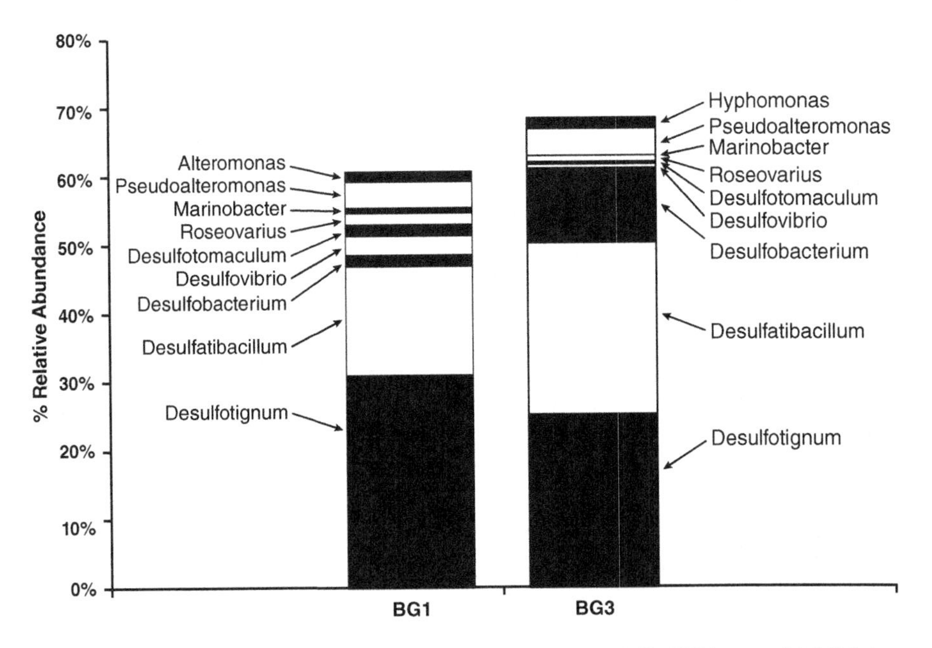

FIGURE 10.5 Associated average taxonomy from bacterial 16S rRNA genes highlighting the top 10 genera amplified from BG1 and BG3 on day 340 from anoxic effluent water.

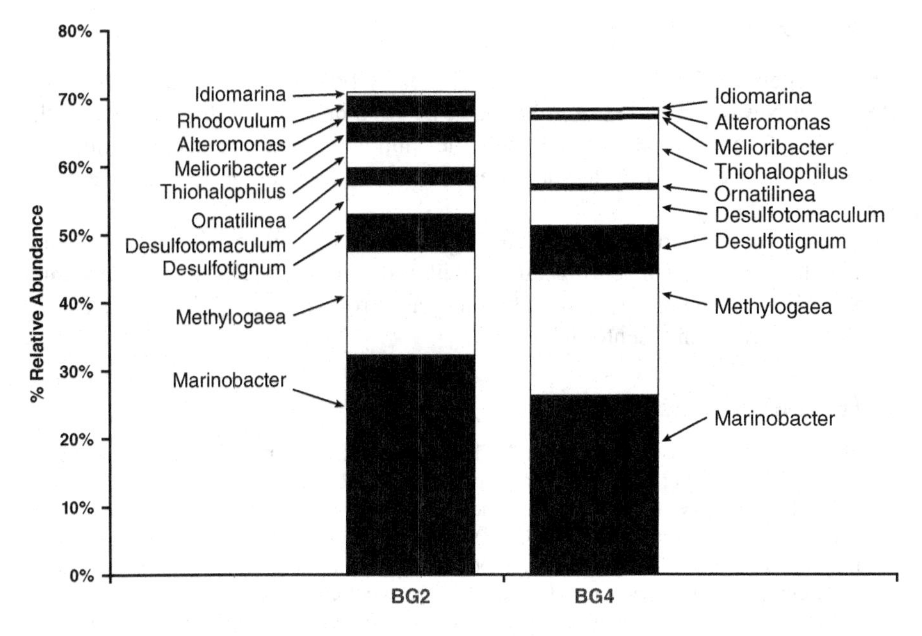

FIGURE 10.6 Associated average taxonomy from bacterial 16S rRNA genes highlighting the top 10 genera amplified from BG2 and BG4 on day 340 from anoxic effluent water.

Accurately calculating nitrate dosing concentrations which are bespoke to I/P pairs within an asset has been shown to be challenging. The optimization of nitrate dosing can only be achieved through robust souring forecasting work, laboratory simulation studies and suitably frequent field water analysis, encompassing microbiological monitoring and chemistry analysis.

10.4 CONCLUSION

The diverse spectrum of microflora present in injection seawater is shifted toward that commonly observed in production fluids due to the selective environmental conditions downhole, promoting the growth of both mesophilic and thermophilic sulfate-reducing microorganisms and inhibition of other species. It is therefore essential that field waters are sampled appropriately and that laboratory simulation studies replicate the downhole environment to ensure that any analysis of souring risk or testing of mitigation treatments are performed against a realistic downhole microbiological community and not that of the injection water community.

These lab-based analyses are further supported by MMM in both simulation studies and in the field, as metagenomic analysis of production fluids can give an early warning to operators of the potential for reservoir souring and MIC. These analytical methods are also able to demonstrate the successful application of remediation treatments, for example nitrate injection, as the most abundant genera can be observed to shift from predominately SRM communities toward those that reduce nitrate, while simultaneously decreasing the diversity of the microbiological community.

ACKNOWLEDGMENTS

The work described in this chapter was supported by Tullow Oil plc, Yara International ASA, and Newcastle University. We thank Dr Julia de Rezende for conducting the metagenomic analysis associated with the laboratory pressure range study, and Professor Ian Head and Professor Casey Hubert for facilitating the pressure range study collaboration.

Finally, we are particularly grateful to Professor Bob Eden who has fully supported all of the internal souring studies within the Scientific Division at Rawwater Engineering Company Limited, and to the late Professor David Auckland for his valuable guidance and mentoring.

REFERENCES

Bødtker, G., T. Thorstenson, B. Lillebø, B. Thorbjørnsen, R. Ulvøen, E. Sunde and T. Torsvik. 2008. The effect of long-term nitrate treatment on SRB activity, corrosion rate and bacterial community composition in offshore water injection systems. *Journal of Industrial Microbiology & Biotechnology*, 35(12), 1625–1636.

Bonifay, V., B. Wawrik, J. Sunner, E. C. Snodgrass, E. Aydin, K. E. Duncan, A. V. Callaghan, A. Oldham, T. Liengen and I. Beech. 2017. Metabolomic and metagenomic analysis of two crude oil production pipelines experiencing differential rates of corrosion. *Frontiers Microbiology*, 8, 99.

Chung, J., K. Amin, S. Kim, S. Yoon, K. Kwon and W. Bae. 2014. Autotrophic denitrification of nitrate and nitrite using thiosulfate as an electron donor. *Water Research*, 58, 169–178.

Crook, B., G. Hill, I. Vance and R. D. Eden. 2015. *Health and Safety Issues in the Oil Industry Related to Sudden or Unexpected Generation of H2S by Micro-Organisms*. 1st Edn. Energy Institute, UK.

Dahle, H., F. Garshol, M. Madsen and N. Birkeland. 2008. Microbial community structure analysis of produced water from a high-temperature North Sea oil-field. *Antonie van Leeuwenhoek*, 93(1–2), 37–49.

Dalsgaard, T. and F. Bak. 1994. Nitrate reduction in a sulfate-reducing bacterium, Desulfovibrio desulfuricans, isolated from rice paddy soil: Sulfide inhibition, kinetics, and regulation. *Applied and Environmental Microbiology*, 60(1), 291–297.

de Rezende, J., M. D. T. Streets, L. Walker, I. Head, C. Hubert and R. D. Eden. 2016. *How representative are your microbial community data? Poster Presented at the 22nd Reservoir Microbiology Forum*, London, UK.

Dunsmore, B., J. Youldon, D. Thrasher and I. Vance. 2006. Effects of nitrate treatment on a mixed species, oil field microbial biofilm. *Journal of Industrial Microbiology and Biotechnology*, 33(6), 454–462.

Eckford, R. and P. Fedorak. 2002. Planktonic nitrate-reducing bacteria and sulfate-reducing bacteria in some western Canadian oil field waters. *Journal of Industrial Microbiology and Biotechnology*, 29(2), 83–92.

Eden, R. D., P. J. Laycock and M. Fielder. 1993. Oilfield reservoir souring. Offshore technology report OTH 92 385. *Health and Safety Executive, UK.*

Folarin, Y., D. An, S. Caffrey, J. Soh, C. W. Sensen, J. Voordouw, T. Jack and G. Voordouw. 2013. Contribution of make-up water to the microbial community in an oilfield from which oil is produced by produced water re-injection. *International Biodeterioration and Biodegradation*, 81, 44–50.

Fuhrman, J. A., J. A. Cram and D. M. Needham. 2015. Marine microbial community dynamics and their ecological interpretation. *Nature Reviews Microbiology*, 13, 133–146.

Gevertz, D., A. Telang, G. Voordouw and G. Jenneman. 2000. Isolation and characterization of Strains CVO and FWKO B, two novel nitrate-reducing, sulfide-oxidizing bacteria isolated from oil field Brine. *Applied and Environmental Microbiology*, 66(6), 2491–2501.

Gieg, L. M., T. R. Jack and J. M. Foght. 2011. Biological souring and mitigation in oil reservoirs. *Applied Microbiology and Biotechnology*, 92(2), 263.

Glaser, P., A. Danchin, F. Kunst, P. Zuber and M. Nakano. 1995. Identification and isolation of a gene required for nitrate assimilation and anaerobic growth of Bacillus subtilis. *The Journal of Bacteriology*, 177(4), 1112–1115.

Heider, J., A. Spormann, H. Beller and F. Widdel. 1998. Anaerobic bacterial metabolism of hydrocarbons. *FEMS Microbiology Reviews*, 22(5), 459–473.

Heukelekian, H. 1943. Effect of the addition of sodium nitrate to sewage on hydrogen sulfide production and B. O.D. reduction. *Sewage Works Journal*, 15(2), 255–261.

Jenneman, G. E., P. D. Moffitt, G. A. Bala and R. H. Webb. 1997. *Field demonstration of sulfide removal in reservoir brine by bacteria indigenous to a Canadian reservoir. SPE 38768*. In *Proceedings of the SPE Annual Technical Conference*. Richardson, TX: Society of Petroleum Engineers.

Johnson, R. J., B. D. Folwell, A. Wirekoh, M. Frenzel and T. L. Skovhus. 2017. Reservoir souring–latest developments for application and mitigation. *Journal of Biotechnology*, 256, 57–67.

Juhler, S., T. L. Skovhus, I. Vance. 2012. *A Practical Evaulation of 21st Century Microbiological Techniques for the Upstream Oil and Gas Industry*. 1st edn. Energy Institute, UK.

Klindworth, A., E. Pruesse, T. Schweer, J. Peplies, C. Quast, M. Horn and F. O. Glockner. 2013. Evaluation of general 16S ribosomal RNA gene PCR primers for classical and next-generation sequencing-based diversity studies. *Nucleic Acids Research*, 41(1), e1.

Larsen, J. 2004. *Downhole nitrate applications to control sulfate reducing bacteria activity and reservoir souring*. In *Proceedings of Corrosion 2004*, New Orleans, paper no. 02025.

Lovley, D. R. and F. H. Chapelle. 1995. Deep subsurface microbial processes. *Reviews of Geophysics*, 33(3), 365–381.

Maier, R. M. 2009. Chapter 14—Biogeochemical cycling. In *Environmental Microbiology*, ed. R. M. Maier, I. L. Pepper, C. P. Gerba, 2nd ed., 287–318. San Diego, CA: Academic Press.

Marietou, A. 2016. Nitrate reduction in sulfate-reducing bacteria. *FEMS Microbiology Letters*, 363(15), fnw155.

Miranda-Tello, E., M. Fardeau, P. Thomas, F. Ramirez, L. Casalot, J. Cayol, J. Garcia and B. Ollivier. 2004. *Petrotoga mexicana sp. nov.*, a novel thermophilic, anaerobic and xylanolytic bacterium isolated from an oil-producing well in the Gulf of Mexico. *International Journal of Systematic and Evolutionary Microbiology*, 54(Pt 1), 169–174.

Mitchell, A. F., I. Skjevrak and J. Waage. 2017. *A re-evaluation of reservoir souring patterns and effect of mitigation in a mature North Sea field*. In *Proceedings of SPE International Conference on Oilfield Chemistry*, Montgomery, TX.

Moest, R. R. 1975. Hydrogen sulfide determination by the methylene blue method. *Analytical Chemistry*, 47(7), 1204–1205.

Myhr, S., B. L. P. Lillebø, J. Beeder, E. Sunde and T. Torsvik. 2002. Inhibition of microbial H_2S production in an oil reservoir model column by nitrate injection. *Applied Microbiology and Biotechnology*, 58(3), 400–408.

National Association of Corrosion Engineers. 2004. Standard test method field monitoring of bacterial growth in oilfield systems. TM0194-2004. *NACE, USA*.

Ommedal, H. and T. Torsvik. 2007. Desulfotignum toluenicum sp. nov., a novel toluene-degrading, sulphate-reducing bacterium isolated from an oil-reservoir model column. *International Journal of Systematic and Evolutionary Microbiology*, 57, 2865–2869.

Reinsel, M., J. Sears, P. Stewart and M. McInerney. 1996. Control of microbial souring by nitrate, nitrite or glutaraldehyde injection in a sandstone column. *Journal of Industrial Microbiology*, 17(2), 128–136.

Ribicic, D., R. Netzer, A. Winkler and O. G. Brakstad. 2018. Microbial communities in seawater from an arctic and a temperate norwegian fjord and their potentials for biodegradation of chemically dispersed oil at low seawater temperatures. *Marine Pollution Bulletin*, 129(1), 308–317.

Sanders, P. F. and P. J. Sturman. 2005. Reservoir souring: Mechanisms and prevention. In *Petroleum Microbiology*, ed. B. Ollivier, and M. Magot, 171–198. Washington, DC: ASM.

Schürmann, A., M. Schroth, M. Saurer, S. Bernasconi and J. Zeyer. 2003. Nitrate-consuming processes in a petroleum-contaminated aquifer quantified using push–pull tests combined with 15N isotope and acetylene-inhibition methods. *Journal of Contaminant Hydrology*, 66(1–2), 59–77.

Stevenson, B. S., H. S. Drilling, P. A. Lawson, K. E. Duncan, V. A. Parisi and J. M. Suflita. 2011. Microbial communities in bulk fluids and biofilms of an oil facility have similar composition but different structure. *Environmental Microbiology*, 13(4), 1078–1090.

Streets, M. D. T., R. D. Eden and J. Hughes. 2014. *Using sand-packed, bioreactors in order to demonstrate the significant impact of downhole pressure on a sulphate-reducing community. Oral Presentation Delivered at the 20th Reservoir Microbiology Forum*, London, UK.

Streets, M. D. T., J. Hilmers, V. Rasmussen, L. Walker and R. D. Eden. 2019. *The use of pressurised, sand-packed bioreactors to improve nitrate treatment calculations within a souring forecasting model.* Poster Presented at the *7th International Symposium on Applied Microbiology and Molecular Biology in Oil Systems*, Halifax, NS.

Streets, M. D. T., L. Walker, J. Wilson, R. D. Eden, I. Head, C. Hubert and J. de Rezende. 2015. *Using pressurized, sand-packed, bioreactors to demonstrate the impact of downhole pressure on the sulphate-reducing microbial community. Oral Presentation Delivered at the 21st Reservoir Microbiology Forum*, London, UK.

Sunde, E., B. L. P. Lillebø, G. Bødtker, T. Torsvik and T. Thorstenson. 2004. H_2S inhibition by nitrate injection on the Gullfaks field. In *Corrosion 2004*. Paper 04760. New Orleans. LA: NACE International.

Terrisse, F., C. Cravo-Laureau, C. Noël, C. Cagnon, A. Dumbrell, T. Mcgenity and R. Duran. 2017. Variation of oxygenation conditions on a hydrocarbonoclastic microbial community reveals alcanivorax and cycloclasticus ecotypes. *Frontiers in Microbiology*, 8, 1549.

van den Berg, E., M. Boleij, J. Kuenen, R. Kleerebezem and M. van Loosdrecht. 2016. DNRA and denitrification coexist over a broad range of acetate/N-NO_3^- ratios, in a chemostat enrichment culture. *Frontiers in Microbiology*, 7, 1842.

Vance, I. and D. Thrasher. 2005. Reservoir souring: Mechanisms and prevention. In *Petroleum Microbiology*, ed. B. Ollivier, and M. Magot, 123–142. Washington, DC: ASM Press.

Vigneron, A., E. B. Alsop, B. P. Lomans, N. C. Kyrpides, I. M. Head and N. Tsesmetzis. 2017. Succession in the petroleum reservoir microbiome through an oil field production lifecycle. *ISME Journal*, 11(9), 2141–2154.

Voordouw, G., A. Grigoryan, A. Lambo, S. Lin, H. Park, T. Jack, D. Coombe, B. Clay, F. Zhang, R. Ertmoed, K. Miner, J. Arensdorf and G. Voordouw. 2009. Sulfide remediation by pulsed injection of nitrate into a low temperature Canadian heavy oil reservoir. *Environmental Science & Technology*, 43(24), 9512–9518.

Voordouw, G. and A. J. Telang. 1999. A genome probe survey of the microbial community in oil fields. In *Microbial Biosystems: New Frontiers. Proceedings of the 8th International Symposium on Microbial Ecology*, ed. C.R. Bell, M. Brylinsky, and P. Johnson-Green. Halifax, Canada: Atlantic Canada Society for Microbial Ecology.

Wu, Y., Y. Cheng, C. G. Hubbard, S. Hubbard and J. B. Ajo-Franklin. 2018. Biogenic sulfide control by nitrate and (per) chlorate–a monitoring and modeling investigation. *Chemical Geology*, 476, 180–190.

Xue, Y. and G. Voordouw. 2015. Control of microbial sulfide production with biocides and nitrate in oil reservoir simulating bioreactors. *Frontiers Microbiology*, 8(6), 1387.

Zhou, X., J. Zhang and C. Wen. 2017. Community composition and abundance of anammox bacteria in cattail rhizosphere sediments at three phenological stages. *Current Microbiology*, 74(11), 1349–1357.

11 Quantitative PCR Approaches for Predicting Anaerobic Hydrocarbon Biodegradation

Courtney R. A. Toth, Gurpreet Kharey, and Lisa M. Gieg

CONTENTS

11.1 Introduction 227
11.2 Overview of Quantitative PCR and Best Practices 229
11.3 Quantitative Gene Markers for Monitoring Anaerobic Hydrocarbon Biodegradation 230
11.3.1 Quantitative PCR Assays for Catabolic Gene Markers 231
11.3.2 Quantitative PCR Assays for Specific Hydrocarbon-Biodegrading Taxa 237
11.3.3 Other Useful Gene Markers 238
11.4 Case Studies 239
11.4.1 Case Study 1: Determining the Potential for Monitored Natural Attenuation by Biodegradation in Fuel-Contaminated Groundwater 239
11.4.2 Case Study 2: Demonstrating Anaerobic Benzene Biodegradation at Canadian Forces Base (CFB) Borden 240
11.5 Conclusion 242
Acknowledgments 243
References 243

11.1 INTRODUCTION

Hydrocarbon-based fuels currently provide for the majority of society's energy demands (IEA 2020). However, accidental spills or leaks that may occur during hydrocarbon recovery, transport, refining processes, and storage can contaminate pristine marine and terrestrial environments. While a number of chemical or physical remedial approaches can be used to restore such contaminated environments, these can be costly or impractical. For example, techniques that rely on active remediation such as "pump-and-treat" have often resulted in poor results because the mass transfer from

NAPL (non-aqueous phase liquid) to the aqueous phase in a subsurface groundwater environment can be a slow process (Keely 1989). Microorganisms that are naturally present in marine or terrestrial ecosystems are known to have the ability to utilize hydrocarbons as carbon and energy sources, converting these contaminants to innocuous respiration end products (such as CO_2 and H_2O). Bioremediation is a remedial strategy that relies on microbial communities to biodegrade contaminants, offering a cost-effective approach for restoring contaminated sites such as hydrocarbon-impacted marine sediments and terrestrial groundwater aquifers. However, the universal application of bioremediation has often been called into question due to its unreliable efficacy. While hydrocarbon-degrading microbes appear to be near-ubiquitous in nature, their necessary catabolic capacity may not always be expressed, or the organisms themselves may be in too low abundance to result in meaningful rates of degradation.

Relying on bioremediation processes for site clean-up requires ongoing monitoring and the collection of multiple lines of evidence to ensure that contaminants such as hydrocarbons are being removed via biological metabolism. Almost three decades ago, the National Research Council in the United States (1993) recommended that three types of evidence be obtained in order to deem bioremediation as an effective clean-up strategy at a given site: 1) demonstrating mass loss of contaminants at a given site over time; 2) demonstrated potential for microorganisms present at the site to biodegrade the contaminants in question (e.g. via lab-based biodegradation tests); and 3) evidence demonstrating that *in situ* biodegradation is actually occurring. Although the latter recommendation is perhaps the most challenging to demonstrate, research investigating the microbial metabolism of hydrocarbons, especially under anoxic conditions, has led to the development of many monitoring tools that can be used to indicate that *in situ* biodegradation is occurring (Bombach et al. 2010). Analytical methods include, among others, monitoring for isotopic changes in hydrocarbons themselves (CSIA, compound-specific isotopic analysis), and seeking unique and diagnostic chemicals (metabolites) indicative of anaerobic hydrocarbon biodegradation pathways within contaminant plumes (Bombach et al. 2010, Gieg and Toth 2020). Biochemical characterization of key hydrocarbon-metabolizing enzymes has led to the development of several functional gene marker assays that can detect and track the growth of hydrocarbon-degrading populations. Given the overwhelming diversity of anaerobic fuel degraders in nature (spanning across the Proteobacteria, the Firmicutes, as well as several bacterial candidate phyla), these "biomarker" assays can effectively screen numerous organisms for evidence of hydrocarbon biodegradation potential in a single assay. Alternately, for hydrocarbons such as benzene wherein the anaerobic biodegradation pathway(s) remain unclear, long-term enrichments and/or molecular analyses derived from multiple contaminated sites are now starting to reveal highly specific taxa that are associated with benzene metabolism (e.g., *Thermincola* spp. and Sva0485 clade Proteobacteria), resulting in the development of taxonomic-based molecular probes for these biodegradative organisms (Toth et al. 2021). We also remark that the recent integration of bioinformatics tools with biodegradation studies is helping to glean new insights into the degradation of hydrocarbons, including and the discovery of novel hydrocarbon-degrading organisms (Hu et al. 2016, Liu et al. 2019) and degradation pathways (Sierra-Garcia et al. 2014).

As new taxa and/or genes encoding for key enzymes involved in the anaerobic biodegradation of hydrocarbons become identified, they can be effectively used as biomarkers to assess and/or monitor contaminated sites for evidence of *in situ* hydrocarbon degradation. They can also be used to explain the overall prevalence of anaerobic hydrocarbon biodegradation in nature. For example, studies are starting to demonstrate that microorganisms in non-contaminated environments harbor basal levels of known hydrocarbon-degrading genes (such as *bssA* or *assA*, encoding for the catalytic subunits of benzylsuccinate synthase and alkylsuccinate synthase, respectively), which helps to explain why pristine environments naturally harbor hydrocarbon-degrading microorganisms that can respond to hydrocarbon input should it occur (Gittel et al. 2015, Kharey et al. 2020). Given this, it is critical to have methods in place for quantifying these diagnostic genes in order to demonstrate their increased abundance over background levels as a measure of increased biodegradation activity, thus providing evidence in support of *in situ* hydrocarbon biodegradation.

In this chapter, we will overview the biomarkers most commonly used to search for evidence of anaerobic hydrocarbon biodegradation *in situ*. We highlight two case studies wherein quantitative polymerase chain reaction (qPCR) approaches were successfully employed to monitor for hydrocarbon biodegradation genes or taxa in fuel-contaminated groundwater in order to provide evidence in support of anaerobic *in situ* hydrocarbon biodegradation. In addition, we begin by briefly overviewing the advantages, caveats, and advances in qPCR-based assays for monitoring anaerobic hydrocarbon biodegradation.

11.2 OVERVIEW OF QUANTITATIVE PCR AND BEST PRACTICES

Quantitative PCR is a well-established molecular method that allows for the simultaneous detection and quantification of target DNA. In conventional PCR, the amplified DNA product, also known as an amplicon, is detected in an end-point analysis. In qPCR, the accumulation of amplification product is measured as the reaction progresses, in real time, with product quantification after each cycle. Although the concept is relatively simple, there are specific issues in qPCR that developers and users of this technology should bear in mind.

Inherent to the successful application of a qPCR assay is the design of primers that effectively target a suitable region of a gene such that appropriate coverage and specificity is attained (de Rezende 2016). A suitable gene should only be present in the target organism(s) and its metabolic function should be clearly established. Databases of anaerobic hydrocarbon biodegradation genes are publicly available (Callaghan and Wawrik 2016), which can help users find genes of interest. Searching in broader nucleotide databases (such as NCBI) can also be helpful, but note that these resources are less curated. Once a gene has been identified, one can then search for existing primer sets and reaction conditions that can target and amplify the gene of interest, or develop their own assay. The latter should be done with the upmost care, as poor primer design and amplification reaction conditions can readily lead to inaccurate results. Several resources are currently available to help ensure the development of a quality qPCR assay. For example, Bustin and Huggett (2017) recently outlined a step-by-step workflow to ensure the development of a robust qPCR assay that includes proper target identification, defining assay properties (such as annealing temperature, polymerase, master mixes), and assay optimization. Following such guides is highly recommended.

As will be discussed later in this chapter, there are a growing number of PCR assays intended for the qualitative detection of hydrocarbon-degrading functional genes and taxa. However, most of these are not suitable for biomarker quantification. qPCR primers, reagent concentrations, and reaction conditions are specifically designed to amplify target gene sequences as efficiently as possible, doubling the number of replicated molecules during each amplification cycle (Bustin and Huggett 2017); this level of precision is less important for qualitative PCR assays. Typically, desired amplification efficiencies for qPCR range from 90% to 110%. Short PCR products (75 – 200 base pairs, bp) are typically amplified with higher efficiency than longer ones, thus amplification of large gene regions should be avoided if possible. With a few exceptions (e.g., Toth and Gieg 2018, Winderl et al. 2007), most qualitative PCR assays for hydrocarbon-degrading genes and taxa have not been validated for use in qPCR.

Understanding the limitations and pitfalls involved in qPCR primer design allows for the careful interpretation and communication of results. For example, it is well known that conventional DNA extraction techniques recover DNA from both living and dead cells, thus can lead to overestimations of active organisms; this seems to be particularly problematic in field samples where dead cell DNA can persist for months or years (Willerslev and Cooper 2005). Some organisms may also harbor multiple copies of the same target gene contributing to overestimates (Kembel et al. 2012). One solution is to extract a series of time course DNA samples and monitor the overall changes in abundance of the target gene(s) (Toth and Gieg 2018, Toth et al. 2021). Increases in target gene copies over time, combined with other lines of evidence of hydrocarbon degradation, can help pinpoint the metabolic processes and microorganisms driving bioremediation at a given site. Another option is to extract and quantify total RNA or messenger RNA (mRNA) using a PCR technique known as reverse transcription qPCR (RT-qPCR). This technology amplifies genes that are in the process of being expressed into proteins and enzymes, thus providing unequivocal evidence of active microbial processes (Fowler et al. 2014, Wei et al. 2018). As the extent of gene expression can vary, careful selection of reference genes is required for reliable performance of RT-qPCR. Users should also be aware that RNA is notoriously difficult to work with, as it is single-stranded and highly unstable.

11.3 QUANTITATIVE GENE MARKERS FOR MONITORING ANAEROBIC HYDROCARBON BIODEGRADATION

Studies of different hydrocarbon activation mechanisms utilized by anaerobic microorganisms have led to the discovery of unique metabolites and catabolic gene biomarkers that can be used as indicators for monitoring and demonstrating *in situ* hydrocarbon biodegradation. Though not the focus of this chapter, the use of metabolites as diagnostic indicators of *in situ* anaerobic hydrocarbon biodegradation (such as the benzyl- and alkylsuccinates produced via fumarate addition) has been reported in dozens of site monitoring studies in the last ~ 25 years, as recently reviewed by

Gieg and Toth (2020). The use of qPCR for field monitoring of bioremediation has only gained attention in the last decade, which we focus on here. Table 11.1 provides a list of qPCR primers that have been used for the detection and quantification of anaerobic hydrocarbon-degrading genes and taxa in groundwater and/or in laboratory enrichments prepared from aquifer materials.

11.3.1 Quantitative PCR Assays for Catabolic Gene Markers

Several microbial reaction mechanisms catalyzing the anaerobic activation of hydrocarbons have been described in literature, including addition to fumarate (also known as fumarate addition), hydroxylation, and carboxylation. Each of these mechanisms have been reviewed extensively in recent articles (Heider et al. 2016a, Heider et al. 2016b, Meckenstock et al. 2016), thus will be only briefly highlighted here. As most anaerobic hydrocarbon activation genes have a single, clearly defined metabolic function, their detection *in situ* can be considered diagnostic of a site's catabolic potential for bioremediation, especially when increases in gene abundance are detected over time.

Fumarate addition is the most widely reported hydrocarbon activation mechanism and is used by diverse anaerobic taxa to activate alkylbenzenes, alkylated polycyclic aromatic hydrocarbons (PAH), and alkanes (Figure 11.1a). The enzymes associated with hydrocarbon activation and downstream metabolic steps have also been at least partially characterized. All fumarate addition enzymes (FAE) are radical-forming, enabling C-C, C-O, and C-N bond breaking and formation steps that are otherwise challenging for non-radical enzymes. The catalytic moieties are conserved among FAE, making them good target regions for PCR. Benzylsuccinate synthase (BSS) subunit A, encoded by the *bssA* gene, catalyzes fumarate addition to toluene (Heider et al. 2016a, Leuthner et al. 1998), yielding benzylsuccinate as an initial intermediate product. Subsequent conversion of benzylsuccinate to benzoyl-CoA, a central intermediate in the anaerobic decomposition of aromatic compounds, has also been extensively characterized (Heider et al. 2016a, Leutwein and Heider 1999). Addition to fumarate has also been observed for xylenes (Beller and Spormann 1997, Morasch et al. 2004), trimethylbenzenes (Parisi et al. 2009), and ethylbenzene under sulfate-reducing conditions (Elshahed et al. 2001, Kniemeyer et al. 2003). Detailed mechanistic studies of the fumarate addition reaction have not been done with most of these other substrates, thus the similarity of the toluene-activating BSS versus other BSS analogs is not well understood. Fumarate addition of alkyl-substituted PAHs such as 2-methylnaphthalene is catalyzed by naphthyl-2-methyl-succinate synthase (NMS) subunit A, encoded by *nmsA* (Annweiler et al. 2000, Meckenstock et al. 2016), and undergoes subsequent conversion to 2-naphthoyl-CoA in a manner similar to toluene conversion to benzoyl-CoA (Meckenstock et al. 2016, Selesi et al. 2010). Alkylsuccinate synthase (ASS) subunit A (also known as (1-methylalkyl) succinate synthase, or MAS), encoded by *assA* (also denoted as *masD*), produces alkylsuccinates ([1-methylalkyl]) succinate) from *n*-alkanes ranging in length from C_2 to at least C_{40} (Kropp et al. 2000, Wilkes et al. 2016). This compound is subsequently metabolized via a carbon skeleton rearrangement (yet to be characterized), followed by a series of modified

TABLE 11.1
List of Established Primers Used for Quantification of Targeted Hydrocarbon Functional Genes and Taxon-Specific (16S rRNA) Genes

Name	Target Gene	Design Reference	Sequence (5' to 3')[a]	Amplicon Size (bp)	Reference
Fumarate Addition					
	bssA	Denitrifying	ACGACGGYGGCATTTCTC	130	Beller et al. 2002
		Betaproteobacteria	GCATGATSGGYACCGACA		
7772f	*bssA*	Beta- and Deltaproteobacteria,	GACATGACCGACGCSATYCT	800	Winderl et al. 2007
8546r		Clostridia	TCGTCGTCRTTGCCCCAYTT		
SRBf	*bssA*	Sulfate-reducing	GTSCCCATGATGCGCAGC	97	Beller et al. 2008
SRBr		Deltaproteobacteria and Firmicutes	CGACATTGAACTGCACGTGRTCG		
bssA3f	*bssA*	Denitrifying and ¨iron-reducing	TCGAYGAYGGSTGCATGGA	500	Staats et al. 2011
bssAr		Betaproteobacteria	TTCTGGTTYTTCTGCAC		
MBssA1F	*bssA*	*Desulfosporosinus* spp.	ATGCCCTTTGTTGCCAGTAT	223	Fowler et al. 2014
MBssA1R			GCTGCATTTCTTCGAAACCT		
MHGC_bssAf	*bssA*	*Desulfotomaculum* spp.	GACGACGGCTGCATGGA	708	Toth and Gieg 2018
MHGC_bssAr			GCCTTCCCAGTTGGCGTA		
682F	*bssA*	*o*-Xylene-degrading	GTCCGGACTGAGGATATGCG	120	Rossmassler et al. 2019
801R		*Peptococcaceae* spp.	TGCATCCTGAGCCATCTTGG		
2294F	*bssA*	*o*-Xylene-degrading *Peptococcaceae* spp.	ACATCGACCACGTCCAGTTC	201	Rossmassler et al. 2019
2494R			CGTCCAGAAACTCCAGGTCC		
971F	*bssA*	*o*-Xylene-degrading	ACCGGCTTTTATGGGAGTGG	140	Rossmassler et al. 2019
1029R		*Peptococcaceae* spp.	GACCACCAAACCGAGTAGCA		
2080F	*bssA*	*o*-Xylene-degrading	CGCAAGGGCTATCCGGTATT	234	Rossmassler et al. 2019
2313R		*Peptococcaceae* spp.	AGGATCAAGGCGTTCGTTGT		

(*Continued*)

TABLE 11.1 (Continued)
List of Established Primers Used for Quantification of Targeted Hydrocarbon Functional Genes and Taxon-Specific (16S rRNA) Genes

Name	Target Gene	Design Reference	Sequence (5' to 3')[a]	Amplicon Size (bp)	Reference
bssOil (forward)	*bssA*	Alpha-, Betaproteobaceria and Clostridia	GAATCCCTGGTTACAGGTCCAC	141	Kharey et al. 2020
bssMys (forward)	*bssA*	Alpha-, Betaproteobaceria and Clostridia	CAATCCGTGGCACAACTGCATG	141	
bssSuf (forward)	*bssA*	*Desulfotomaculum* spp.	GAATACGTGGAGCGACCCGCTC	141	
bssWin (forward)	*bssA*	Alpha-, Betaproteobaceria and Clostridia	CAATCCGTGGCTTCAGGTTCAT	141	
bssHitr (forward)			TCCTCGTAGCCTTCCCAGTT		
assA2Fq	*assA*	Deltaproteobacteria	ATGTACTGGCACGGACA	442	Aitken et al. 2013
assA2Rq			GCGTTTTCAACCCATGTA		
assA3Fq	*assA*	Deltaproteobacteria	CGCACCTGGGTTCATCA	467	Aitken et al. 2013
assA3Rq			GGCCATGGTGTACTTCTT		
assAqF	*assA*	*Smithella* spp.	CACTTGAGCTGCTCTGCCCAGG	430	Oberding and Gieg 2018
assAqR			AGATGGGGCCTCAAACGGG		
assOri (forward)	*assA*	*Desulfatibacillum alkanivorans* AK-01	CTCCGCCACGGCCAACTG	486	Kharey et al. 2020
assMsd (forward)	*assA*	Deltaproteobacteria	CTCAGCCACCGCCAACTG	486	
assEx (forward)	*assA*	*Desulfoglaeba alkanexedens*	CTCTGCGACCGCGAATTG	486	
assSml (forward)	*assA*	*Smithella* spp. SCADC	TAGCGCCACGGCCAACTG	486	
8543r (reverse)			TCGTCRTTGCCCCAYTTNGG		
Carboxylation					
bc_F	*abcA*	Benzene-degrading *Thermincola*	GCGGTGAGGTATTGACCACT	175	van der Waals et al. 2017
bc_R			TTCGGGCTGACATATCCTTC		
abcA_1005f	*abcA*	Benzene-degrading *Thermincola*	GCCGACGGAAATGGTTATGC	287	Shayan et al. 2017
abcA_1291r			ATGCCTTGCTCCAGGTTCTC		
abcA_254f	*abcA*	Benzene-degrading *Thermincola*	GGCGCGAAATCCAGGATACA	119	Toth et al. 2021
abcA_373r			GGTCGAACAGGTTGACGTCT		
Hydroxylation					
ebdA 2433F	*ebdA*	*Azoarcus aromaticum* strain EbN1	TGCCCAGTTCTACCTTGAC	496	Kühner et al. 2005
ebdA 2928R			TGCTTTCTTGSTGCTTSCC		

(Continued)

TABLE 11.1 (Continued)
List of Established Primers Used for Quantification of Targeted Hydrocarbon Functional Genes and Taxon-Specific (16S rRNA) Genes

Name	Target Gene	Design Reference	Sequence (5' to 3')[a]	Amplicon Size (bp)	Reference
ebdA 300F	*ebdA*	*Azoarcus aromaticum* strain EbN1	CTCCGCGGCGTCCTTGCT	186	Wöhlbrand et al. 2013
ebdA 486R			CGCGCCGTGCCCAGTTCTACC		
ebdA2 2666F	*ebdA*	*Azoarcus aromaticum* strain EbN1	GGCGGAGCTCGGTATCAA	190	Wöhlbrand et al. 2013
ebdA2 2856R			GGGCTTCCATTCAGGTAGTA		
Downstream pathways					
bcr-1f	*bcrA*	*Thauera aromatica*, *Azoarcus evansii*, *Rhodopseudomonas palustris*	GTYGGMACCGGCTACGGCCG	480	Hosoda et al. 2005
bcr-2r			TTCTKVGCIACICCDCCGG		
bamA-SP9-f	*bamA*	Alpha-, Beta- and Deltaproteobacteria	CAGTACAAYTCCTACACVACBG	300	Kuntze et al. 2008
bamA-ASP1-r			CMATGCCGATYTCCTGRC		
oah_f	*bamA*	Iron-reducing *Rhodocyclaceae*, *Geobacteraceae*	GCAGTACAAYTCCTACACSACYGABATGGT	350	Staats et al. 2011
oah_r			CCRTGCTTSGGRCCVGCCTGVCCGAA		
bzdNf	*bzdN*	Denitrifying *Azoarcus*	GAGCCGCACATCTTCGGCAT	700	Kuntze et al. 2011
bzdNr			TRTGVRCCGGRTARTCCTTSGTCGG		
Ncr_for	*Ncr*	Deltaproteobacteria	TGGACAAAYAAAMGYACVGAT	320	Morris et al. 2014
Ncr_rev			GATTCCGGCTTTTTTCCAAVT		
Taxon-specific primers					
Bac_1055f	16S rRNA	Total Bacteria	ATGGCTGTCGTCAGCT	338	Amann et al. 1995; Ferris et al. 1996
Bac_1392r			ACGGGCGGTGTGTAC		
Arch_787f	16S rRNA	Total Archaea	ATTAGATACCCGBGTAGTCC	273	Yu et al. 2004
Arch_1059r			GCCATGCACCWCCTCT		
ORM2_168f	16S rRNA	Benzene-degrading candidate clade Sva0485	GAGGGAATAGCCAAAGGTGA	274	Toth et al. 2021
ORM2_422r			GAGCTTTACGACCCGAAGAC		
Thermincola_634f	16S rRNA	*Thermincola* (benzene and non-benzene-degrading spp.)	GRATYTCTTGAGGGTATGAG	109	Magnuson 2017; Toth et al. 2021
Thermincola_743r			CAGTTGTATGCCAGAAAGCC		

[a] Nucleotide codes for positions with 2 or more possible bases are as followed: R = A or G, Y = C or T, S = G or C, W = A or T, K = G or T, M = A or C, D = A or G or T,

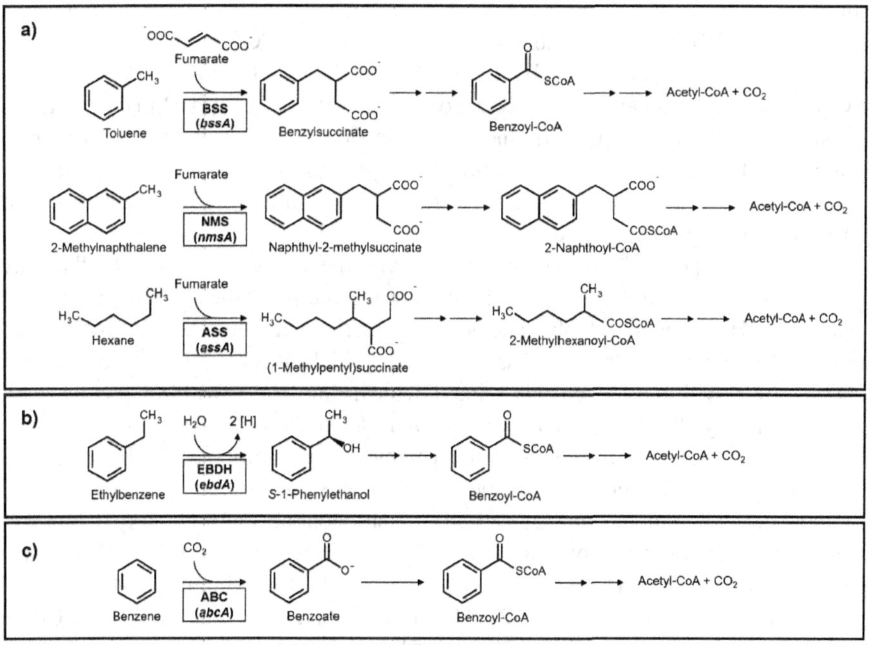

FIGURE 11.1 Overview of the major hydrocarbon activation mechanisms under anaerobic conditions and associated catalytic enzymes described in literature. The activation mechanisms are as follows: a) fumarate addition, b) hydroxylation, and c) carboxylation. Select downstream biodegradation steps and intermediate products are also shown. Multiple *arrows* represent more than one enzymatic step.

beta-oxidation reactions leading to short-chain fatty acids that can enter into central metabolic pathways (Wilkes et al. 2016). Fumarate addition of branched and cyclic alkanes (Agrawal and Gieg 2013), as well as alkenes (Grossi et al. 2007), is also possible under strictly anoxic conditions.

Beller (2002) was the first to design a qPCR primer set targeting *bssA*, with the primer design based on *bssA* gene sequences found in three toluene-degrading denitrifiers. A subsequent qPCR primer set was developed that was more specific for the *bssA* gene found in sulfate-reducing Deltaproteobacteria and Firmicutes (Beller et al. 2008) as the original qPCR primers (targeting nitrate-reducers) did not successfully capture this gene under these other electron-accepting conditions. In fact, as *bssA* was sought further in numerous anaerobic cultures or samples following these initial reports, it became apparent that there was enormous diversity in this gene sequence (von Netzer et al. 2016). As of September 2020, there were over 36,000 nucleotide entries for fumarate addition enzymes (based on a search for "*bssA*" OR "*assA*" OR "*nmsA*"). In addition, genes encoding for fumarate addition enzymes, and in particular for *bssA*, have been identified in many physiological groups of microorganisms, including within the Betaproteobacteria, Deltaproteobacteria, and Firmicutes (von Netzer et al. 2016). The fact that these genes are now reported so widely and across multiple phyla presents a significant challenge for accurate *in situ* quantification of fumarate addition genes as no single primer set can capture all sequence variants. von Netzer et al. (2013) sought to address this by developing PCR assays that captured a much broader diversity of taxa, but these were not designed or intended for quantitative use. Recently, Kharey et al. (2020) designed mixtures of qPCR primer sets that consisted of 4 forward primers and 1 reverse primer for each of *bssA* and *assA* (Table 11.1), which successfully captured and quantified a broader suite of fumarate addition genes in a single assay. We present highlights of this study later in this chapter as one of our two case studies.

Oxygen-independent hydroxylation involves the addition of H_2O to the alkyl group of some substituted aromatic hydrocarbons and *n*-alkanes (Figure 11.1b, Ball et al. 1996, Heider et al. 2016b). The best characterized enzyme facilitating this reaction is ethylbenzene dehydrogenase (EBDH), which was initially discovered and characterized within a few closely related denitrifying strains within the betaproteobacterial family *Rhodocyclaceae*, with *Aromatoleum aromaticum* strain EbN1 serving as the paradigm for this mechanism (Kniemeyer and Heider 2001). Ethylbenzene dehydrogenase adds water to the methylene group on ethylbenzene forming (*S*)-1-phenylethanol, which is then converted to acetophenone and ultimately to benzoyl-CoA. Genetic, genomic, proteomic, and enzymatic analyses of strain EbN1 have also helped to design a handful of qPCR assays for the catalytic subunit (*ebhA*) of EBDH, as seen in Table 11.1 (Heider et al. 2016b). In addition, paralogous enzymes to EBDH have been proposed to catalyze the hydroxylation of propylbenzene, *p*-cymene, and *p*-ethylphenol (Rabus and Widdel 1995, Strijkstra et al. 2014). An analogous hydroxylation mechanism has also been proposed for alkane-degrading *Desulfococcus oleovorans* strain Hxd3 that does not utilize fumarate addition, as genes encoding an enzyme closely related to EBDH have been identified in this organism (Callaghan et al. 2009, So et al. 2003).

For unsubstituted hydrocarbons such as benzene, naphthalene, and phenanthrene, hydroxylation, methylation, and carboxylation have been proposed as initial activation mechanisms under anoxic conditions (Foght 2008). While one than one activation mechanism likely exists (Mancini et al. 2008), carboxylation has emerged in the literature as the most frequently reported mechanism of activation for unsubstituted aromatic compounds (Figure 11.1c), based on initial reports demonstrating the formation of benzoate from benzene (Abu Laban et al. 2010, Phelps et al. 2001), naphthoic acid from naphthalene (Mouttaki et al. 2012, Zhang and Young 1997), and phenanthrene carboxylic acid from phenanthrene (Davidova et al. 2007, Zhang and Young 1997). These findings were later heavily supported using proteomic (Koelschbach et al. 2019), transcriptomic (Luo et al. 2014) and metabolic reconstructions (Kraiselburd et al. 2019) of cultured microorganisms. A particular emphasis has been placed on elucidating the metabolic pathway for anaerobic benzene biodegradation as benzene typically persists in hydrocarbon-contaminated aquifers and is the key regulatory target compound that must be mitigated in any remediation approach. Abu Laban et al. (2010) were the first to identify the catalytic subunit (*abcA*) of anaerobic benzene carboxylase (ABC), predicted to be involved in the direct carboxylation of benzene and leading to the formation of benzoate and subsequently benzoyl-CoA. A handful of qPCR assays targeting *abcA* were later established (see Table 11.1) and were used to screen several other enrichment cultures, bioreactors, and hydrocarbon-contaminated groundwater site for evidence of anaerobic benzene degraders (van der Waals et al. 2017, Wei et al. 2018). To date, *abcA* has only been identified in one bacterial genus (*Thermincola*), though other anaerobic benzene-degrading taxa have been described in literature (Luo et al. 2016, Zhang et al. 2012).

11.3.2 Quantitative PCR Assays for Specific Hydrocarbon-Biodegrading Taxa

Gene marker assays based on catabolic genes indicative of anaerobic hydrocarbon activation have been most commonly used for pinpointing anaerobic hydrocarbon degraders because of their clear functional affiliation and presumed widespread occurrence in nature (Callaghan et al. 2010, von Netzer et al. 2016). However, several hydrocarbons have poorly defined metabolic pathways and/or may become activated by more than one reaction mechanism, as may be the case for benzene (Luo et al. 2016, Mancini et al. 2008). Therefore, another approach is to screen for the presence and abundance of microbial clades closely linked to anaerobic hydrocarbon biodegradation, usually by amplifying portions of the 16S rRNA gene sequence unique to these particular taxa.

Sequencing of ribosomal ribonucleic acid (rRNA) subunits is a powerful tool used in phylogenetic, evolutionary, and taxonomic studies of both prokaryotes and eukaryotes. Involved in protein synthesis, rRNA genes are conserved nucleotide sequences within a given species. In most cases, the 16S small subunit of a prokaryotic ribosome can be used for microbial identification because 1) it is present in almost all bacteria and archaea, 2) the function of the 16S rRNA gene has not changed over evolutionary time, and 3) the 16S rRNA gene (1,500 base pairs, bp) is large enough for informatics purposes as a reliable molecular clock. The 16S rRNA gene contains several conserved and variable regions that allow for the design of assays targeting different taxonomic levels. Bioinformaticians also employ this technology to estimate microbial diversity in a given environment without the need for culturing (Gieg and Toth 2016).

With respect to benzene, distinct clades within the Firmicutes and Deltaproteobacteria have emerged as widespread benzene-degrading specialists. As previously stated, *Thermincola* (belonging to the Firmicutes) has also been identified as a key benzene biodegrading organism, and are typically enriched under iron and sulfate-reducing conditions (Kunapuli et al. 2007, Ulrich and Edwards 2003, van der Zaan et al. 2012). 16S rRNA gene-based qPCR primers targeting *Thermincola* have been developed and can be used alongside *abcA* primers for tracking microbial growth (Toth et al. 2021). A second cluster of predicted anaerobic benzene degraders belong to a candidate clade (Sva0485) within the Deltaproteobacteria, whose 16S rRNA gene sequences have been retrieved from methanogenic and sulfate-reducing, benzene-reducing enrichment cultures across Canada, the US, and Japan (Sakai et al. 2009, Ulrich and Edwards 2003). These organisms do not harbor any known or predicted hydrocarbon activation genes, thus their benzene reaction mechanism appears to be novel (Toth et al., unpublished data). Additionally, benzene is the only known growth substrate of Sva0485 clade Deltaproteobacteria, suggesting that their metabolism is specialized for this compound (Toth et al. 2021). 16S rRNA-based investigations of a representative taxon from this clade, known as ORM2, have shown that the abundance of this organism increases as benzene is biodegraded (Da Silva and Alvarez 2007, Luo et al. 2016). Similarly, the abundance of ORM2 decreases dramatically upon depletion of benzene substate. Thus, 16S rRNA-based qPCR can be considered a robust method of monitoring the growth and abundance of benzene-degrading Sva0485 clade Deltaproteobacteria. Our second case study demonstrates the application of 16S rRNA gene primers for tracking anaerobic benzene degraders.

11.3.3 Other Useful Gene Markers

In addition to seeking and quantifying initial hydrocarbon activation genes (such as *bssA*, *assA*, *abcA*) and 16S rRNA genes targeting specific hydrocarbon-degrading taxa, other genes known to be part of "downstream" metabolic pathways and electron-accepting processes can also be useful for providing evidence in support of hydrocarbon biodegradation in contaminated aquifers. For example, the metabolic pathways for the anaerobic biodegradation of aromatic hydrocarbons are known to converge on benzoyl-CoA or an analog (Boll et al. 2020). The metabolic steps involved in the further breakdown of benzoate are also known and can vary in different microorganisms. Anaerobic benzoate metabolism occurs via distinct reductive dearomatization by benzoyl-CoA reductases (BCR) encoded by *bcrA*, *bcrC*, *bzdN* (class I, ATP-dependent), or *bamB* (class II, ATP-independent) reactions. Some qPCR assays have been designed to quantify these genes in pure cultures and in field-derived samples (Kuntze et al. 2011). An additional gene that encodes the ring opening step further down in the benzoate degradation pathway, *bamA*, has also been used to detect and/or quantify aromatic compound degraders in field sites or samples (Porter and Young 2013, Staats et al. 2011). Although less strictly targeted toward anaerobic pollutant degraders, these assays allow for insights into the diversity and identity of intrinsic aromatics degrader populations and could prove particularly useful at sites where fumarate addition is not involved.

Finally, genes diagnostic of different anaerobic electron-accepting processes, such as for dissimilatory nitrate reduction, iron reduction, sulfate reduction, and methanogenesis can also be useful in providing evidence in support of *in situ* anaerobic hydrocarbon biodegradation. Although not reflective of degradation itself, examining contaminated areas for increased abundances of these genes compared to corresponding uncontaminated areas can indicate microbial growth or stimulation in response to hydrocarbon input (e.g., as a carbon and energy source). Genes indicative of these different electron-accepting processes include the *nirS* and *nirK* genes for nitrate-reducers (Bonilla-Rosso et al. 2016), the *dsrAB* gene for sulfate-reducers (Müller et al. 2015), and the *mcr*A gene for methanogens (Luton et al. 2002, Morris et al. 2014). Several unique iron-reducing genes have been described in the literature (e.g,. Cyc2, *mtoAB*, and *mtrCAB*) for which qPCR primers are available (Shi et al. 2012). Unsurprisingly, genes indicative of these electron-accepting processes can be very diverse; thus, it can be challenging to select appropriate primer sets. For example, Müller et al. (2015) assessed the diversity of the *dsrAB* gene that encodes (bi)sulfite reductase, the enzyme catabolizing the final step in the dissimilatory sulfate-reducing pathway. They analyzed thousands of *dsrAB* genes available in genomic databases to determine, as part of their study, whether consensus primer sets could be created that would capture sulfate-reducer diversity that spans multiple clades in the Bacteria and Archaea. Ultimately, they found nine primer sets that could reasonably capture the majority of sulfate-reducing microorganisms. Vigneron et al. (2017) recently designed qPCR primer mixtures based on primer sets recommended by Müller et al. (2015) to successfully quantify sulfate-reducers in a variety of different environments, including those impacted by hydrocarbons.

11.4 CASE STUDIES

Recent review articles have highlighted a number of reports describing the use of catabolic gene markers (including *bssA*, *bamA*, and *bcrA*) to characterize and/or monitor hydrocarbon-contaminated groundwater aquifers for evidence of *in situ* hydrocarbon biodegradation (Kleinsteuber et al. 2012, Lueders 2017, von Netzer et al. 2016). Here, we highlight two recent case studies illustrating the use of qPCR assays for targeting key hydrocarbon-biodegrading taxa and/or biodegradation genes to offer evidence in support of anaerobic hydrocarbon biodegradation.

11.4.1 Case Study 1: Determining the Potential for Monitored Natural Attenuation by Biodegradation in Fuel-Contaminated Groundwater

The widespread use of gasoline and diesel to power transportation vehicles has occasionally resulted in fuel storage tank leakage into the subsurface environment underlying fuel dispensing depots. As fuel mixtures such as gasoline and diesel are abundant in low molecular weight hydrocarbons such as BTEX, cyclic alkanes, and *n*-alkanes that are comparatively water soluble, leaked fuels that enter underlying groundwater and be transported away from the leak site via groundwater flow, forming a plume of contamination. Monitored natural attenuation (also known as intrinsic bioremediation) is one option for remediating such plumes. Although passive, this approach requires ongoing monitoring within and beyond a contaminant plume to ensure that hydrocarbon concentrations are decreasing over time due to their metabolism by indigenous microbial communities (Bombach et al. 2010). As such, a qPCR approach targeting biodegradation genes can help site owners to determine whether a given site is amenable to remediation via monitored natural attenuation.

Kharey et al. (2020) assessed two former fuel depot sites (Site A and Site B) for their potential to be remediated using a monitored natural attenuation approach by using a variety of monitoring tools including qPCR assays targeting anaerobic hydrocarbon biodegradation (fumarate addition) genes. At both sites, underlying groundwater had become contaminated with mixtures of gasoline and diesel. Both sites were believed to be undergoing natural attenuation as the hydrocarbon plumes were stable, but prior to the study, no evidence in support of this phenomenon was available. Groundwater was collected from sampling wells installed at each site from both non-contaminated and hydrocarbon-contaminated areas across a two-year period. Site A was found to contain hydrocarbon concentrations ranging from 0.1 to

10 ppm, while Site B exhibited higher hydrocarbon concentrations, ranging from 2 to 35 ppm. Each groundwater sample was analyzed for various electron-accepting processes (e.g., by measuring nitrate, Fe(II), and sulfate), microbial community composition (based on 16S rRNA gene analysis), diagnostic fumarate addition metabolites, and for *assA* and *bssA* genes using both qualitative and qPCR assays. New primer sets were designed for qPCR analyses of both *bssA* and *assA* in an attempt to capture a broad diversity of these genes (Kharey et al. 2020; Table 11.1). Although measurements targeting electron accepting conditions and microbial community composition indicated that anaerobic processes and putative hydrocarbon-degrading microorganisms were prevalent at the sites, discernable patterns definitively indicating *in situ* hydrocarbon biodegradation could not be observed from these measurements due to site heterogeneity. In contrast, the detection of fumarate addition metabolites from both alkylbenzenes (benzylsuccinates) and *n*-alkanes/cyclic alkanes (alkylsuccinates) along with the detection and quantification of *assA* and *bssA* genes at multiple locations across each site unequivocally indicated that microbial communities indigenous to each site harbored the metabolic ability to anaerobically biodegrade hydrocarbons. At both sites, qPCR assays indicated that groundwater samples containing measurable levels of alkanes harbored 10^7 to 10^8 copies of the *assA* gene per L groundwater. In non-contaminated wells, the *assA* gene could either not be quantified, or contained 10^5–10^6 copies of the gene per L groundwater. qPCR assays targeting the *bssA* gene showed the contaminated wells harbored 10^6–10^{10} *bssA* copies per L groundwater, while this gene could not be quantified in uncontaminated wells. While *bssA* quantification in contaminated groundwater has been performed previously (e.g., Beller et al. 2008; Winderl et al., 2007; Table 11.1), this was the first report of *assA* gene abundances in this environment. Notably, qualitative PCR assays did detect the *bssA* and *assA* genes in many of the uncontaminated groundwater samples, suggesting that microorganisms indigenous to pristine groundwater environments may possess the natural ability to biodegrade hydrocarbons upon exposure, consistent with findings by other authors (Gittel et al. 2015). Overall, Kharey et al. (2020) demonstrated that the two sites were indeed amenable to monitored natural attenuation and that using qPCR assays targeting known biodegradation genes can be used effectively to assess contaminated sites for management by bioremediation approaches.

11.4.2 Case Study 2: Demonstrating Anaerobic Benzene Biodegradation at Canadian Forces Base (CFB) Borden

For over 40 years, an aquifer underlying a large military base situated 80 km northwest of Toronto, Ontario, Canada has served as an experimental site for groundwater transport and remediation of contaminants including petroleum hydrocarbons (Cherry et al. 1996, Sudicky and Illman 2011). The water table at this location is shallow (~ 1 m below ground surface), fluctuating seasonally, and the groundwater flow rate is approximately 9 cm/day (Mackay et al. 1986). Groundwater sulfate concentrations have ranged between 10 to 30 mg/L and are supplied from a nearby historical leachate plume (MacFarlane et al. 1983). Repeated geochemical experiments and modeling analyses suggested that oxygen was the major electron acceptor driving groundwater hydrocarbon bioremediation at this site (Barbaro and Barker 2000, Barker et al. 1987), yet the amount of organics lost could not be accounted for by aerobic mineralization or partial mineralization alone (Fraser et al. 2008).

While early biodegradation column studies had proposed the occurrence of anaerobic hydrocarbon biodegradation at CFB Borden (Acton and Barker 1992), convincing *in situ* evidence was first obtained using a combination of RT-qPCR and other monitoring methods such as CSIA and signature metabolite analysis (Wei et al. 2018). In a controlled field test, a defined petroleum mixture (containing benzene, toluene, and xylene; BTX) was released into the aquifer. Multiple transects of depth-discrete sampling wells were installed along the nominal groundwater flow path to capture the spatial and temporal behaviors of target compounds and microorganisms during field testing. The authors detected temporal enrichment of ^{13}C and ^{2}H in BTX compounds in sampling wells demonstrating active sulfate reduction, indicative of anaerobic biodegradation. This was supported by matching temporal patterns of benzylsuccinate and 2-methylbenzylsuccinate metabolites (diagnostic of anaerobic toluene and *o*-xylene degradation, respectively) were reported, as well as mRNA transcripts (10^2 – 10^3 copies/L) diagnostic of sulfate-reducing alkylbenzene metabolism (*bssA*-SRB; see Table 11.1), benzene metabolism (*abcA*), and sulfate respiration (*dsrB*). The use of RT-qPCR was particularly notable in this study, as it validated analytical evidence of sulfate-reducing BTX degradation at CFB Borden (Wei et al. 2018). Indeed, the identification of mRNA transcripts provided a direct line of evidence that the anaerobic BTX-degrading microorganisms present at this site were metabolically active.

Following up on this study, Toth et al. (2021) collected groundwater aquifer sediments from CFB Borden to further assess the value in using qPCR targeting benzene-degrading taxa (16S rRNA gene-based assays) and functional genes (*abcA*) for monitoring anaerobic benzene biodegradation. These sediments were used to construct anoxic microcosms that simulated different bioremediation approaches, including natural attenuation (no treatment), biostimulation with 2 mM nitrate, biostimulation with 2 mM sulfate, and bioaugmentation with a known methanogenic benzene-degrading consortium containing ORM2 (~ 10^7 copies per mL of culture), the aforementioned representative taxon of the Sva0485 clade Deltaproteobacteria. Hydrocarbon and biogeochemical measurements (e.g., nitrate, sulfate, and methane analyses), along with 16S rRNA gene sequencing, were used in combination with qPCR assays targeting benzene-biodegrading taxa (*Thermincola* spp. and Sva0485 clade Deltaproteobacteria) and *abcA* were performed over time to assess the effectiveness of each bioremediation approach for benzene biodegradation, and to search for evidence of enrichment of either known benzene-degrading clade.

After two years of incubation, benzene was not consumed in the "natural attenuation" or in the sulfate-biostimulated incubations; accordingly, neither the *abcA* gene nor the benzene-degrading taxa were detected above assay detection limits ($< 10^3$ copies/mL). However, benzene biodegradation occurred under all other test conditions, correlating with an increase in abundance of the *abcA* gene and/or benzene-degrading taxa. Under nitrate-reducing conditions, benzene loss was observed after three to five months and coincided with an increase (> 10^5 copies/mL) in the abundance of *Thermincola* spp. and *abcA* gene copies. Bioaugmented bottles received ~ 10^6 copies/mL of ORM2 and the onset of benzene degradation occurred within one month of incubation, attributed to the higher starting concentration of benzene-degrading organisms. 16S rRNA gene copies of ORM2 increased by an additional order of magnitude after 109 days of incubation; no enrichment of *abcA* or *Thermincola* was detected. As increases in the targeted gene abundances positively correlated with benzene decreases, the results of this study clearly demonstrated that using qPCR assays to enumerate benzene-degrading microorganisms or benzene biodegradation genes has high value for monitoring the effectiveness of benzene biodegradation (Toth et al. 2021). Further, the study showed that biostimulation with nitrate or bioaugmentation with benzene-biodegrading consortia are promising approaches for increasing rates of anaerobic benzene biodegradation in fuel-contaminated aquifers, approaches that will be field-tested in the future to treat benzene-contaminated sites.

11.5 CONCLUSION

Enormous progress made in understanding the activation methods and biodegradation pathways of different classes of hydrocarbons, and many qPCR primers targeting key catabolic genes and/or biodegrading taxa (such as for benzene) are now available for tracking this metabolism in hydrocarbon-contaminated field sites (Table 11.1). Although we focused here on the use of such assays for tracking biodegradation in groundwater aquifers, these qPCR assays can be used to interrogate any hydrocarbon-associated systems including subsurface petroliferous reservoirs, as well as marine environments (Gieg and Toth 2016). Further, while some of the qPCR assays highlighted here have been applied to groundwater systems, such as those targeting the *bssA*, *assA*, *bam/bcr*, and the *abcA* genes, many of the assays/genes have not yet been sought in environmental samples for assessing *in situ* biodegradation of hydrocarbons. This is a clear gap in knowledge, and we call for field site assessments to more frequently apply molecular approaches such as targeted qPCR assays summarized here to both evaluate site potential for *in situ* bioremediation and to track ongoing hydrocarbon biodegradation.

We also advocate for further integration of bioinformatics with biodegradation studies, both in laboratories and in the field, as they will enable greater access to information related to the biochemistry and prevalence of microbial degradation of hydrocarbons. These include the curation of additional databases, biodegradation pathway prediction systems, bioremediation modelling programs, and next-generation sequencing pipelines. Many of these computational tools had previously only existed for aerobic microbial processes (Arora and Bae 2014) but recently have been in development for anaerobic hydrocarbon biodegradation. As we learn more about specific mechanisms, genes, and taxa capable of hydrocarbon metabolism, this information should be translated into new qPCR assays that can be used to validate computational data, as well as improve monitoring efforts for anaerobic *in situ* hydrocarbon bioremediation.

ACKNOWLEDGMENTS

The authors are grateful to Dr. Elizabeth Edwards and her research team at the University of Toronto sharing their ongoing research of anaerobic benzene biodegradation, some results of which were shared in this chapter.

REFERENCES

Abu Laban, N., Selesi, D., Rattei, T., Tischler, P. and Meckenstock, R.U. (2010) Identification of enzymes involved in anaerobic benzene degradation by a strictly anaerobic iron-reducing enrichment culture. *Environ Microbiol* 12(10), 2783–2796.

Acton, D.W. and Barker, J.F. (1992) In situ biodegradation potential of aromatic hydrocarbons in anaerobic groundwaters. *J Contam Hydrol* 9, 325–352.

Agrawal, A. and Gieg, L.M. (2013) *In situ* detection of anaerobic alkane metabolites in subsurface environments. *Front Microbiol* 4, 140.

Aitken, C.M., Jones, D.M., Maguire, M.J., Gray, N.D., Sherry, A., Bowler, B.F.J., Ditchfield, A.K., Larter, S.R. and Head, I.M. (2013) Evidence that crude oil alkane activation proceeds by different mechanisms under sulfate-reducing and methanogenic conditions. *Geochim Cosmochim Acta* 109,162–174.

Amann, R.I., Ludwig, W. and Schleifer, K.H. (1995) Phylogenetic identification and in situ detection of individual microbial cells without cultivation. *Microbiol Rev* 59, 143–169.

Annweiler, E., Materna, A., Safinowski, M., Kappler, A., Richnow, H.H., Michaelis, W. and Meckenstock, R.U. (2000) Anaerobic degradation of 2-methylnaphthalene by a sulfate-reducing enrichment culture. *App Environ Microbiol* 66(12), 5329–5333.

Arora, P.K. and Bae, H. (2014) Integration of bioinformatics to biodegradation. *Biol Proced Online* 16, 8.

Ball, H.A., Johnson, H.A., Reinhard, M. and Spormann, A.M. (1996) Initial reactions in anaerobic ethylbenzene oxidation by a denitrifying bacterium, strain EB1. *J. Bacteriol* 178(19), 5755–5761.

Barbaro, J.R. and Barker, J.F. (2000) Controlled field study on the use of nitrate and oxygen for bioremediation of a gasoline source zone. *Biorem J* 4(4), 259–270.

Barker, J.F., Patrick, G.C. and Major, D. (1987) Natural attenuation of aromatic hydrocarbons in a shallow sand aquifer. *Groundwater Monit Rem* 7(1), 64–71.

Beller, H.R. (2002) Analysis of benzylsuccinates in groundwater by liquid chromatography tandem mass spectrometry and its use for monitoring in situ BTEX biodegradation. *Environ Sci Technol* 36(12), 2724–2728.

Beller, H.R. and Spormann, A.M. (1997) Anaerobic activation of toluene and *o*-xylene by addition to fumarate in denitrifying strain T. *J Bacteriol* 179(3), 670–676.

Beller, H.R., Kane, S.R., Legler, T.C., and Alvarez, P.J. (2002) A real-time polymerase chain reaction method for monitoring anaerobic, hydrocarbon-degrading bacteria based on a catabolic gene. *Environ Sci Technol* 36, 3977–3984.

Beller, H.R., Kane, S.R., Legler, T.C., McKelvie, J.R., Sherwood Lollar, B., Pearson, F., Balser, L. and Mackay, D.M. (2008) Comparative assessment of benzene, toluene, and xylene natural attenuation by quantitative polymerase chain reaction analysis of a catabolic gene, signature metabolites, and compound-specific isotope analysis. *Environ Sci Technol* 42, 6065–6072.

Boll, M., Estelmann, S. and Heider, J. (2020) Anaerobic degradation of hydrocarbons: Mechanisms of hydrocarbon activation in the absence of oxygen. In: *Anaerobic utilization of hydrocarbons, oils, and lipids*. Boll, M. (eds), pp. 213–277, Springer, Cham.

Bombach, P., Richnow, H.H., Kästner, M. and Fischer, A. (2010) Current approaches for the assessment of in situ biodegradation. *Appl Microbiol Biotechnol* 86(3), 839–852.

Bonilla-Rosso, G., Wittorf, L., Jones, C.M. and Hallin, S. (2016) Design and evaluation of primers targeting genes encoding NO-forming nitrite reductases: Implications for ecological inference of denitrifying communities. *Sci Rep* 6, 39208.

Bustin, S. and Huggett, J. (2017) qPCR primer design revisited. *Biomol Detect Quantif* 14, 19–28.

Callaghan, A.V. and Wawrik, B. (2016) AnHyDeg: A curated database of anaerobic hydrocarbon degradation genes. https://github.com/AnaerobesRock/AnHyDeg.

Callaghan, A.V., Tierney, M., Phelps, C.D. and Young, L.Y. (2009) Anaerobic biodegradation of *n*-hexadecane by a nitrate-reducing consortium. *Appl Environ Microbiol* 75(5), 1339–1344.

Callaghan, A.V., Davidova, I.A., Savage-Ashlock, K., Parisi, V.A., Gieg, L.M., Suflita, J.M., Kukor, J.J. and Wawrik, B. (2010) Diversity of benzyl- and alkylsuccinate synthase genes in hydrocarbon-impacted environments and enrichment cultures. *Environ Sci Technol* 44(19), 7287–7294.

Cherry, J.A., Barker, J.F., Feenstra, S., Gillham, R.W., Mackay, D.M. and Smyth, D.J.A. (1996) *Groundwater and subsurface remediation*. Kobus, H., Barczewski, B. and Koschitzky, H.P. (eds), Springer-Verlag, Berlin Heidelberg.

da Silva, M.L. and Alvarez, P.J. (2007) Assessment of anaerobic benzene degradation potential using 16S rRNA gene-targeted real-time PCR. *Environ Microbiol* 9(1), 72–80.

Davidova, I.A., Gieg, L.M., Duncan, K.E. and Suflita, J.M. (2007) Anaerobic phenanthrene mineralization by a carboxylating sulfate-reducing bacterial enrichment. *ISME J* 1(5), 436–442.

de Rezende, J. (2016) Quantification of sulfate-reducing microorganisms by quantitative PCR: Current challenges and developments. In *Microbially influenced corrosion in the upstream oil and gas industry*. Skovhus, T.L., Enning, D. and Lee, J.S. (eds), pp. 213–227, CRC Press, Boca Raton.

Elshahed, M.S., Gieg, L.M., McInerney, M.J. and Suflita, J.M. (2001) Signature metabolites attesting to the in situ attenuation of alkylbenzenes in anaerobic environments. *Environ Sci Technol* 35, 682–689.

Ferris, M.J., Muyzer, G. and Ward, D.M. (1996) Denaturing gradient gel electrophoresis profiles of 16S rRNA-defined populations inhabiting a hot spring microbial mat community. *Appl Environ Microbiol* 62(2), 340–346.

Foght, J. (2008) Anaerobic biodegradation of aromatic hydrocarbons: Pathways and prospects. *J Mol Microbiol Biotechnol* 15(2–3), 93–120.

Fowler, S.J., Gutierrez-Zamora, M.L., Manefield, M. and Gieg, L.M. (2014) Identification of toluene degraders in a methanogenic enrichment culture. *FEMS Microbiol Ecol* 89(3), 625–636.

Fraser, M., Barker, J.F., Butler, B., Blaine, F., Joseph, S. and Cooke, C. (2008) Natural attenuation of a plume from an emplaced coal tar creosote source over 14 years. *J Contam Hydrol* 100(3–4), 101–115.

Gieg, L.M. and Toth, C.R.A. (2016) Anaerobic biodegradation of hydrocarbons: Metagenomics and metabolomics. In: *Consequences of microbial interactions with hydrocarbons, oils and lipids*. Steffan, R. (eds), pp. 1–42, Springer, Cham.

Gieg, L.M. and Toth, C.R.A. (2020) Signature metabolite analysis to determine in situ anaerobic hydrocarbon biodegradation. In: *Anaerobic utilization of hydrocarbons, oils, and lipids*. Boll, M. (eds), pp. 361–390, Springer, Cham.

Gittel, A., Donhauser, J., Røy, H., Girguis, P.R., Jørgensen, B.B. and Kjeldsen, K.U. (2015) Ubiquitous presence and novel diversity of anaerobic alkane degraders in cold marine sediments. *Front Microbiol* 6, 1414.

Grossi, V., Cravo-Laureau, C., Méou, A., Raphel, D., Garzino, F. and Hirschler-Réa, A. (2007) Anaerobic 1-alkene metabolism by the alkane- and alkene-degrading sulfate reducer *Desulfatibacillum aliphaticivorans* strain CV2803T. *Appl Environ Microbiol* 73(24), 7882–7890.

Heider, J., Szaleniec, M., Martins, B.M., Seyhan, D., Buckel, W. and Golding, B.T. (2016a) Structure and function of benzylsuccinate synthase and related fumarate-adding glycyl radical enzymes. *J Mol Microbiol Biotechnol* 26(1–3), 29–44.

Heider, J., Szaleniec, M., Sünwoldt, K. and Boll, M. (2016b) Ethylbenzene dehydrogenase and related molybdenum enzymes involved in oxygen-independent alkyl chain hydroxylation. *J Mol Microbiol Biotechnol* 26(1–3), 45–62.

Hosoda, A., Kasai Y., Hamamura, N., Takahata, Y. and Watanabe, K. (2005) Development of a PCR method for the detection and quantification of benzoyl-CoA reductase genes and its application to monitored natural attenuation. *Biodegradation* 16, 591–601.

Hu, P., Tom, L., Singh, A., Thomas, B.C., Baker, B.J., Piceno, Y.M., Andersen, G.L. and Banfield, J.F. (2016) Genome-resolved metagenomic analysis reveals roles for candidate phyla and other microbial community members in biogeochemical transformations in oil reservoirs. *MBio* 7(1), e01669–e01615.

IEA (2020) *Global energy review 2020*. IEA, Paris. https://www.iea.org/reports/global-energy-review-2020.

Keely, J.F. (1989) *Performance evaluations of pump-and-treat remediations*. EPA/540/4-89/005 (ed), EPA, Ada, Oklahoma.

Kembel, S.W., Wu, M., Eisen, J.A. and Green, J.L. (2012) Incorporating 16S gene copy number information improves estimates of microbial diversity and abundance. *PLoS Comput Biol* 8(10), e1002743.

Kharey, G., Scheffer, G. and Gieg, L.M. (2020) Combined use of diagnostic fumarate addition metabolites and genes provides evidence for anaerobic hydrocarbon biodegradation in contaminated groundwater. *Microorganisms* 8, 1532.

Kleinsteuber, S., Schleinitz, K.M. and Vogt, C. (2012) Key players and team play: Anaerobic microbial communities in hydrocarbon-contaminated aquifers. *Appl Microbiol Biotechnol* 94(4), 851–873.

Kniemeyer, O. and Heider, J. (2001) Ethylbenzene dehydrogenase, a novel hydrocarbon-oxidizing molybdenum/iron-sulfur/heme enzyme. *J Biol Chem* 276(24), 21381–21386.

Kniemeyer, O., Fischer, T., Wilkes, H., Glöckner, F.O. and Widdel, F. (2003) Anaerobic degradation of ethylbenzene by a new type of marine sulfate-reducing bacterium. *Appl Environ Microbiol* 69(2), 760–768.

Koelschbach, J.S., Mouttaki, H., Merl-Pham, J., Arnold, M.E. and Meckenstock, R.U. (2019) Identification of naphthalene carboxylase subunits of the sulfate-reducing culture N47. *Biodegradation* 30(2–3), 147–160.

Kraiselburd, I., Brüls, T., Heilmann, G., Kaschani, F., Kaiser, M. and Meckenstock, R.U. (2019) Metabolic reconstruction of the genome of candidate *Desulfatiglans* TRIP_1 and identification of key candidate enzymes for anaerobic phenanthrene degradation. *Environ Microbiol* 21(4), 1267–1286.

Kropp, K.G., Davidova, I.A. and Suflita, J.M. (2000) Anaerobic oxidation of *n*-dodecane by an addition reaction in a sulfate-reducing bacterial enrichment culture. *Appl Environ Microbiol* 66(12), 5393–5398.

Kühner, S., Wöhlbrand, L., Fritz, I., Wruck, W., Hultschig, C., Hufnagel, P., Kube, M., Reinhardt, R. and Rabus, R. (2005) Substrate-dependent regulation of anaerobic degradation pathways for toluene and ethylbenzene in a denitrifying bacterium, strain EbN1. *J Bacteriol* 187(4), 1493–1503.

Kunapuli, U., Lueders, T. and Meckenstock, R.U. (2007) The use of stable isotope probing to identify key iron-reducing microorganisms involved in anaerobic benzene degradation. *ISME J* 1(7), 643–653.

Kuntze, K., Shinoda, Y., Moutakki, H., McInerney, M.J., Vogt, C., Richnow, H.H. and Boll, M. (2008) Hosoda, A., Kasai Y., Hamamura, N., Takahata, Y. and Watanabe, K. (2005) 6-Oxycyclohex-1-ene-1-carbonyl-coenzyme A hydrolases from obligately anaerobic bacteria: Characterization and identification of its gene as a functional marker for aromatic compounds degrades anaerobes. *Environ Microbiol* 10(6), 1547–1556.

Kuntze, K., Vogt, C., Richnow, H.H. and Boll, M. (2011) Combined application of PCR-based functional assays for the detection of aromatic-compound-degrading anaerobes. *Appl Environ Microbiol* 77(14), 5056–5061.

Leuthner, B., Leutwein, C., Schulz, H., Hörth, P., Haehnel, W., Schiltz, E., Schägger, H. and Heider, J. (1998) Biochemical and genetic characterization of benzylsuccinate synthase from *Thauera aromatica*: A new glycyl radical enzyme catalysing the first step in anaerobic toluene metabolism. *Mol Microbiol* 28(3), 615–628.

Leutwein, C. and Heider, J. (1999) Anaerobic toluene-catabolic pathway in denitrifying *Thauera aromatica*: Activation and β-oxidation of the first intermediate, (*R*)-(+)-benzylsuccinate. *Microbiology* 145, 3265–3271.

Liu, Y.F., Qi, Z.Z., Shou, L.B., Liu, J.F., Yang, S.Z., Gu, J.D. and Mu, B.Z. (2019) Anaerobic hydrocarbon degradation in candidate phylum 'Atribacteria' (JS1) inferred from genomics. *ISME J* 13(9), 2377–2390.

Lueders, T. (2017) The ecology of anaerobic degraders of BTEX hydrocarbons in aquifers. *FEMS Microbiol Ecol* 93(1), fiw220.

Luo, F., Gitiafroz, R., Devine, C.E., Gong, Y., Hug, L.A., Raskin, L. and Edwards, E.A. (2014) Metatranscriptome of an anaerobic benzene-degrading, nitrate-reducing enrichment culture reveals involvement of carboxylation in benzene ring activation. *Appl Environ Microbiol* 80(14), 4095–4107.

Luo, F., Devine, C.E. and Edwards, E.A. (2016) Cultivating microbial dark matter in benzene-degrading methanogenic consortia. *Environ Microbiol* 18(9), 2923–2936.

Luton, P.E., Wayne, J.M., Sharp, R.J. and Riley, P.W. (2002) The *mcrA* gene as an alternative to 16S rRNA in the phylogenetic analysis of methanogen populations in landfill. *Microbiology (Reading)* 148(Pt 11), 3521–3530.

MacFarlane, D.S., Cherry, J.A., Gillham, R.W. and Sudicky, E.A. (1983) Migration of contaminants in groundwater at a landfill: A case study: 1. Groundwater flow and plume delineation. *J Hydrol* 63, 1–29.

Mackay, D.M., Freyberg, D.L., Roberts, P.V. and Cherry, J.A. (1986) A natural gradient experiment on solute transport in a sand aquifer: 1. Approach and overview of plume movement. *Water Resour Res* 22(13), 2017–2029.

Magnuson, E.N. (2017) Characterization of the microbial community and activity of nitrate-reducing, benzene-degrading enrichment cultures. MASc Thesis, Toronto, Canada: University of Toronto. Chem Eng and Appl Chem, http://hdl.handle.net/1807/79143.

Mancini, S.A., Devine, C.E., Elsner, M., Nandi, M.E., Ulrich, A.C., Edwards, E.A. and Sherwood Lollar, B. (2008) Isotopic evidence suggests different initial reaction mechanisms for anaerobic benzene biodegradation. *Environ Sci Technol* 42, 8290–8296.

Meckenstock, R.U., Boll, M., Mouttaki, H., Koelschbach, J.S., Cunha Tarouco, P., Weyrauch, P., Dong, X. and Himmelberg, A.M. (2016) Anaerobic degradation of benzene and polycyclic aromatic hydrocarbons. *J Mol Microbiol Biotechnol* 26(1–3), 92–118.

Morasch, B., Schink, B., Tebbe, C.C. and Meckenstock, R.U. (2004) Degradation of *o*-xylene and *m*-xylene by a novel sulfate-reducer belonging to the genus *Desulfotomaculum*. *Arch Microbiol* 181(6), 407–417.

Morris, B.E.L., Gissibl, A., Kümmel, S., Richnow, H.H., and Boll, M. (2014) A PCR-based assay for the detection of anaerobic naphthalene degradation. *FEMS Microbiol Lett* 354, 55–59.

Mouttaki, H., Johannes, J. and Meckenstock, R.U. (2012) Identification of naphthalene carboxylase as a prototype for the anaerobic activation of non-substituted aromatic hydrocarbons. *Environ Microbiol* 14(10), 2770–2774.

Müller, A.L., Kjeldsen, K.U., Rattei, T., Pester, M. and Loy, A. (2015) Phylogenetic and environmental diversity of DsrAB-type dissimilatory (bi)sulfite reductases. *ISME J* 9(5), 1152–1165.

National Research Council(NRC), N.R.C. (1993) *In situ bioremediation: When does it work?* National Academy Press, Washington, DC.

Oberding, L.K. and Gieg, L.M. (2018). Methanogenic paraffin biodegradation: Alkylsuccinate synthase gene quantification and dicarboxylic acid production. *Appl Environ Microbiol* 84(1), e01773–e01717.

Parisi, V.A., Brubaker, G.R., Zenker, M.J., Prince, R.C., Gieg, L.M., da Silva, M.L., Alvarez, P.J. and Suflita, J.M. (2009) Field metabolomics and laboratory assessments of anaerobic intrinsic bioremediation of hydrocarbons at a petroleum-contaminated site. *Microb Biotechnol* 2(2), 202–212.

Phelps, C.D., Zhang, X. and Young, L.Y. (2001) Use of stable isotopes to identify benzoate as a metabolite of benzene degradation in a sulphidogenic consortium. *Environ Microbiol* 3(9), 600–603.

Porter, A.W. and Young, L.Y. (2013) The *bamA* gene for anaerobic ring fission is widely distributed in the environment. *Front Microbiol* 4, 302.

Rabus, R. and Widdel, F. (1995) Anaerobic degradation of ethylbenzene and other aromatic hydrocarbons by new denitrifying bacteria. *Arch Microbiol* 163, 96–103.

Rossmassler, K., Snow, C.D., Taggart, D., Brown, C. and De Long, S.K. (2019) Advancing biomarkers for anaerobic *o*-xylene biodegradation via metagenomic analysis of a methanogenic consortium. *Appl Microbiol Biotechnol* 103, 4177–4192.

Sakai, N., Kurisu, F., Yagi, O., Nakajima, F. and Yamamoto, K. (2009) Identification of putative benzene-degrading bacteria in methanogenic enrichment cultures. *J Biosci Bioeng* 108(6), 501–507.

Selesi, D., Jehmlich, N., von Bergen, M., Schmidt, F., Rattei, T., Tischler, P., Lueders, T. and Meckenstock, R.U. (2010) Combined genomic and proteomic approaches identify gene clusters involved in anaerobic 2-methylnaphthalene degradation in the sulfate-reducing enrichment culture N47. *J Bacteriol* 192(1), 295–306.

Shayan, M., Thomson, N.R., Aravena, R., Barker, J.F., Madsen, E.L., Massimo, M., DeRito, C.M., Bouchard, D., Buscheck, T., Kolhatkar, R. and Daniels, E.J. (2017) Integrated plume treatment using persulfate coupled with microbial sulfate reduction. *Ground Water Monit Remed* 38(4), 45–61.

Shi, L., Rosso, K.M., Clarke, T.A., Richardson, D.J., Zachara, J.M. and Fredrickson, J.K. (2012) Molecular underpinnings of Fe(III) oxide reduction by *Shewanella oneidensis* MR-1. *Front Microbiol* 3, 50.

Sierra-Garcia, I.N., Correa Alvarez, J., de Vasconcellos, S.P., Pereira de Souza, A., dos Santos Neto, E.V. and de Oliveira, V.M. (2014) New hydrocarbon degradation pathways in the microbial metagenome from Brazilian petroleum reservoirs. *PLoS One* 9(2), e90087.

So, C.M., Phelps, C.D. and Young, L.Y. (2003) Anaerobic transformation of alkanes to fatty acids by a sulfate-reducing bacterium, strain Hxd3. *Appl Environ Microbiol* 69(7), 3892–3900.

Staats, M., Braster, M. and Röling, W.F.M. (2011) Molecular diversity and distribution of aromatic hydrocarbon-degrading anaerobes across a landfill leachate plume. *Environ Microbiol* 13(5), 1216–1227.

Strijkstra, A., Trautwein, K., Jarling, R., Wöhlbrand, L., Dörries, M., Reinhardt, R., Drozdowska, M., Golding, B.T., Wilkes, H. and Rabus, R. (2014) Anaerobic activation of *p*-cymene in denitrifying betaproteobacteria: Methyl group hydroxylation versus addition to fumarate. *Appl Environ Microbiol* 80(24), 7592–7603.

Sudicky, E.A. and Illman, W.A. (2011) Lessons learned from a suite of CFB Borden experiments. *Ground Water* 49(5), 630–648.

Toth, C.R.A. and Gieg, L.M. (2018) Time course-dependent methanogenic crude oil biodegradation: Dynamics of fumarate addition metabolites, biodegradative genes, and microbial community composition. *Front Microbiol* 8, 2610.

Toth, C.R.A., Luo, F., Bawa, N., Webb, J., Guo, S., Dworatzek, S. and Edwards, E.A. (2021) Anaerobic benzene biodegradation linked to growth of highly specific bacterial clades. *BioRxiv*, 2021.01.23.427911.

Ulrich, A.C. and Edwards, E.A. (2003) Physiological and molecular characterization of anaerobic benzene-degrading mixed cultures. *Environ Microbiol* 5(2), 92–102.

van der Zaan, B.M., Saia, F.T., Stams, A.J., Plugge, C.M., de Vos, W.M., Smidt, H., Langenhoff, A.A. and Gerritse, J. (2012) Anaerobic benzene degradation under denitrifying conditions: *Peptococcaceae* as dominant benzene degraders and evidence for a syntrophic process. *Environ Microbiol* 14(5), 1171–1181.

van der Waals, M.J., Atashgahi, S., da Rocha, U.N., van der Zaan, B.M., Smidt, H. and Gerritse, J. (2017) Benzene degradation in a denitrifying biofilm reactor: Activity and microbial community composition. *Appl Microbiol Biotechnol* 101(12), 5175–5188.

Vigneron, A., Alsop, E.B., Cruaud, P., Philibert, G., King, B., Baksmaty, L., Lavallée, D., Lomans, B.P., Kyrpides, N.C., Head, I.M. and Tsesmetzis, N. (2017) Comparative metagenomics of hydrocarbon and methane seeps of the Gulf of Mexico. *Sci Rep* 7(1), 16015.

von Netzer, F., Pilloni, G., Kleindienst, S., Krüger, M., Knittel, K., Gründger, F. and Lueders, T. (2013) Enhanced gene detection assays for fumarate-adding enzymes allow uncovering of anaerobic hydrocarbon degraders in terrestrial and marine systems. *Appl Environ Microbiol* 79(2), 543–552.

von Netzer, F., Kuntze, K., Vogt, C., Richnow, H.H., Boll, M. and Lueders, T. (2016) Functional gene markers for fumarate-adding and dearomatizing key enzymes in anaerobic aromatic hydrocarbon degradation in terrestrial environments. *J Mol Microbiol Biotechnol* 26(1–3), 180–194.

Wei, Y., Thomson, N.R., Aravena, R., Marchesi, M., Barker, J.F., Madsen, E.L., Kolhatkar, R., Buscheck, T., Hunkeler, D. and Derito, C.M. (2018) Infiltration of sulfate to enhance sulfate-reducing biodegradation of petroleum hydrocarbons. *Ground Water Monit Remediat* 38(4), 73–87.

Wilkes, H., Buckel, W., Golding, B.T. and Rabus, R. (2016) Metabolism of hydrocarbons in *n*-alkane-utilizing anaerobic bacteria. *J Mol Microbiol Biotechnol* 26(1–3), 138–151.

Willerslev, E. and Cooper, A. (2005) Ancient DNA. *Proc Biol Sci* 272(1558), 3–16.

Winderl, C., Schaefer, S. and Lueders, T. (2007) Detection of anaerobic toluene and hydrocarbon degraders in contaminated aquifers using benzylsuccinate synthase (*bssA*) genes as a functional marker. *Environ Microbiol* 9(4), 1035–1046.

Wöhlbrand, L., Jacob, J.H., Kube, M., Mussmann, M., Jarling, R., Beck, A., Amann, R., Wilkes, H., Reinhardt, R. and Rabus, R. (2013) Complete genome, catabolic sub-proteomes and key-metabolites of *Desulfobacula toluolica* Tol2, a marine, aromatic compound-degrading, sulfate-reducing bacterium. *Environ Microbiol* 15(5), 1334–1355.

Yu, Y., Lee, C., Kim, J. and Hwang, S. (2004) Group-specific primer and probe sets to detect methanogenic communities using quantitative real-time polymerase chain reaction. *Biotechnol Bioeng* 89(6), 670–679.

Zhang, X. and Young, L.Y. (1997) Carboxylation as an initial reaction in the anaerobic metabolism of naphthalene and phenanthrene by sulfidogenic consortia. *Appl Environ Microbiol* 63(12), 4759–4764.

Zhang, T., Bain, T.S., Nevin, K.P., Barlett, M.A. and Lovley, D.R. (2012) Anaerobic benzene oxidation by *Geobacter* species. *Appl Environ Microbiol* 78(23), 8304–8310.

12 Leveraging Bioinformatics to Elucidate the Genetic and Physiological Adaptation of *Pseudomonas aeruginosa* to Hydrocarbon-Rich Jet Fuel

Thusitha S. Gunasekera

CONTENTS

12.1 Introduction 249
12.2 Genomics 251
 12.2.1 Core and Accessory Genome Analysis 251
12.3 Single Nucleotide Polymorphism (SNPs) and Alk Gene Expression 253
12.4 Transcriptomics 254
 12.4.1 *P. aeruginosa* Response to Alkanes and Jet Fuel 254
12.5 Physiological Adaptations 264
 12.5.1 Biofilm Formation 264
 12.5.2 Solvent Resistance 266
 12.5.3 Trace Elements and Iron Acquisition 266
 12.5.4 Stress Response 267
12.6 Conclusion 269
Acknowledgment 270
References 270

12.1 INTRODUCTION

Bacteria adaptation to different environments has been studied extensively. Microorganisms that have the ability to survive and proliferate in hydrocarbon-rich environments have a wide-range of adaptive mechanisms (Gunasekera et al., 2013).

Alkanes and aromatic compounds in fuel can serve as rich carbon sources for microbial proliferation but toxic compounds in fuel have adverse effects on microorganisms. Microorganisms can use hydrocarbons as a carbon source by aerobic or anaerobic oxidation by preferentially degrading specific fuel constituents; some microorganisms have a high capacity to degrade alkanes while others prefer to grow by consuming aromatic components (Striebich et al., 2014). However, in order to survive and degrade hydrocarbons, microorganisms should be well adapted to counteract and decrease the toxic effects of fuel constituents. An opportunistic human pathogen, *Pseudomonas aeruginosa*, can colonize in different habitats, including jet fuels and hydrocarbon-contaminated environments (Gunasekera et al., 2013). Genomes cover all chromosomal and extra chromosomal genetic materials including plasmids DNA. The characterization of microbial genomes helps us to elucidate how genetic material contributes to their characteristics, functions, mechanisms, as well as their evolutionary relationships. On the other hand, transcriptomes, the sum of all of its RNA transcripts, identify the subset of genes that induce under certain conditions. Therefore, transcriptomics uncovers coordinated cell responses to particular conditions. Transcriptomics also can be used to characterize how gene expression changes in different organisms to different environments. With the advancement of next-generation sequencing technologies, genomics and transcriptomics have become widespread tools applied to microbiology. Genomics in combination with transcriptomics can be used to study how bacteria adapt to a fuel environment by changing their gene expression profile and in turn change their proteome and enzyme levels in the cell. This quick change in gene expression provides correct cell physiology, allowing them to adapt and proliferate in a jet fuel system rapidly.

P. aeruginosa genome has a highly conserved core genome that represents 88% of its total genome. (Valot et al., 2015). Both environmental and pathogenic strains share highly conserved core genomes (Alonso et al., 1999) and possess key genes important for hydrocarbon degradation. For example, *P. aeruginosa* isolated from human tissue, the type strain PAO1 and the strain isolated from a fuel tank, ATCC 33988 (Brown et al., 2014) have alkane monooxygenase genes *alkB1*and *alkB2* (Smits et al., 2003). However, their growth rates in jet fuel are different (Figure 12.1) because the environmental strain (ATCC 33988) is well adapted to the hydrocarbon-rich fuel environment (Gunasekera et al., 2017). *P. aeruginosa* ATCC 33988 activates multiple metabolic pathways and adaptations to overcome fuel stress (Gunasekera et al., 2013; Gunasekera et al., 2017). Genome sequencing and comparative genomics of closely related individuals provided additional information about genes and gene structure as well as genome-wide association analyses of signal nucleotide polymorphisms (SNPs) (Gunasekera et al., 2017). SNPs are a substitution of a single nucleotide at a specific position in the genome. SNPs help to understand fine scale evolutionary processes involved with gene expression and genetic and physiological adaptation to jet fuel. A large amount of genomic diversity of two closely related *P. aeruginosa* strains is shown in relation to hydrocarbon degradation (Gunasekera et al., 2017). SNPs analysis also showed a large amount of genetic heterogeneity between these two strains that contributed to the development of adaptive mechanisms for their specific niche.

An integrated approach combining genomics and transcriptomics along with functional genetic studies provide valuable insights into complex biological systems.

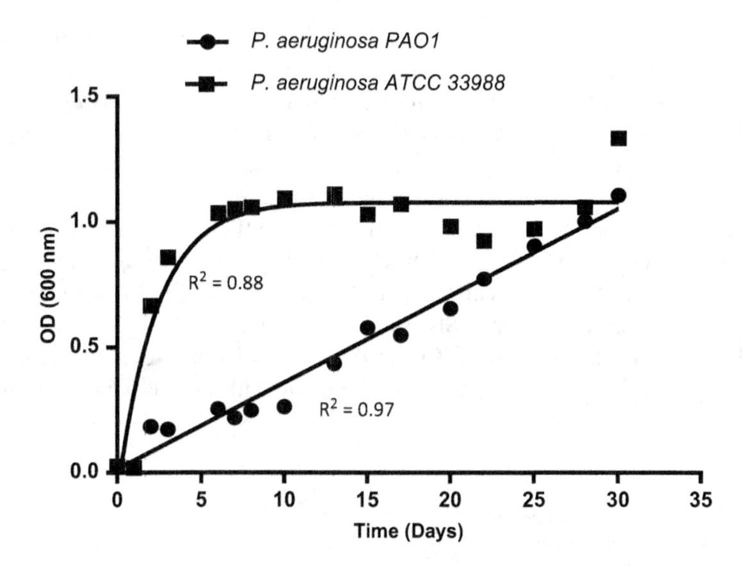

FIGURE 12.1 Growth characterization *P. aeruginosa* strains. *P. aeruginosa* strains were grown in M9 minimal media (M9 minimal media was prepared according to the directions described by Gunasekera & Paliy, 2006) with filter-sterilized Jet-A fuel at 28 °C with aeration. Each data point is an average of three replicates. R Squared = regression coefficient. *P. aeruginosa* ATCC 33988 bacterium isolated from a fuel tank, whereas *P. aeruginosa* PAO1 (the type strain) isolated from human tissue.

To advance cellular microbiology, multi-omics technologies are essential for elucidating key genetic pathways and uncovering novel regulatory mechanisms. Although RNA sequencing is a rapidly evolving technology, microarrays are still a powerful method to study genome-wide expression studies of well-defined bacteria such as *P. aeruginosa*, which are often found in hydrocarbon-contaminated sites and contaminated fuel tanks. In contrast to RNA-seq, c-DNA microarrays do not require depleting ribosomal RNA from the sample. However, RNA sequencing has several advantages over hybridization methods with one key advantage that it can be performed for non-model organisms, whose genomes are not yet available. RNA-seq gives unprecedented details such as non-coding RNAs, SNPs, and regulatory sequences. As the field moves forward with RNA sequencing as a key technology for functional genomics, c-DNA microarrays still provide high-throughput statistically robust data for a handful of model organisms. Throughout this chapter, examples will be provided of how transcriptomics helps to discover novel adaptive mechanisms of *P. aeruginosa* and how sequence-based data support experimental data and vice versa.

12.2 GENOMICS

12.2.1 Core and Accessory Genome Analysis

The genome of *P. aeruginosa* ATCC 33988 bacterium isolated from a fuel tank was sequenced (Brown et al., 2014) and its genome and transcriptome was compared with the bacterium *P. aeruginosa* PAO1, which is considered an opportunistic human pathogen. This life style transition from pathogenic level to environmental level or vice versa allows adaptation to a specific niche. *P. aeruginosa* ATCC 33988 is well adapted to a hydrocarbon-rich environment and is capable of degrading normal straight chain alkanes efficiently and faster than the PAO1 strain. Additionally, *P. aeruginosa* PAO1 strain cannot degrade n-C_8 or n-C_{10}, while ATCC 33988 strain degrades these short-chain alkanes efficiently (Gunasekera et al., 2017). *P. aeruginosa* ATCC 33988 has a relatively larger genome (6.41 Mb) as compared to that of the clinical strain PAO1, which has a 6.26 Mb size genome (Figure 12.2). A large number of SNPs and a slightly larger genome of the environmental organism *P. aeruginosa* ATCC 33988 suggest a micro-evolutionary process has taken place. *P. aeruginosa* PAO1 has a genome with a total of 5708 genes while

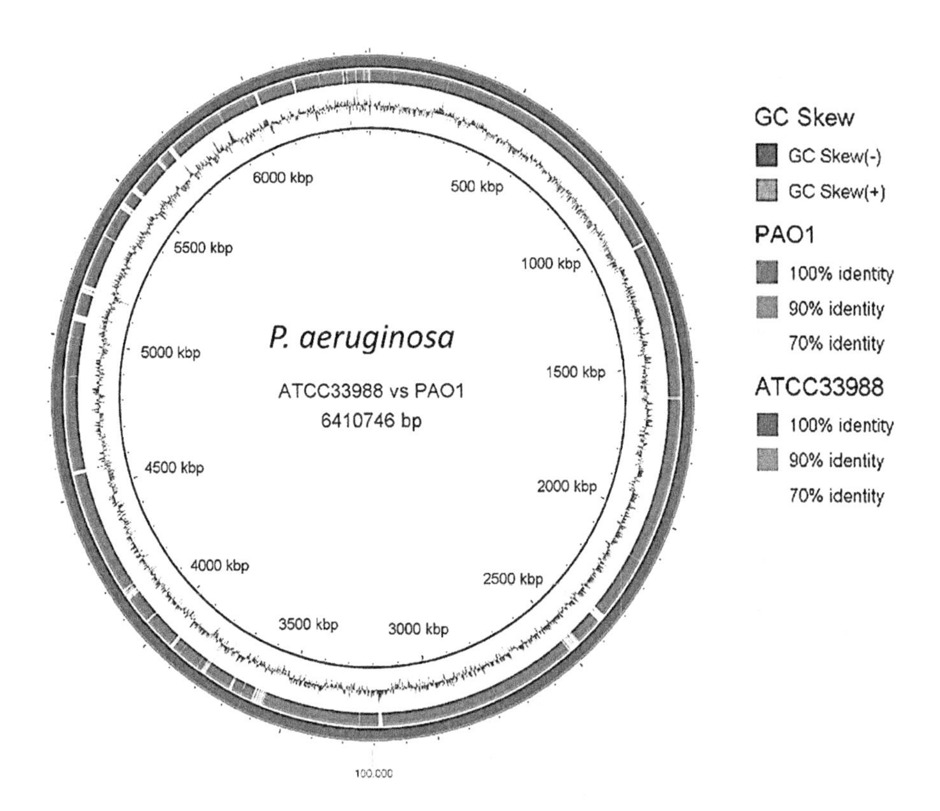

FIGURE 12.2 Schematic representation of *P. aeruginosa* ATCC 33988 strain Genome (Blue). The inner circle (Purple) represents the genome of *P. aeruginosa* PAO1 strain. GC-skew ((G-C)/(G+C)), purple indicates (-) GC skew, green indicates (+) GC skew. *P. aeruginosa* ATCC genome was sequenced by the whole-genome shotgun project and has been deposited in DDBJ/EMBL/GenBank under the accession no. JPQQ00000000 (Brown et al., 2014). The size of the *P. aeruginosa* ATCC genome is 6.4-Mb and has 5983 genes. In contrast, the genome of *P. aeruginosa* PAO1, an important opportunistic human pathogen, is 6.26 Mb in size, with 5708 genes (https://www.pseudomonas.com)).

P. aeruginosa ATCC 33988 has 5983 genes, an additional 275 genes as compared to the PAO1 strain (https://www.pseudomonas.com). Several genetic variations have been identified in these two organisms including SNPs insertions, and deletions (Gunasekera et al., 2017). The core genome is estimated at about 88% of the whole genome and is comprised of genes involved in aerobic and anaerobic respiration, transcription and translation regulation, DNA and protein repair, hydrocarbon degradation, and multidrug efflux systems. Both strains have functional *alkB1* and *alkB2* genes in their core genome and these genes are essential for alkane degradation. However, *P. aeruginosa* ATCC 33988 degrades jet fuel faster than the PAO1 strain.

12.3 SINGLE NUCLEOTIDE POLYMORPHISM (SNPs) AND ALK GENE EXPRESSION

SNPs at a specific base position in the regulatory region or in the coding sequence can alter gene expression. *P. aeruginosa* ATCC 33988 adaptation to a hydrocarbon environment was primarily due to fuel-specific adaptive mechanisms and highly inducible *alk* genes (Gunasekera et al., 2017). Using microarray and RT-qPCR, it was confirmed that both ATCC 33988 and PAO1 have active alkane monooxygenase enzymes (Gunasekera et al., 2017). These genes are highly induced in *P. aeruginosa* ATCC 33988 compared to the PAO1 strain. For example, *alkB1* and *alkB2* genes, coding alkane monooxygenase, are conserved, presenting only a few synonymous-SNPs. However, both *alkB1* and *alkB2* genes were significantly induced in the presence of fuel in *P. aeruginosa* ATCC 33988 as compared to PAO1 strain. Results indicated two nucleotide polymorphisms in the *alkB1* genes and a single-nucleotide polymorphism in the *alkB2* gene. A promoter polymorphism was also found in *alkB1* and *alkB2* genes which affects the expression of these genes. The expression of Green Fluorescent Protein (GFP) under the control of *alk* promoters confirmed the *alk* gene promoter polymorphism affects *alk* gene expression significantly. Overexpression of *alk* genes using a native promoter from ATCC 33988 stimulated the growth of PAO1 in fuel. A single SNP at the promoter of *alkB2* gene significantly increased the expression of GFP, when the promoters were swapped from the ATCC 33988 strain to PAO1 (Gunasekera et al., 2017). This indicates the promoter contains functionally relevant polymorphisms although the gene contains silent synonymous mutations. These beneficial mutations may have enhanced the binding of transcriptional factors to regulatory regions of the gene. Increased expression of the habitat-associated genes offer fitness for better adapting to their related environments. Of the 4298 *P. aeruginosa* genomes in Gene Bank, 3860 strains show 100% sequence similarity (at the amino acid level) to the *alkB2* gene of PAO1 type strain. The rest of the strain shows 1–5 missmatches to the 377 total alignment length of the protein. All these strains carry multiple SNPs in the promoter region of the gene. Genome analysis revealed a large number of strains contain polymorphisms in the *alkB2* promoter region. Of the 130 genomes analyses revealed, 63 strains carry 1–7 different SNPs in the *alkB2* promoters and 42 strains carried 1–5 SNPs in the AlkB2 protein. Of 130 strains, 32 strains carried no mutations either in the promoter or in the amino acid sequence of the mature protein. *P. aeruginosa* environmental strain CR1 (Sood et al.,

2019) contains 7 SNPs in the promoter regions and 5 amino acid SNPS in the AlkB2 proteins. Another taxonomic outlier, PA7, carried 7 SNPs in the promoter region and 4 amino acid SNPs in the AlkB2 protein (Klockgether et al., 2011). Therefore, on occasions, there are more SNPs in the promoter region than there are non-synonymous SNPs in the coding region of the gene. Further studies are needed into how these *alk* gene promoter genotypes and SNPs in regulatory sequences affect protein expression and how it relates to alkane degradation efficiency. Other researchers have shown evidence that single-nucleotide polymorphisms in the promoter region can influence gene expression and pathogen phenotype (Chauhan et al., 2010; de Greeff et al., 2014).

12.4 TRANSCRIPTOMICS

12.4.1 *P. aeruginosa* Response to Alkanes and Jet Fuel

Although jet fuel and alkanes are energy-rich molecules, they are often toxic to any type of cells, therefore microbial cells exert various adaptive mechanisms to overcome hydrocarbon toxicity. Gene expression brings new insights into the bacterial response to hydrocarbons (Gunasekera et al., 2013, 2017; Grady et al., 2017). Degradation of hydrocarbon fuels by microrganisms is well known, but crucial details about genetic programming used by bacteria to survive and proliferate in hydrocarbon environments are not completely known. Utilizing cDNA microarrays (Gunasekera et al., 2013, Grady et al., 2017; Gunasekera et al., 2017) and RNA sequencing (Grady et al., 2017) revealed the transcriptional architecture, especially the strategies used by bacteria to overcome stress in response to hydrocarbon exposure. Microarray or RNAseq generates large data sets and profiling gene expression data using various computational tools to find co-expressed genes is a useful tool to generate insights into the mechanisms involved in hydrocarbon degradation. In addition, constructing a molecular interaction network and finding bottlenecks in new or existing pathways are useful.

Microarray data mining was performed by a combination of K-mean clustering, principal component analysis (PCA), and hierarchical clustering analysis to identify differentially expressed genes in the presence of alkanes. Microarray or RNAseq gene expression data often shows high dimensionality. Therefore, extracting important features such as co-regulated genes or novel gene subtypes is difficult from a large gene expression data set. Clustering algorithm techniques extract valuable information from large and highly variable transcriptomic data sets by reducing dimensionality. Cluster analysis is a popular approach and classical algorithms such as K-mean clustering, hierarchical clustering, self-organization, principal component analysis (PCA) are widely used. However, background-corrected normalized data are always preferable for cluster analysis. Bio-informatics research shows K-mean and hierarchical clustering performed significantly better than other cluster methods when they are used to analyze large gene expression data sets (Freyhult et al., 2010). K-mean clustering is a useful analytical tool for transcriptomic data analysis, especially extracting co-regulated genes under specific conditions (Gunasekera et al., 2008).

Further analysis of previously published (Grady et al., 2017) microarray data using K-mean clustering enables the extraction of subsets of genes that are important for alkane degradation. K-mean divides the transcriptome into 16 discrete clusters (Figure 12.3) and clusters 1, 13, 15 represent subset genes that are significantly induced in the presence of alkanes (Figure 12.4). Cluster 13 is a unique cluster comprised of genes responsible for alkane degradation and, in this subset, include genes for membrane-bound alkane monooxygenase 2 (*alkB2*), *laoABC* operon, cytochrome oxidase subunits I, II, & III, PA3427 (probable short-chain dehydrogenase), acetyl co-synthase, and dehydrogenase. Panasiaa and Philipp (2018)

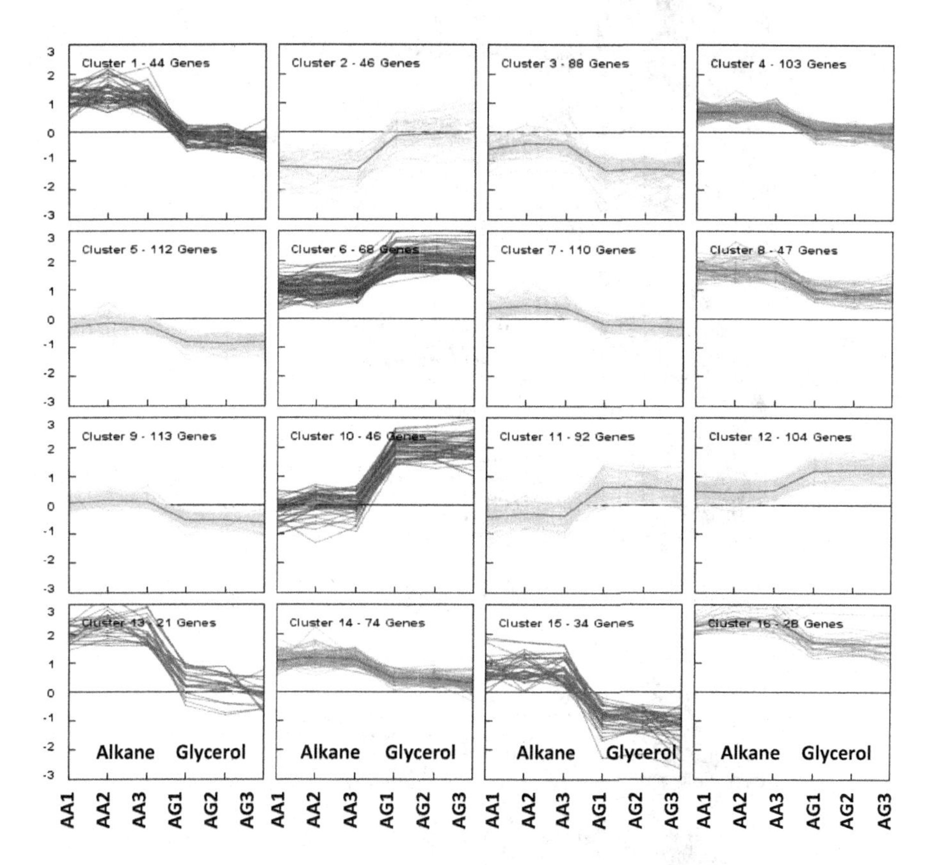

FIGURE 12.3 K-mean clustering of differentially expressed genes in the presence of alkanes as compared to glycerol. *P. aeruginosa* cells were grown in alkane and glycerol to the early log phase and transcriptome was assessed using DNA microarray. Subsets of alkane-induced genes were identified via K-means clustering (K = 16) of the complete transcriptome of *P. aeruginosa* ATCC 33988 in response to alkane mix. Cluster 1, 13, and 15 comprised of highly induced genes in response to alkanes (genes are color-coded in blue). Clusters 4, 8, 14, 15 contained marginally induced genes. Significantly down-regulated genes are in clusters 6 and 10 (color-coded in red). AA1, AA2, AA3 indicate Alkane mix and AG1, AG2, AG3 indicate Glycerol). Normalized cell intensity data were taken as input for K-means cluster analysis.

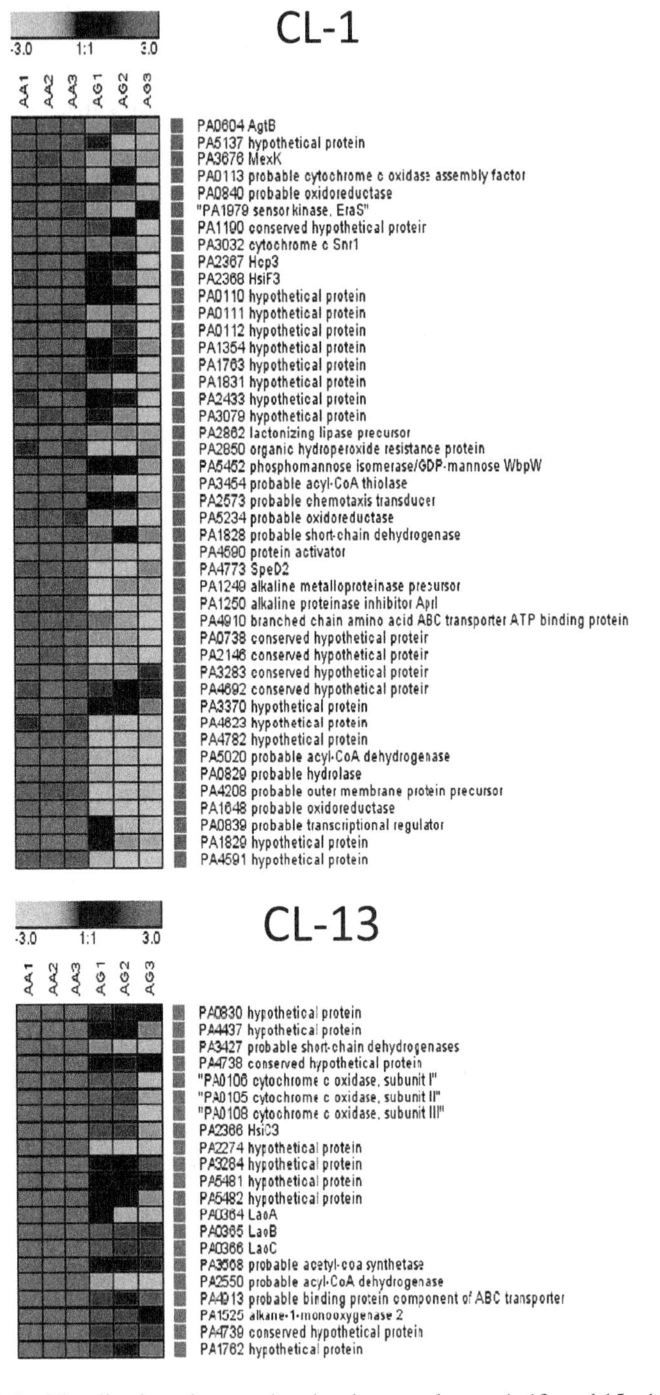

FIGURE 12.4 Visualization of expression data in gene clusters 1, 13, and 15 with different intensities. AA1, AA2, AA3 (Alkane mix) and AG1, AG2, AG3 (Glycerol). Normalized cell intensity data were taken as input for K-means cluster analysis. Red and green represents genes up regulation and down regulation, respectively. *(Continued)*

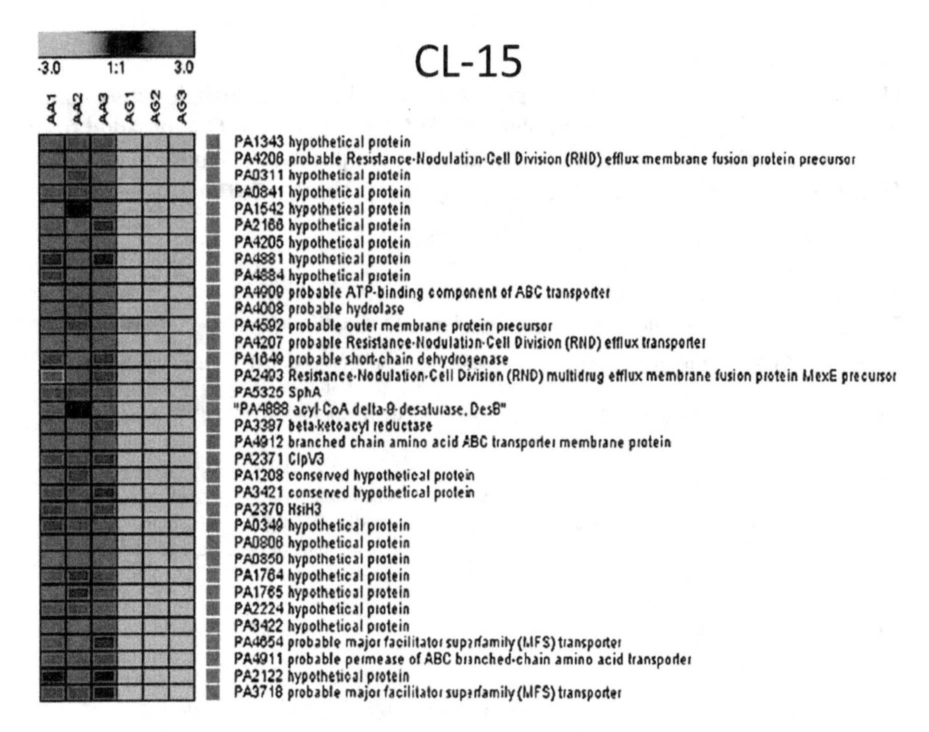

FIGURE 12.4 (Continued) Visualization of expression data in gene clusters 1, 13, and 15 with different intensities. AA1, AA2, AA3 (Alkane mix) and AG1, AG2, AG3 (Glycerol). Normalized cell intensity data were taken as input for K-means cluster analysis. Red and green represents genes up regulation and down regulation, respectively.

recently demonstrated LaoA (PA0364) and LaoB (PA0365) both have alcohol dehydrogenase activity and LaoC (PA0366) has been identified as a presumable aldehyde dehydrogenase. Further, using deletion mutants, the authors revealed that LaoA and LaoB are involved in the degradation of primary long-chain alcohols, which are the primary products of degradation of alkanes by alkane monooxygenase. Interestingly all three genes are highly induced in the presence of alkanes, (Table 12.1) suggesting these genes have a presumptive role in alkane degradation. *P. aeruginosa* has a highly branched respiratory chain with multiple terminal oxidases including bo_3-type quinol oxidase (Cyo), cyanide-insensitive oxidase (CIO), aa_3-type cytochrome *c* oxidase (aa_3), and two cbb_3-type cytochrome *c* oxidases (cbb_3-1 and cbb_3-2) (Arai, 2011). These oxidases are encoded by the *cox*, *cco1*, *cco2*, *cyo*, and *cio* gene clusters, respectively. Generally, these terminal oxidases are differentially expressed under various growth condition and it is believed that differential expression of these terminal oxidases contribute toward survival in different environments. Interestingly the cox gene cluster (PA0105-0108), coxB (PA0105), coxA (PA0106), and coxC (coIII) (PA0108) encode subunits II, I, and III of the A-type cytochrome c oxidase, respectively, and these three genes are highly upregulated in alkanes and found in the same K-mean cluster (cluster 13). Osamura et al., (2017) demonstrated aa3-type cytochrome c oxidase is tightly regulated to be expressed only under starvation

TABLE 12.1
Highly Induced Gene in the Presence of Alkanes. Clusters of Orthologous Groups of Proteins (COG) and Gene Ontology (GO) Categories Were Downloaded from the *Pseudomonas* Genome Database (http://www.pseudomonas.com). Differentially Expressed Genes Have an Adjusted *p*-Value < 0.01

Locus Tag	Cluster #	Function	Fold Change (Change in the Expression)	COG Category	GO
PA0829	CL#1	probable hydrolase	56.62	lipid transport and metabolism	catalytic activity
PA0839	CL#1	probable transcriptional regulator	33.90	transcription	DNA binding
PA1648	CL#1	probable oxidoreductase	29.88	general function prediction only	oxidoreductase activity
PA2146	CL#1	conserved hypothetical protein	27.63	general function prediction only	unknown
PA5020	CL#1	probable acyl-CoA dehydrogenase	26.66	lipid transport and metabolism	oxidoreductase activity, acting on the CH-CH group of donors
PA4590	CL#1	protein activator	23.70	unknown	transport
PA4782	CL#1	hypothetical protein	23.30	unknown	unknown
PA3454	CL#1	probable acyl-CoA thiolase	19.16	lipid transport and metabolism	transferase activity, transferring acyl groups other than amino-acyl groups
PA1829	CL#1	hypothetical protein	17.42	general function prediction only	unknown
PA4773	CL#1	SpeD2	14.52	amino acid transport and metabolism	spermidine biosynthetic process
PA2862	CL#1	lactonizing lipase precursor	13.08	general function prediction only	carbon utilization
PA1190	CL#1	conserved hypothetical protein	12.43	unknown	unknown
PA0738	CL#1	conserved hypothetical protein	12.21	function unknown	unknown
PA3032	CL#1	cytochrome c Snr1	11.85	inorganic ion transport and metabolism	electron transfer activity
PA1828	CL#1	probable short-chain dehydrogenase	11.42	general function prediction only	oxidoreductase activity
PA4591	CL#1	hypothetical protein	11.15	cell wall/membrane/envelope biogenesis	unknown

(Continued)

TABLE 12.1 (Continued)

Highly Induced Gene in the Presence of Alkanes. Clusters of Orthologous Groups of Proteins (COG) and Gene Ontology (GO) Categories Were Downloaded from the *Pseudomonas* Genome Database (http://www.pseudomonas.com). Differentially Expressed Genes Have an Adjusted *p*-Value < 0.01

Locus Tag	Cluster #	Function	Fold Change (Change in the Expression)	COG Category	GO
PA4208	CL#1	probable outer membrane protein precursor	11.06	intracellular trafficking, secretion, and vesicular transport	efflux transmembrane transporter activity
PA1249	CL#1	alkaline metalloproteinase precursor	9.76	secondary metabolites biosynthesis, transport and catabolism	metalloendopeptidase activity
PA0111	CL#1	hypothetical protein	9.10	unknown	unknown
PA3079	CL#1	hypothetical protein	9.08	general function prediction only	membrane
PA3283	CL#1	conserved hypothetical protein	8.97	function unknown	unknown
PA2368	CL#1	HsiF3	8.77	function unknown	unknown
PA5234	CL#1	probable oxidoreductase	8.64	energy production and conversion	oxidoreductase activity
PA4910	CL#1	branched chain amino acid ABC transporter ATP binding protein	8.59	amino acid transport and metabolism	ATP binding
PA0112	CL#1	hypothetical protein	8.57	posttranslational modification, protein turnover, chaperones	heme a biosynthetic process
PA5137	CL#1	hypothetical protein	8.46	amino acid transport and metabolism	unknown
PA1250	CL#1	alkaline proteinase inhibitor AprI	8.43	unknown	endopeptidase inhibitor activity
PA1354	CL#1	hypothetical protein	7.93	function unknown	unknown
PA0110	CL#1	hypothetical protein	7.84	function unknown	membrane
PA1831	CL#1	hypothetical protein	7.83	carbohydrate transport and metabolism	unknown
PA2850	CL#1	organic hydroperoxide resistance protein	7.64	posttranslational modification, protein turnover, chaperones	response to oxidative stress

(*Continued*)

TABLE 12.1 (Continued)
Highly Induced Gene in the Presence of Alkanes. Clusters of Orthologous Groups of Proteins (COG) and Gene Ontology (GO) Categories Were Downloaded from the *Pseudomonas* Genome Database (http://www.pseudomonas.com). Differentially Expressed Genes Have an Adjusted *p*-Value < 0.01

Locus Tag	Cluster #	Function	Fold Change (Change in the Expression)	COG Category	GO
PA2573	CL#1	probable chemotaxis transducer	7.45	signal transduction mechanisms	signal transduction
PA0113	CL#1	probable cytochrome c oxidase assembly factor	7.36	posttranslational modification, protein turnover, chaperones	integral component of membrane
PA3676	CL#1	MexK	7.28	defense mechanisms	membrane
PA4623	CL#1	hypothetical protein	7.16	unknown	unknown
PA1763	CL#1	hypothetical protein	7.06	unknown	unknown
PA0604	CL#1	AgtB	6.57	amino acid transport and metabolism	unknown
PA3370	CL#1	hypothetical protein	6.33	unknown	unknown
PA4692	CL#1	conserved hypothetical protein	6.10	general function prediction only	oxidoreductase activity, acting on a sulfur group of donors
PA2433	CL#1	hypothetical protein	6.10	unknown	unknown
PA0840	CL#1	probable oxidoreductase	5.82	energy production and conversion	FMN binding
PA5452	CL#1	phosphomannose isomerase/ GDP-mannose WbpW	5.65	cell wall/membrane/ envelope biogenesis	polysaccharide metabolic process
PA1979	CL#1	sensor kinase, EraS"	5.45	signal transduction mechanisms	Phosphor-relay signal transduction system
PA2367	CL#1	Hcp3	5.03	function unknown	unknown
PA2274	CL#13	hypothetical protein	138.34	unknown	unknown
PA0364	CL#13	LaoA	86.11	amino acid transport and metabolism	oxidoreductase activity, acting on CH-OH group of donors
PA2550	CL#13	probable acyl-CoA dehydrogenase	49.86	lipid transport and metabolism	oxidoreductase activity, acting on the CH-CH group of donors

(Continued)

TABLE 12.1 (Continued)
Highly Induced Gene in the Presence of Alkanes. Clusters of Orthologous Groups of Proteins (COG) and Gene Ontology (GO) Categories Were Downloaded from the *Pseudomonas* Genome Database (http://www.pseudomonas.com). Differentially Expressed Genes Have an Adjusted *p*-Value < 0.01

Locus Tag	Cluster #	Function	Fold Change (Change in the Expression)	COG Category	GO
PA4437	CL#13	hypothetical protein	41.95	unknown	unknown
PA3427	CL#13	probable short-chain dehydrogenases	41.83	general function prediction only	oxidoreductase activity
PA4738	CL#13	conserved hypothetical protein	33.03	function unknown	unknown
PA0830	CL#13	hypothetical protein	31.09	general function prediction only	unknown
PA5482	CL#13	hypothetical protein	29.81	function unknown	plasma membrane
PA0366	CL#13	LaoC	27.15	energy production and conversion	cellular aldehyde metabolic process
PA0365	CL#13	LaoB	23.83	unknown	unknown
PA1525	CL#13	alkane-1-monooxygenase 2	21.48	lipid transport and metabolism	oxidoreductase activity
PA4739	CL#13	conserved hypothetical protein	20.66	general function prediction only	unknown
PA4913	CL#13	probable binding protein component of ABC transporter	14.05	amino acid transport and metabolism	amino acid transport
PA5481	CL#13	hypothetical protein	11.85	unknown	periplasmic space
PA3284	CL#13	hypothetical protein	11.20	function unknown	unknown
PA3568	CL#13	probable acetyl-coa synthetase	10.34	lipid transport and metabolism	catalytic activity
PA0105	CL#13	cytochrome c oxidase, subunit II"	9.21	energy production and conversion	cytochrome-c oxidase activity
PA1762	CL#13	hypothetical protein	9.00	general function prediction only	ATP binding
PA0108	CL#13	cytochrome c oxidase, subunit III"	5.93	energy production and conversion	heme-copper terminal oxidase activity
PA2366	CL#13	HsiC3	5.83	function unknown	unknown
PA0106	CL#13	cytochrome c oxidase, subunit I"	4.41	energy production and conversion	oxidation-reduction process
PA3422	CL# 15	hypothetical protein	153.03	unknown	unknown

(*Continued*)

TABLE 12.1 (Continued)
Highly Induced Gene in the Presence of Alkanes. Clusters of Orthologous Groups of Proteins (COG) and Gene Ontology (GO) Categories Were Downloaded from the *Pseudomonas* Genome Database (http://www.pseudomonas.com). Differentially Expressed Genes Have an Adjusted *p*-Value < 0.01

Locus Tag	Cluster #	Function	Fold Change (Change in the Expression)	COG Category	GO
PA4206	CL# 15	probable Resistance-Nodulation-Cell Division (RND) efflux membrane fusion protein precursor	117.15	cell wall/membrane/envelope biogenesis	membrane
PA4207	CL# 15	probable Resistance-Nodulation-Cell Division (RND) efflux transporter	95.59	defense mechanisms	transmembrane transporter activity
PA4205	CL# 15	hypothetical protein	48.87	function unknown	quorum sensing
PA0850	CL# 15	hypothetical protein	40.02	unknown	unknown
PA3387	CL# 15	beta-ketoacyl reductase	35.07	secondary metabolites biosynthesis, transport and catabolism	Gram-negative-bacterium-type cell wall
PA4912	CL# 15	branched chain amino acid ABC transporter membrane protein	29.06	amino acid transport and metabolism	membrane
PA4884	CL# 15	hypothetical protein	28.66	unknown	unknown
PA4654	CL# 15	probable major facilitator superfamily (MFS) transporter	27.83	carbohydrate transport and metabolism	integral component of plasma membrane
PA3718	CL# 15	probable major facilitator superfamily (MFS) transporter	25.20	carbohydrate transport and metabolism	integral component of plasma membrane
PA4909	CL# 15	probable ATP-binding component of ABC transporter	22.90	amino acid transport and metabolism	ATP binding
PA4008	CL# 15	probable hydrolase	20.08	lipid transport and metabolism	unknown
PA3421	CL# 15	conserved hypothetical protein	19.53	unknown	unknown

(Continued)

TABLE 12.1 (Continued)
Highly Induced Gene in the Presence of Alkanes. Clusters of Orthologous Groups of Proteins (COG) and Gene Ontology (GO) Categories Were Downloaded from the *Pseudomonas* Genome Database (http://www.pseudomonas.com). Differentially Expressed Genes Have an Adjusted *p*-Value < 0.01

Locus Tag	Cluster #	Function	Fold Change (Change in the Expression)	COG Category	GO
PA4888	CL# 15	acyl-CoA delta-9-desaturase, DesB"	18.79	lipid transport and metabolism	lipid metabolic process
PA1542	CL# 15	hypothetical protein	17.53	general function prediction only	unknown
PA0349	CL# 15	hypothetical protein	17.16	unknown	unknown
PA0806	CL# 15	hypothetical protein	15.75	function unknown	unknown
PA0311	CL# 15	hypothetical protein	13.90	unknown	unknown
PA2166	CL# 15	hypothetical protein	13.81	unknown	unknown
PA1208	CL# 15	conserved hypothetical protein	13.21	inorganic ion transport and metabolism	N,N-dimethylaniline monooxygenase activity
PA2493	CL# 15	Resistance-Nodulation-Cell Division (RND) multidrug efflux membrane fusion protein MexE precursor	11.46	cell wall/membrane/envelope biogenesis	membrane
PA1764	CL# 15	hypothetical protein	11.33	lipid transport and metabolism	unknown
PA1765	CL# 15	hypothetical protein	10.30	unknown	carbohydrate transport
PA4911	CL# 15	probable permease of ABC branched-chain amino acid transporter	10.11	amino acid transport and metabolism	membrane
PA2224	CL# 15	hypothetical protein	10.10	unknown	unknown
PA0841	CL# 15	hypothetical protein	9.06	function unknown	unknown
PA5325	CL# 15	SphA	8.93	energy production and conversion	pathogenesis
PA2122	CL# 15	hypothetical protein	8.23	unkown	iron ion binding
PA2371	CL# 15	ClpV3	8.22	posttranslational modification, protein turnover, chaperones	ATP binding

(*Continued*)

TABLE 12.1 (Continued)
Highly Induced Gene in the Presence of Alkanes. Clusters of Orthologous Groups of Proteins (COG) and Gene Ontology (GO) Categories Were Downloaded from the *Pseudomonas* Genome Database (http://www.pseudomonas.com). Differentially Expressed Genes Have an Adjusted *p*-Value < 0.01

Locus Tag	Cluster #	Function	Fold Change (Change in the Expression)	COG Category	GO
PA4592	CL# 15	probable outer membrane protein precursor	7.25	intracellular trafficking, secretion, and vesicular transport	efflux transmembrane transporter activity
PA1343	CL# 15	hypothetical protein	7.03	unknown	unknown
PA2370	CL# 15	HsiH3	5.75	function unknown	unknown
PA4881	CL# 15	hypothetical protein	4.98	unknown	unknown
PA1649	CL# 15	probable short-chain dehydrogenase	4.40	general function prediction only	oxidoreductase activity

conditions in cooperation with other terminal oxidases to facilitate survival in nutrient starvation conditions and this gene cluster is upregulated in the presence of alkanes. The genes located downstream of the cox genes, PA0112 and PA0113, encode enzymes that are probably involved in the production of heme and are also highly induced as they are part of cluster #1. The *alkB1* and *alkB2* were upregulated 6.95- and 21.48-fold, respectively. These two crucial genes encode AlkB1 and AlkB2 enzymes, important for alkane degradation, and are split into two separate clusters. The less induced *alkB1* gene is part of cluster 8 and the highly induced *alkB2* is part of cluster #13 as mentioned previously.

Furthermore, the genes in these clusters were assigned to different Clusters of Orthologous Groups of proteins (COG) categories (Table 12.1). The COG has been designed to classify proteins based on orthology concepts by grouping them with direct evolutionary counterparts. While COG categorized genes evolved by the event of speciation, gene ontology (GO) focuses on the function of the genes and gene products. Both the COGs and GO categories were downloaded from the Pseudomonas Genome Database (Winsor et al., 2011; http://www.pseudomonas.com) and assigned (Table 12.1).

12.5 PHYSIOLOGICAL ADAPTATIONS

12.5.1 Biofilm Formation

Generally, biofilm formation is considered an adaptive strategy as the planktonic cells switch to a sessile form of growth to produce cell aggregates. For the formation of biofilms, microbial communities often produce exopolysaccharides, proteins or use extracellular DNA for self-assembly of individual cells to form cell aggregates. Research into *P. aeruginosa* ATCC 33988 indicates they are well adapted to a fuel environment and have the ability to form thick biofilms (Gunasekera et al., 2013). The ability to form biofilms is critical for its survival in a fuel environment (Passman, 2003). In fuel environments, biofilms enable bacteria to avoid direct contact with toxic solvents and *P. aeruginosa* produce visible biofilms at the fuel and growth media interface (Gunasekera et al., 2013). Experimental data shows this bacterium induce both alginate production, a linear polymer of β-1,4-linked D-mannuronic acid and the Pel pathway which produces a glucose-rich, cellulose-like polymer. In a whole genome expression study, it was found, with the exception of *algC*, genes involved in biosynthesis and secretion of alginate were induced in the presence of jet fuel as compared to glycerol (Gunasekera et al., 2013). Alginate is known to be over produced in cystic fibrosis patients and *P. aeruginosa* alginate has been shown to contribute to antibiotic resistance and can evade human antibacterial defense mechanisms. Hydrocarbons are, in general, toxic to living cells including bacteria and there is enough evidence that bacteria produce biofilms to overcome these recalcitrant toxic compounds in fuel (Gunasekera et al., 2013). However, the Psl polysaccharide, which is rich in mannose and galactose, is not highly induced in the presence of jet fuel from cells isolated in mid-log phase indicating alginate and Pel are key polysaccharides involved in biofilm production when cells are growing logarithmically. It is also possible that the Psl polysaccharide is playing a role in forming biofilms during the initial attachment phase in early log phase or during lag phase. Ma et al. (2006) showed Psl polysaccharide plays an important role in initial cell adhesion to initiate formation of biofilms.

The formation of a biofilm in *P. aeruginosa* is regulated by numerous regulators including Quorum Sensing (QS), secondary regulatory system - bis-(3′-5′)-cyclic diguanosine monophosphate (c-di-GMP) molecules and small RNA (sRNA). Quorum sensing plays a key role in modulating the biofilm expression genes of polycyclic aromatic hydrocarbons (PAHs) in *P. aeruginosa* N6P6 (Mangwani et al., 2015). The degradation of the PAH compounds, phenanthrene and pyrene, was directly affected by biofilm growth and expression of *lasI* genes, which plays a role in synthesis and the use of N-(3-oxo-dodecanoyl)-L-homoserine lactone (3OC12-HL). Cyclic di-GMP is central to the post-transcriptional regulation of biofilm formation. For example, PelD is a c-di-GMP receptor and binding to c-di-GMP is essential for Pel polysaccharide production (Lee et al., 2007). In addition, the RNA-binding protein RsmA negatively controls biofilm formation. For example, RsmA post-transcriptionally regulated Psl biosynthesis by binding to *psl* mRNA and inhibits translation (Irie et al., 2010). However, two small non-coding RNAs (ncRNAs), RsmY and RsmZ, sequester the translational repressor RsmA and de-repress biofilm formation by increasing cyclic di-GMP level (Jimenez et al., 2012). Biofilm formation in fuel tanks are a significant concern because biofilms can clog fuel lines and fuel filters. Scientists are developing approaches to inhibit microbial biofilm formation and keep bacteria in planktonic growth state as single cells, which are more susceptible to biocides or antimicrobial agents.

12.5.2 Solvent Resistance

Bacteria use a wide range of mechanisms to overcome the toxicity of hydrocarbons in fuel. Among them are impermeabilization of solvents through the membrane and active efflux of excess solvents from the cell. The down-regulation of porins such as OprF and OprG in the presence of Jet-A fuel indicates these porins are involved in regulating hydrocarbons entering into the cells and regulating hydrocarbon homeostasis (Gunasekera et al., 2013; Gunasekera et al., 2017). Also, down-regulating porins can act as a permeability barrier for toxic solvents by preventing the uptake of toxic solvents into the cells. In addition to prevention of internalization, *P. aeruginosa* is equipped with multiple efflux pumps to remove toxic solvents from the cells. The MexCD-OprJ and MexEF-OprN efflux pumps were shown to up-regulated in the presence of jet fuel (Gunasekera et al., 2013). In addition to these two efflux pumps, a large number of efflux pumps were up-regulated in the presence of fuel (Gunasekera et al., 2013; Gunasekera et al., 2017). The role of efflux pumps in antibiotic resistance is well understood and for example, MexCD-OprJ is known to export cefsulodin & novobiocin antibiotics while MexEF-OprN can confer triclosan resistance (Chuanchuen et al., 2003; Piddock, 2006). Li et al., 1998 provided experimental evidence that MexA-MexB-OprM, MexC-MexD-OprJ, and MexE-MexF-OprN play a significant role in organic solvent tolerance, including n-hexane and p-xylene in *P. aeruginosa*. By adding the efflux pump inhibitor, Phe-Arg-β-napthylamide (PAβN), a c-capped dipeptide to fuel *P. aeruginosa* growth, was inhibited, indicating blocking efflux pumps caused cell toxicity (Gunasekera et al., 2013; Gunasekera et al., 2017). No growth inhibition for cells in glycerol showed that Phe-Arg-β-napthylamide has no toxicity and efflux pumps do remove toxic solvents from fuel (Gunasekera et al., 2013). Further, blocking of efflux pumps using PAβN was effective when *P. aeruginosa* was growing in n-dodecane and n-hexadecane as a sole carbon source indicating excess alkanes are toxic to the cell (Gunasekera et al., 2017). Grady et al., 2017 demonstrated that by using multi-omics tools that *mexGHI-opmD* efflux pumps were induced in the presence alkanes. Fuel adapted *P. aeruginosa* strain, ATCC 33988, has the ability to degrade shorter chain alkanes, n-C_8 or n-C_{10}, while theses shorter chain alkanes are very toxic to *P. aeruginosa* PAO1 strain (Gunasekera et al., 2017). This suggests ATCC 33988 was better adapted to overcome n-C8 or n-C10 toxicity by either preventing the uptake of these toxic hydrocarbons into the cells or removing them using active efflux mechanisms. Currently, it is not well understood what specific efflux pumps are responsible for removing specific solvents in fuel. Thus, further research is needed to fully understand the role of efflux pumps in solvent resistance in jet fuel and hydrocarbon-rich environments. In addition to the above mechanisms and strategies, bacteria living in organic rich solvents in hostile environments change membrane fluidity by changing fatty acid composition, fatty acid chain length, and phospholipid composition as adaptive mechanisms because organic solvents change the membrane fluidity (Ramos et al., 2002; Murínová & Dercová 2014).

12.5.3 Trace Elements and Iron Acquisition

For efficient degradation of hydrocarbons in fuel, *P. aeruginosa* requires iron in its growth environment. There are a number of iron-dependent enzymes/proteins including integral membrane di-iron alkane hydroxylases (e.g. AlkB), cytochrome P450 enzymes, ferredoxin reductase, and ferredoxin that share a requirement of iron for activity. Also, several iron-containing alcohol dehydrogenases require iron as a co-factor. For aromatic hydrocarbon degradation, Fe II/Fe III activates molecular oxygen and promote aromatic dihydroxylation and oxidative ring-cleavage. These reactions are catalyzed by iron-dependent dioxygenases (Wang et al., 2017). Therefore, there is a major need for iron during hydrocarbon degradation. Gunasekera et al., 2013 showed *P. aeruginosa* upregulates significant number of iron-related genes during the growth in fuel and this indicates a greater demand for iron during alkane degradation. Whole genome gene expression studies revealed that both the high affinity pyroverdine and the lower affinity pyochelin systems were induced during fuel growth (Gunasekera et al., 2013). Both pyoverdine and pyochelin bind extracellular iron ($Fe3+$) and are transported to the inside of the cell. Both siderophores are actively translocated using FptA and FpvA outer membrane receptors for pyochelin and pyoverdine transport, respectively, and these receptors are induced when *P. aeruginosa* is grown in Jet-A fuel. Iron uptake is controlled by several regulators including Fur, which is considered as a negative regulator for the iron uptake system. Down regulation of Fur, activated a large number of iron-related genes including iron-responsive extracytoplasmic function (ECF) σ factors such as PvdS that governs the expression of pyoverdine biosynthesis. In addition to two endogenous siderophores (pyoverdine and pyochelin), *P. aeruginosa* also upregulates a transport system for ferric enterobactin. The *fepBDGC* operon encodes an ABC transporter which is involved in the transport of ferric enterobactin and was induced when *P. aeruginosa* was grown in fuel. In addition to alkane hydrocarbons, iron limitation affects degradation aromatic hydrocarbons. For example, activity of toluene monooxygenase and benzoate-1,2-dioxygenase requires iron as a cofactor (Dinkla et al., 2001). Banin et al., 2005 demonstrated using pyoverdine and pyochelin mutants and showed that a critical level of intracellular iron serve as the signal for biofilm development because there are several iron regulated steps in biofilm development. It is also known that iron-replete condition affects alginate biosynthesis (Kang and Kirienko, 2018), which is an important constituent in the biofilm matrix. The role of iron in *P. aeruginosa* biofilm formation in relation to its pathogenesis has been recently reviewed (Kang & Kirienko, 2018). Therefore, it is not surprising that *P. aeruginosa* has developed an efficient iron uptake mechanism and these pathways are induced in the presence of fuel.

12.5.4 Stress Response

Fuel-adapted *P. aeruginosa* ATCC 33988 strain employs a rapid heat shock response as a mechanism to tolerate stresses produced by the hydrocarbons (Gunasekera et al., 2017). The well-known genes encoding the heat shock proteins, DnaK and GroES, were induced along with *katA*, which encodes a catalase, and *sodB*, which encodes superoxide dismutase. Induction of heat shock proteins is primarily caused by increased cellular levels of the heat shock sigma-factor sigma32 encoded by the *rpoH* gene. Induction of *rpoH* gene in fuel-resistant ATCC 33988 strain further validates the heat shock response during growth in a fuel environment. Upregulation of genes encoding heat shock/chaperone proteins suggest that ATCC 33988 cells have developed efficient adaptive mechanisms to prevent protein mis-folding and to degrade the irreversibly denatured proteins (Gunasekera et al., 2017).

The summary of adaptive mechanisms and common strategies used by *P. aeruginosa* to survive in hydrocarbon-rich environments is listed in Table 12.2.

TABLE 12.2
Summary of Adaptive Mechanisms and Common Strategies Used by *P. aeruginosa* to Survive in Hydrocarbon Rich Environments

Adaptive Mechanism	Strategies	References
Change membrane composition	Membrane fluidity is re-adjusted by altering the composition of the lipid bilayer.	Ramos et al., 2015; Murínová & Dercová, 2014
Hydrocarbon solubilization	Rhamnolipids increase alkane solubility. Modified pathways were observed in well adapted *P. aeruginosa* ATCC33988 strain.	Liu et al., 2017; Grady et al., 2017 Das et al., 2015
Surface tension	Reduce surface tension allows increased contact with hydrocarbons.	Kim & Weber 2003; Song et al., 2006; Hua & Wang, 2012; Grady et al., 2017
Outer membrane porins and membrane permeability	Genes encoding outer membrane porins, OprF and OprG, were down-regulated when *P. aeruginosa* was grown in jet fuel.	Gunasekera et al., 2013
Efflux of toxic compounds	Activation of energy-dependent active efflux of hydrocarbons. A large number of genes encoding RND pumps induced in the presence of hydrocarbons. The MexA-MexB-OprM, MexC-MexD-OprJ, and MexE-MexF-OprN play a significant role in organic solvent such as such as n-hexane and p-xylene tolerance in *P. aeruginosa*.	Gunasekera et al., 2013 Grady et al., 2017 Ramos et al., 2015; Li et al., 1998.
Quorum sensing	Quorum Sensing (QS) modulates the expression of biofilm genes of polycyclic aromatic hydrocarbons (PAHs) in *P. aeruginosa* N6P6. Modified QS pathways found for fuel adapted *P. aeruginosa* ATCC33988.	Mangwani et al., 2015; Grady et al., 2017
Biofilm formation	Jet fuel induces *P. aeruginosa* cell aggregation. In the presence of fuel, alginate and Pel biosynthesis and export mechanisms were induced. QS plays a key role in modulating the expression biofilm genes of polycyclic aromatic hydrocarbons (PAHs) *P. aeruginosa* N6P6.	Gunasekera et al., 2013 Mangwani et al., 2015
Single-Nucleotide Polymorphism	Promoter polymorphism affects the expression of *alk* genes in *P. aeruginosa* ATCC 33988.	Gunasekera et al., 2017
Trace element acquisition	Siderophore-dependent iron acquisition pathways are induced.	Gunasekera et al., 2013
Stress Response	*P. aeruginosa* ATCC 33988 strain employs a rapid heat shock response as a mechanism to tolerate stresses produced by the hydrocarbons.	Gunasekera et al., 2017

12.6 CONCLUSION

P. aeruginosa ATCC 33988 has an efficient alkane degradation machinery compared to its closely related type strain, PAO1. Closely related strains having different hydrocarbon degradation rates suggest genetic mutation driving the different ecotypes and degradation efficiencies. It must not be forgotten that in addition to having an efficient hydrocarbon degradation mechanism, bacterial adaptive mechanisms play a vital role in individual adaptation and strain differentiation. Bacteria use well-known adaptive mechanisms such as biofilm formation, solvent resistance, stress response, and efficient trace element acquisition mechanisms to combat hydrocarbon-rich toxic environments and to obtain essential nutrients and trace elements for their growth. Hydrocarbons are not water soluble and microorganisms have developed mechanisms to bring them inside the cell. Although exact mechanisms of hydrocarbon uptake are not fully understood, in most cases they produce bio-surfactants or specific proteins to solubilize hydrocarbons that facilitate uptake of water insoluble hydrocarbons. The research on microbial degradation of hydrocarbons has been aided by the advent of next generation sequencing (NGS). NGS is now established as a cost-effective technology, enabling sequencing of novel microbial genomes and decoding their genetic information in a timely fashion for microbial research. Simultaneously, RNA sequencing (RNAseq) and microarrays provide key gene expression data. Therefore, a multi-omics approach has potential to provide great insights into comprehension of novel mechanisms exploited by bacteria to survive and proliferate in hydrocarbon-rich environments. Grady et al., 2017 used RNA-seq, microarrays, ribosome foot-printing, proteomics, and small molecule LC-MS experiments to compare gene expression in *P. aeruginosa* and discovered hydrocarbon degrading *P. aeruginosa* strains use different cell density-dependent quorum sensing responses and potentially different approaches to solubilizing hydrophobic fuels. Overall, these mechanisms and pathways can be exploited to control microbial growth in hydrocarbon-rich environments, where bacteria cause issues for users. For example, inhibition of biofilms in fuel tanks can be achieved by inhibiting bacterial quorum sensing (QS) mechanisms or hydrolyzing bacterial extracellular polymeric substances (EPSs). QS is known to regulate more than 10% genes in *P. aeruginosa* genome (Moradali et al., 2017). It was previously demonstrated that the blockage of efflux pumps can be used to stop *P. aeruginosa* growth in fuel and reduce bio deterioration (Gunasekera et al., 2013). Conversely, degradation effective, well-adapted bacteria to recalcitrant hydrophobic compounds are considered as an essential trait for successful bioremediation of contaminated sites.

The mechanism that regulates the *alk* genes in *P. aeruginosa* is not well understood. However, the regulation of *alk* genes in *P. putida* has been studied adequately. It is evident from recent research that small regulatory RNAs (sRNAs) also play a role in the regulation of gene expression by directly binding to target mRNA or binding to the regulatory proteins (Sonnleitner et al., 2012; Sonnleitner & Bläsi, 2014). For example, CrcZ (a noncoding RNA) plays an important role in regulation of alk genes in *P. putida* GpO1 and the small RNA ErsA is involved in biofilm formation by negative post-transcriptional regulation of AlgC in *P. aeruginosa* (Falcone et al., 2018). Therefore, small RNAs have been implicated as important players in transcriptional and post-translational regulation of important biological processes.

Although bacteria have developed various adaptations to individual stress, future research should be focused on interaction between different stresses and cross-regulatory mechanisms. Previous research found *E. coli* used cross-regulatory mechanisms to combat heat and osmoregulatory stresses (Gunasekera et al., 2008). It is equally important to characterize microbial communities that are associated with microbial induced corrosion (MIC), biofouling, and reservoir souring. Genomic and transcriptomic technology together with other –omics techniques is increasingly being used to gain insight to understand these metabolic processes related to microorganisms' native to other petroleum reservoirs.

ACKNOWLEDGMENT

I thank Dr. Oscar Ruiz (Air Force Research Laboratory), Dr. Sarah Grady (Lawrence Livermore National Laboratory), Loryn Bowen (University of Dayton Research Institute), Lisa Brown (University of Dayton Research Institute), and Richard Striebich (University of Dayton Research Institute) for their valued contribution to this research. Research reported in this article is supported by funds from the Air Force Research Laboratory, Fuels and Energy Branch. Approved for Public Release/ Unlimited Distribution; Case File Number 88ABW-2020-2252.

REFERENCES

Alonso, A., Rojo, F., Martínez, J.L. 1999. Environmental and clinical isolates of *Pseudomonas aeruginosa* show pathogenic and biodegradative properties irrespective of their origin. *Environ. Microbiol.* 1:421–430.

Arai, H. 2011. Regulation and function of versatile aerobic and anaerobic respiratory metabolism in *P. aeruginosa*. *Front. Microbiol.* 2:103.

Banin, E., Vasil, M.L., Greenberg, E.P. 2005. Iron and *Pseudomonas aeruginosa* biofilm formation. *Proc. Natl. Acad. Sci. U.S.A.* 102(31):11076–11081.

Brown, L.M., Gunasekera, T.S., Ruiz, O.N. 2014. Draft genome sequence of *Pseudomonas aeruginosa* ATCC 33988, a bacterium highly adapted to fuel-polluted environments. *Genome Announc.* 2(6):e01113–e01114.

Chauhan, S., Singh, A., Tyagi, J.S. 2010. A single-nucleotide mutation in the −10 promoter region inactivates the narK2X promoter in *Mycobacterium bovis* and *Mycobacterium bovis* BCG and has an application in diagnosis. *FEMS Microbiol. Lett.* 303:190–196.

Chuanchuen, R., Karkhoff-Schweizer, R.R., Schweizer, H.P. 2003. High-level triclosan resistance in *Pseudomonas aeruginosa* is solely a result of efflux. *Am. J. Infect. Control* 31:124–127.

Das, D., Baruah, R., Sarma Roy, A., Singh, A.K., Deka Boruah, H.P., Kalita, J., Bora, T.C. 2015. Complete genome sequence analysis of *Pseudomonas aeruginosa* N002 reveals its genetic adaptation for crude oil degradation. *Genomics* 105(3):182–190.

de Greeff, A., Buys, H., Wells, J.M., Smith, H.E. 2014. A naturally occurring nucleotide polymorphism in the orf2/folc promoter is associated with *Streptococcus suis* virulence. *BMC Microbiol.* 14:264. doi:10.1186/s12866-014-0264-9.

Dinkla, I.J., Gabor, E.M., Janssen, D.B. 2001. Effects of iron limitation on the degradation of toluene by *Pseudomonas* strains carrying the tol (pWWO) plasmid. *Appl. Environ. Microbiol.* 67(8):3406–3412.

Falcone, M., Ferrara, S., Rossi, E., Johansen, H.K., Molin, S., Bertoni, G. 2018. The small RNA ErsA of *Pseudomonas aeruginosa* contributes to biofilm development and motility through post-transcriptional modulation of AmrZ. *Front. Microbiol.* 9:238. Published 2018 Feb 15. doi:10.3389/fmicb.2018.00238.

Freyhult, E., Landfors, M., Önskog, J., Hvidsten, T.R., Rydén, P. 2010. Challenges in microarray class discovery: a comprehensive examination of normalization, gene selection and clustering. *BMC Bioinf.* 11:503. PMC3098084.

Grady, S.L., Malfatti, S.A., Gunasekera, T.S., Dalley, B.K., Lyman, M.G., Striebich, R.C., Mayhew, M.B., Zhou, C.L., Ruiz, O.N., Dugan, L.C. 2017. A comprehensive multi-omics approach uncovers adaptations for growth and survival of *Pseudomonas aeruginosa* on n-alkanes. *BMC Genomics* 18(1):334.

Gunasekera, T.S., Bowen, L.L., Zhou, C.E., Howard-Byerly, S.C., Foley, W.S., Striebich, R.C., Dugan, L.C., Ruiz, O.N. 2017. Transcriptomic analyses elucidate adaptive differences of closely related strains of *Pseudomonas aeruginosa* in fuel. *Appl. Environ. Microbiol.* 83:e03249–e03316. doi:10.1128/AEM.03249-16.

Gunasekera, T.S., Csonka, L.N., Paliy, O. 2008. Genome-wide transcriptional responses of Escherichia coli K-12 to continuous osmotic and heat stresses. *J. Bacteriol.* 190: 3712–3720.

Gunasekera, T.S., Paliy, O. 2006. Growth of E. coli BL21 in minimal media with different gluconeogenic carbon sources and salt contents [published correction appears in Appl Microbiol Biotechnol. Dec;73(4):968]. *Appl. Microbiol. Biotechnol.* 2007;73(5): 1169–1172.

Gunasekera, T.S., Striebich, R.C., Strobel, E.M., Mueller, S.S., Ruiz, O.N. 2013. Transcriptional profiling suggests that multiple metabolic adaptations are required for effective proliferation of *Pseudomonas aeruginosa* in jet fuel. *Environ. Sci. Technol.* 47:13449–13458.

Hua F, Wang, H. 2012. Uptake modes of octadecane by *Pseudomonas* sp. *DG17 and synthesis of biosurfactant. J Appl Microbiol.* 112:25–37.

Irie, Y., Starkey, M., Edwards, A. N.,Wozniak, D. J., Romeo, T., Parsek, M. R. 2010. Pseudomonas aeruginosa biofilm matrix polysaccharide Psl is regulated transcriptionally by RpoS and post-transcriptionally by RsmA. *Mol. Microbiol.* 78:158–172.

Jimenez, P.N., Koch, G., Thompson, J.A., Xavier, K.B., Cool, R.H., Quax, W.J. 2012. The multiple signaling systems regulating virulence in *Pseudomonas aeruginosa. Microbiol. Mol. Biol. Rev.* 76:46–65.

Kang, D., Kirienko, N.V. 2018. Interdependence between iron acquisition and biofilm formation in Pseudomonas aeruginosa. *J. Microbiol.* 56(7), 449–457. doi:10.1007/s12275-018-8114-3.

Kim, H.S., Weber, W.J. Jr. 2003. Preferential surfactant utilization by a PAH-degrading strain: effects on micellar solubilization phenomena. *Environ. Sci. Technol.* 37(16):3574–3580. doi:10.1021/es0210493.

Klockgether, J., Cramer, N., Wiehlmann, L., Davenport, C.F., Tümmler, B. 2011. *Pseudomonas aeruginosa* genomic structure and diversity. *Front. Microbiol.* 2:150.

Lee, V.T., Matewish, J.M., Kessler, J.L., Hyodo, M., Hayakawa, Y., Lory, S. 2007. A cyclic-di-GMP receptor required for bacterial exopolysaccharide production. *Mol. Microbiol.* 65:1474–1484.

Li, X.Z., Zhang, L., Poole, K. 1998. Role of the multidrug efflux systems of *Pseudomonas aeruginosa* in organic solvent tolerance. *J. Bacteriol.* 180(11):2987–2991.

Liu, Y., Zeng, G., Zhong, H., Wang, Z., Liu, Z., Cheng, M., Liu, G., Yang, X., Liu, S. 2017. Effect of rhamnolipid solubilization on hexadecane bioavailability: enhancement or reduction? *J. Hazard. Mater.* 322:394–401.

Ma, L., Jackson, K.D., Landry, R.M., Parsek, M.R., Wozniak, D.J. 2006. Analysis of *Pseudomonas aeruginosa* conditional psl variants reveals roles for the psl polysaccharide in adhesion and maintaining biofilm structure post attachment. *J. Bacteriol.* 188:8213–8221.

Mangwani, N., Kumari, S., Das, S.. 2015. Involvement of quorum sensing genes in biofilm development and degradation of polycyclic aromatic hydrocarbons by a marine bacterium *Pseudomonas aeruginosa* N6P6. *Appl. Microbiol. Biotechnol.* 99(23):10283–10297. doi:10.1007/s00253-015-6868-7.

Moradali, M.F., Ghods, S., Rehm, B.H. 2017. *Pseudomonas aeruginosa* lifestyle: A paradigm for adaptation, survival, and persistence. *Front. Cell. Infect. Microbiol.* 7:39. doi:10.3389/fcimb.2017.00039.

Murínová, S., Dercová, K. 2014. Response mechanisms of bacterial degraders to environmental contaminants on the level of cell walls and cytoplasmic membrane. *Int. J. Microbiol.* 873081. doi:10.1155/2014/873081.

Osamura, T., Kawakami, T., Kido, R., Ishii, M., Arai, H. 2017. Specific expression and function of the A-type cytochrome c oxidase under starvation conditions in *Pseudomonas aeruginosa. PLoS One* 12(5):e0177957.

Panasiaa, G., Philipp, B. 2018. LaoABCR, a novel system for oxidation of long-chain alcohols derived from SDS and alkane degradation in *Pseudomonas aeruginosa. Appl. Environ. Microbiol.* 84(13): e00626–e00718.

Passman, F.J. 2003. Standard guide for microbial contamination in fuels and fuel systems. In *Fuel and Fuel System Microbiology: Fundamentals, Diagnosis, and Contamination Control*, Passman, F. J., Ed., West Conshohocken, PA: ASTM International, pp. 81–91.

Piddock, L.J. 2006. Clinically relevant chromosomally encoded multidrug resistance efflux pumps in bacteria. *Clin. Microbiol.* 19(2):382–402.

Ramos, J.L., Sol Cuenca, M., Molina-Santiago, C., Segura, A., Duque, E., Gómez-García, M.R., Udaondo, Z., Roca, A. 2015. Mechanisms of solvent resistance mediated by interplay of cellular factors in *Pseudomonas putida. FEMS Microbiol. Rev.* 39:555–566. doi:10.1093/femsre/fuv006.

Smits, T.H., Witholt, B., van Beilen, J.B. 2003. Functional characterization of genes involved in alkane oxidation by *Pseudomonas aeruginosa. Antonie Van Leeuwenhoek* 84:193–200.

Song, R., Hua, Z., Li, H., Chen, J. 2006. Biodegradation of petroleum hydrocarbons by two *Pseudomonas aeruginosa* strains with different uptake modes. *J Environ Sci Health A Tox Hazard Subst Environ Eng*. 41: 733–748.

Sonnleitner, E, Bläsi, U. 2014. Regulation of Hfq by the RNA CrcZ in *Pseudomonas aeruginosa* carbon catabolite repression. *PLoS Genet.* 10:e1004440. doi:10.1371/journal.pgen.1004440.

Sonnleitner, E., Romeo, A., Bläsi, U. 2012. Small regulatory RNAs in *Pseudomonas aeruginosa. RNA Biol.* 9(4):364–371. doi:10.4161/rna.19231.

Sood, U., Hira, P., Kumar, R., Bajaj, A., Rao, D.L.N., Lal, R., Shakarad, M. 2019. Comparative genomic analyses reveal core-genome-wide genes under positive selection and major regulatory hubs in outlier strains of *Pseudomonas aeruginosa. Front. Microbiol.* 10:53. doi:10.3389/fmicb.2019.00053.

Striebich, R.C., Smart, C.E., Gunasekera, T.S., Mueller, S.S., Strobel, E.M., McNichols, B., Ruiz, O.N. 2014. Characterization of the F-76 diesel and Jet A aviation fuel hydrocarbon degradation profiles of *Pseudomonas aeruginosa* and *Marinobacter hydrocarbonoclasticus. Int. Biodeter. Biodegr.* 93:33–43.

Valot, B., Guyeux, C., Rolland, J.Y., Mazouzi, K., Bertrand, X., Hocquet, D. 2015. What it takes to be a *Pseudomonas aeruginosa*? The core genome of the opportunistic pathogen updated. *PLoS One* 10:e0126468.

Wang, Y., Li, J., Liu, A. 2017. Oxygen activation by mononuclear nonheme iron dioxygenases involved in the degradation of aromatics. *J Biol Inorg Chem.* 22: 395–405.

Winsor, G.L., Lam, D.K.W., Fleming, L., Lo, R., Whiteside, M.D., Yu, N.Y., Hancock, R.E., Brinkman, F.S. 2011. Pseudomonas Genome Database: improved comparative analysis and population genomics capability for *Pseudomonas* genomes. *Nucleic. Acids. Res.* 39:D596–D600. doi:10.1093/nar/gkq869.

Index

A

Acetobacterium, 70
acetoclastic, 147
acetyl co-synthase, 255
acidophilic, 141
acid-producing bacteria (APB), 52, 70, 77, 111, 114, 120, 134, 138
Actinobacteria, 131
adaptive mechanisms, 250
alginate, 268
algorithm, 254
alkaliphic, 141
alkane monooxygenase, 250, 253
alkanes, 231, 240, 250
alkyl dimethyl benzyl ammonium chloride (ADBAC), 109–110, 114
alkylsuccinate synthase gene, catalytic subunit (assA), 229, 242
alpha-diversity metrics, 20
Alphaproteobacteria, 112, 132
ammonium persulfate, 101
amplicon, 20
amplicon sequence variants (ASVs), 7, 20
amplicon sequencing, 175
Anaerobaculum, 51
anaerobic ammonium oxidation (anammox), 216
anaerobic benzene carboxylase gene, catalytic subunit (abcA), 233, 236–238, 242
anaerobic digesters, 29
anaerobic hydrocarbon biodegradation, 227–230, 237–242
anaerobic methane-oxidizing archaea (ANME), 148
anaerobic oxidation of methane (AOM), 148
anaerobic respiration, 215
ANCOM, 21
ANOSIM, 21
Anvi'o, 24
Aquifex, 57
ARAGORN, 23
Archaea, 92–94, 121, 126, 131, 138
Archaeoglobi, 155
Archaeoglobus, 51, 192
Arcobacter, 52, 60, 77, 78, 86, 87, 132
Assemblers, 23
assimilatory nitrate reduction, 215
ATLAS, 24
ATP, 112, 121, 122
  analysis, 55

B

Bacteria, 121, 129–132
Bacteroidetes, 77, 210
Bakken basin, 99
Barrnap, 23
Bathyarchaeota, 147
Benzene, 227, 236–238, 240–242
benzylsuccinate synthase gene, catalytic subunit (bssA), 229–233, 238–242
beta diversity, 21
bicarbonate, 48
Bignorm, 23
Binning, 23
Biocatalysts, 28
Biocide, 28, 70, 79, 82–85, 89–91, 209
  Electrophilic, 108, 110–111
  field evaluation of, 112
  incompatibility, 105
  laboratory evaluation of, 112
  lytic, 108, 111
  non-oxidising, 108, 172
  oxidising, 105
  synergistic effects, 105
biodegradation, 22
bioethanol, 28
biofilm, 82, 90, 173, 269
biofilm formation, 268
biofouling, 15
biofuel, 28
biogas, 29
bioinformatics, 15, 72, 78, 199, 120
biomarker, 228–230
bioremediation, 228, 239–242
biostratigraphic profile, 27
bio-surfactants, 269
black smokers, 44
blackwater, 64
borate salts, 101
bowtie, 22
Bray-Curtis distance, 21
breaker, 101
2-bromo-2-nitro-1, 3-propanediol (bronopol), 109
bug bottles, 119–126, 132, 140
Burroghs Wheeler Transform algorithm, 7

C

Caminicella, 59, 60, 77
Candidatus Methanophagales, 146
CAPEX, 42

cathodic depolarization theory, 140
c-di-GMP, 13
cDNA microarrays, 251
cellulases, 28
cellulosic, 28
cellulytic enzymes, 28
CheckM, 23
chemical treatment, 84
chemically microbiological influenced corrosion (CMIC), 70
chemolithotrophs, 215
chemoorganotrophs, 215
chimeras, 20
chlorine dioxide, 103–107
  generation, 106
  mode of action, 106
1-(3-chloroallyll)-3, 5, 7-triaza-1-azoniaadamantane chloride (CTAC), 110, 111
Chloromethylisothiazolinone, 109
citric acid, 101
clay control, 101
Clostridia/um, 63, 86–89, 192
clusters of orthologous groups (COGs), 24
cold seep, 26
CONCOCT, 23
core genome, 251
corrosion, 70, 78, 82, 92, 119–121, 135
  inhibitor, 101
  rate, 120, 125, 133
  risk, 94
crosslinker, 101
cumulative sum scaling (CSS), 24

D

DADA2, 20
dazomet, 110
DBNPA, 83, 90–92, 111
DDAC, 110
dearation tower, 45, 58
Deblur, 20
Defluvibacter, 77
Deltaproteobacteria, 63, 86, 92, 192
denitrification, 215
Desulfacinum, 64, 189
Desulfitobacter, 64
Desulfobacter, 77
Desulfocaldus, 59–60
Desulfocapsa, 188
Desulfohalobium, 51
Desulfomicrobium, 51, 188
Desulfopila corrodens IS5, 154
Desulfotignum, 214
Desulfotomaculum, 64, 191, 192
Desulfovibrio, 51, 70, 77, 86, 92, 188
  D. ferrophilus, 181
Desulfovibrio ferrophilus IS4, 154
Desulfuromonadales, 78
Desulfuromonas, 51, 77
DFAST, 23
dibromonitrilopropionamide, 109
didecyl dimethyl ammonium chloride (DDAC), 109
differential abundance, 21
digital normalization, 23
dimethyloxazolidinone (DMO), 109, 111
disposal well, 47
dissimilatory nitrate reduction to ammonium (DNRA), 215
DNA, 86, 119–126, 131
DNA/RNA Shield, 176
DNA sequencing (Sanger), 2, 15
DNA sequencing cost, 4
DNAzol, 176
dsrAB gene, 220

E

ecosystem, oilfield, 43, 44
efflux mechanisms, 266
efflux pumps, 266
electrical microbiological influenced corrosion (EMIC), 70
electron acceptor, common, 48, 49, 50
electron donors, common, 49, 50
end member mixing analysis, 27
endoxylanase, 28
enzymatic hydrolysis, 28
Epsilonproteobacteria, 60, 63, 86, 93, 132, 189
ethylbenzene, 231, 236
Euryarchaeota, 146
extracellular polymeric substances (EPSs), 269

F

Faith's PD, 21
Fayetteville shale, 114
Ferroglobus, 63
Fervidobacterium, 51
FeS, 42, 64, 65
FeS_2, 42
fine filter, 45
Firmicutes, 77, 210
Flexistipes, 192
floating production storage and offloading (FPSO), 58, 217
flowback water, 100, 101
fluorescence in situ hybridization (FISH), 72
454 pyrosequencing, 16, 53
friction reducer, 101, 104
Fusibacter, 77

G

Galaxy, 17
Gammaproteobacteria, 63, 87, 90, 112, 211
ganon, 23
gallant, 101
gene expression, 250
gene ontology (GO), 258
genomes, 250
genomics, 250
genotypes, 254
Geobacillus, 191
glutaraldehyde, 65, 83, 84, 90, 103, 109–111, 114, 209
Gneiss, 21
green fluorescent protein (GFP), 253
GreenGenes, 19, 53
guar gum, 101

H

H_2S, 70, 71, 82–84, 92, 102, 124, 134
  oxidation, 64
Halanaerobium, 51, 57, 64, 86, 89, 101, 193
halophile, 44
halophilic, 141
Haynesville shale, 102
health, safety and environment (HSE), 26
hierarchical clustering, 254
horizontal drilling, 99
horizontal well, 99, 100
Human Genome Project, 16, 26
Human Microbiome Project, 6, 26
hydraulic fracturing, 99–101, 192
  water components, 101
hydrocarbon, 227–232, 234–242, 250
  seep sediment, 102
  solubilization, 268
hydrocarbon/fuel-contaminated groundwater, 227, 240
hydrochloric acid, 101
hydrogenotrophic, 147
hydrogen peroxide, 107
hydrogen sulfide, 18
hydroxyethyl cellulose, 101
hypochlorous acid, 105

I

Illumina, 53
Illumina platform, 22
IMG/M, 24
IMP, 26
injection plant (SIP), 123, 125–127, 129, 134
injectivity, 82, 84, 92
injector well, 47
in silico PCR primer design, 20
internal transcribed spacer (ITS), 5, 19
InterPro, 23
iron control, 101
iron-reducing bacteria, 70
isothiazolin, 101

J

Jaccard distance, 21
Jet-A, 266
jet fuels, 250

K

Kaiju, 23
Kbase, 17
KEGG, 23
KEGG Orthology Groups (KO), 24
key performance indicator (KPI), 72–74, 104, 114
khmer, 23
K-mean clustering, 254
k-mer classifiers, 23
Korarchaeota, 147
Kosmotoga, 64
KRAKEN2, 23
Krona plot, 156

L

lateral gene transfer (LTG), 144
lithotrophy, 155
long-read technologies, 10
Luteolibacter, 60

M

machine learning, 26
Marinobacter, 57, 60, 86, 87
Marinobacterium, 57, 86, 87
MataPhlAn, 7
MaxBin 2.0, 23
mcrA gene, 148
MEGAHIT, 23
MEGAN6, 23
MetaBAT2, 23
metabolomics, 175
metadata, 16
MetaErg, 23
metagenome assembled genomes (MAGs), 22, 23
metagenomics, 6, 16, 22, 175, 185
MetaPhlAn2, 23
metaproteomics, 175, 187
MetaSPAdes, 23
metatranscriptome, 19, 26

metatranscriptomics, 6, 175
methane, 141
Methanobacteriales, 53, 146
Methanocalculus, 52, 188
Methanocellales, 146
Methanococcales, 146
Methanococcus maripaludis, 154, 181
methanogenesis, 50, 141
methanogenic archaea, 52, 70, 102, 138
methanogenic corrosion, 140
methanogens, 138
Methanohalophilus, 63
Methanolobus, 188
Methanomassiliicoccales, 146
Methanomicrobiales, 146
Methanopyrales, 146
Methanosarcinales, 52, 146
Methanosarcineae, 101
Methanothermococcus, 77
Methermicoccus, 63
methyl-coenzyme M reductase (Mcr), 148
methylisothiozolinone, 109
Methylophaga, 59, 60
methylotrophic, 147
MG-RAST, 7, 24
mic factor, 153
mic hydrogenase (micH), 154
mic island, 153
MIC management, 71
MIC risk, 71, 125
Microarray, 251
microbial control, frac operations, 103
microbial mat, 3
microbial status, 58
microbiologically influenced corrosion (MIC), 15, 27, 42, 52, 64, 69, 70–72, 77, 82, 84, 92, 102, 119–123, 132–134, 138, 170, 172
microbiome, 16
Micrococcineae, 131
microorganisms, classification, 51
microseepage, 27
mineral oil, 101
Minox process, 45
mock communities, 18
molecular biological methods (MMM), 173, 209
monitoring, 119–122
monitoring program, 52
most probable number (MPN), 16, 55, 72, 77, 209
MOTHUR, 7
MOTHUR platform, 20
Multi-omics techniques, 186, 189
MUSiCC, 24

N

N,n-dimethyl formamide, 101
naïve Bayesian classifier, 21
NCBI, 23
Nebula Genomics, 16
next generation sequencing (NGS), 3, 16, 53, 64, 70, 72, 79, 86, 91, 119–122, 133–135, 269
nitrate, 18
  addition of, 61
  reduction, 52
  treatment of souring, 190
nitrate-reducing bacteria (NRB), 70, 209
nitrate-reducing, sulfate-reducing microorganisms (NRSRM), 216
nitrate-reducing, sulfide-oxidizing microorganisms (NRSOM), 215
Nitrospirae, 51
non-coding RNAs, 251
normalization, 23
NSERC IRC Petroleum microbiology, 17
nucleic acid extraction, 176
nucleic acid preservation buffer (NAP), 18

O

oil reservoir, temperature, 46
oil water separator, 47
open ponds, 47
operational taxonomic units (OTUs), 7, 20, 210
OPEX, 42, 65
ORNA, 23
orthologous groups of proteins (COG), 258
orthology groups, 24
osmoprotectants, 141
OUT clustering, 20
Oxford Nanotechnology, 5
oxygen scavenger, 56
ozone, 105, 107
  generation, 108

P

Pacific Biosciences, 5
PCR, 18
Pelobacter, 57, 70
peracetic acid, 103, 105, 107, 109, 112
  mode of action, 107
perchlorate-reducing microorganims (PRB), 209
PERMANOVA, 21
Permian basin, 99, 114
Petrobacter, 191
Pfam, 23
pH, limit of life, 44
Phe-Arg-β-napthylamide, 266
PhiX, 22
Phosphonate, 101
PhyloFlash, 23
Phylogenetic, 21
phyloseq platform, 20

Picrust2, 21
Pielou's Evenness, 21
pigging debris, 19, 160
pipeline deposits, 157
Piphillin, 21
polyacrylamide, 101
polymerase chain reaction (PCR)
group-specific primers, 178–181
universal primers, 177
Postgate medium, 140
predictive models, 71
preservation of samples, 174, 176
pressurized bioreactor, 208
primers, 229–238, 242
principal component analysis (PCA), 254
principle coordinate analysis (PCoA), 214
Prodigal, 23
produced water (PW), 76–78
produced water re-injection system (PWRI), 76–78
production allocation, 27
production well, 46
PROKA, 23
PROMMenade, 23
promoter polymorphism, 253
proppant, 101, 105
protein coding genes (CDS), 23
Proteus, 51
Pseudomonas, 52, 63, 188
P. aeruginosa, 250

Q

QAC, 83, 84, 90, 111
Qiita, 17
qPCR primers, 182–183
quantitative insights Into microbial ecology (QIIME), 7, 17, 210
quantitative polymerase chain reaction (qPCR), 19, 54, 58, 65, 72, 86, 92, 119–129, 134, 140, 210, 229–231, 235–242
quarternary ammonium compounds (QACs), 103, 108
mode of action, 110
quorum sensing (QS), 265

R

RDP, 53
RDP database, 19
redox reactions, 47
relative rog expression (RLE), 24
reservoir souring, 27, 42, 102, 207
REVEAL, 27
reverse methanogenesis, 148
risk model, microbial control, 104
RNA analysis, 184
RNA sequencing, 251
RNAlater, 18, 176
RNAmmer, 23
Robert Hungate, 140
Roseobacter, 132
rRNA, 23
RT-qPCR, 253

S

sanger sequencing, 16
scale inhibitor, 101
schmoo, 42
seawater (SW), 76–78, 123, 125–127, 134
injection, 69, 73, 74
mechanical dearation, 46
nitrate addition, 46
temperature, 45
seawater treatment plant (STP), 123, 125–127, 130, 131, 135
Sediminibacterium, 86, 87
separator, 59
sessile, 51
shale reservoir, 102
Shannon, 16, 21
shotgun metagenomics, 6
silica sand, 101
SILVA, 53
SILVA database, 19
single marker gene sequencing, 18, 19
single nucleotide polymorphism (SNPs), 251
16s rRNA, 19, 51, 86, 123, 129, 140
16S rRNA metagenomics, 210
soda lake, 44
sodium hypochlorite, 101, 103, 105, 109
mode of action, 106
solvent resistance, 266
souring control, 27
souring potential, 27
souring prediction, 27
souring risk in hydraulic fracturing, 102
souring simulation, 27
souring, control of, 170
SourSimRL, 27
South America, 82
SqeezeMeta, 24
stress response, 267
sulfate-reducing bacteria (SRB), 72, 75, 78, 89, 114, 120, 134, 138
sulfate-reducing microorganisms (SRM), 208, 271
sulfate reduction, 51
sulfide, 207
sulfide-oxidizing bacterium (SOB), 77–79, 92
sulfur reduction, 51
Sulfurimonas, 52, 78, 138
Sulfurospirillum, 52, 78
Sulfurovum, 52, 132

sulphate reducing prokarytoes (SRP), 18, 69–71, 77
surface tension, 268
Sva0485 clade Deltaproteobacteria, 228, 241
sweet sorghum, 28
Synergista, 63
Syntrophy, 151

T

TACK Group, 157
Tax4Fun2, 21
tetrakis hydroxymethyl phosphonium sulfate (THPS), 83–85, 90, 103, 104, 109, 111, 155, 209
Thauera, 191
thermal viability shell (TVS), 214
Thermincola, 228, 242
Thermoacetogenium, 102, 189
Thermoanaerobacter, 102, 189
Thermodesulfobacteria, 51
Thermodesulfobacterium, 184, 189
Thermodesulforhabidus, 64, 184
Thermodesulfovibrio, 51, 57, 184
thermogenic methane, 28
thermophilic, 141
thermophilic SRB, 102
Thermosipho, 64
thermospores, 26
Thermotoga, 51, 64
Thermotogaea, 63
Thermovirga, 102
Thiomicrospira, 52, 60
thiosulfate, 210
thiosulfate-reducing bacteria, 70, 89
thiosulfate reduction, 51
toluene, 231, 235, 241
total dissolved salts (TDS), 100
trace elements and iron acquisition, 266
transcriptomics, 250
transmission pipelines, 85
tributyl tetradecyl phosphonium chloride (TTPC), 103, 109
trimethyloxizolidinone, 110
trimmed mean of M-values (TMM), 24
trimming, 22
Trimmomatic, 22
tris(hydroxymethyl)nitromethane (THNM), 65, 110, 111, 112, 114
tRNA, 23
tRNAscan, 23
trunk line, 46
tubing failures, 158

U

unconventional well, 100
UniFrac matrices, 21
UniProt, 23

V

vacuum stripping, seawater, 45
VAMB, 23
Verstraetearchaeota, 147

W

water activity, Aw, 45
West Texas, 99
white-light interferometry (WLI), 160
whole metagenome sequencing, 22
Wolfe cycle, 149

X

xylene, 231, 241

Printed in the United States
by Baker & Taylor Publisher Services